Steps toward a Philosophy of Engineering

Steps toward a Philosophy of Engineering

Historico-Philosophical and Critical Essays

Carl Mitcham

London • New York

Published by Rowman & Littlefield International, Ltd.
6 Tinworth Street, London SE11 5AL, United Kingdom
www.rowmaninternational.com

Rowman & Littlefield International, Ltd., is an affiliate of Rowman & Littlefield
4501 Forbes Boulevard, Suite 200, Lanham, Maryland 20706, USA
With additional offices in Boulder, New York, Toronto (Canada), and Plymouth (UK)
www.rowman.com

Copyright © 2020 Carl Mitcham

All rights reserved. No part of this book may be reproduced in any form or by any electronic or mechanical means, including information storage and retrieval systems, without written permission from the publisher, except by a reviewer who may quote passages in a review.

British Library Cataloguing in Publication Data
A catalogue record for this book is available from the British Library

ISBN: HB 978-1-78661-126-0
       PB 978-1-78661-127-7

Library of Congress Control Number: 2019949259

ISBN 978-1-78661-126-0 (cloth: alk. paper)
ISBN 978-1-78661-127-7 (pbk: alk. paper)
ISBN 978-1-78661-128-4 (electronic)

*In gratitude for and to the students
especially in China
who have thought along with me
about more than what is included here*

Contents

Preface   xiii

Acknowledgments   xvii

Fragments in Search of an Introduction:
Remarks on Engineering as a Theme in Philosophy   1
    Emergence   1
    Complementary Contexts: West and East   6
    Engineering Studies   9
    North American Initiatives   11
    Engineering in Words   14
    Philosophy and Engineering   18
    Engineering Is Everywhere   21
    Aspirations   22

PART ONE   25

1 Science, Technology, Engineering, and the Military   27
    Observations from History   27
    Historico-Philosophical Background   29
    After World War II in the United States   32
    Military Embeds with Philosophy of Engineering   35
    Conclusion   36
    Addendum: Reengineering Warfare   37

2 Ethics into Design   39
    On the Existence of Design   40
    On the Social Dimensions of Modern Design   43
    On the Ethics of Designing   44

viii   Contents

   Two Versions of an Ethics in Design   46
   Notes toward an Inner Ethics of Design   48
   Notes   50

3 **The Importance of Philosophy to Engineering**   53
   1. Self-Defense and Philosophy   54
   2. Self-Interest and Philosophy   56
   3. Excursus: Three Questions   60
   4. Engineering and Ethics   63
   5. Beyond Applied Ethics: Self-Knowledge and Philosophy   67
   Notes   69

4 **From *Dasein* to Design: The Problematics of Turning
   Making into Thinking (*with J. Britt Holbrook*)**   73
   The Etymology of "Design"   75
   Technological Design History   77
   Engineering Design as the Turning of Making into Thinking   79
   The Problematics of Engineering Design   81
   A Duty *Plus Respicere* and Its Discontents   82
   The Metaphysics of Engineering Design   84
   Authenticity in Engineering Design   86

5 **Professional Idealism among Scientists and Engineers:
   A Neglected Tradition in STS Studies**   89
   1. FAS, the *Bulletin*, Pugwash, and UCS   91
      The Federation of American Scientists   93
      Bulletin of the Atomic Scientists   94
      The Pugwash Movement   95
      The Union of Concerned Scientists   96
   2. Committee for Scientific Freedom and Responsibility   98
   Implications   101

6 **Can Engineering Be Philosophical?**   105
   Oppositions   106
   Obligations   109
   Options   111
   Conclusion   116
   Philosophical Engineering: Five Theses   117

7 **Convivial Software: An End-User Perspective
   on Free and Open Source Software**   119
   Technological Invention in a Social Context   121
   The Engineering Ideal   125
   The Convivial Technology Ideal   128

Conclusion and Implications   133
Addendum: The Speed Trap   136

8 Comparing Approaches to Philosophy of Engineering: Including the Linguistic Philosophical Approach (*with Robert Mackey*)   139
1. Introduction   139
2. Six Basic Types   141
3. Toward a Linguistic Philosophy of Engineering   144
4. Conclusion   148

PART TWO   151

9 A Spectrum of Ideals in Engineering Ethics, Simplified   153
1. Historical Dialectics of Ethics and Engineering   155
First Thesis: Obedience to Authority and Company Loyalty   156
The Principle of Loyal Obedience   160
Second Thesis: Technocratic Efficiency   161
The Principle of Efficiency   163
Third Thesis: Public Safety, Health, and Welfare   163
   In the ECPD-ABET-AAES   164
   In the NSPE   165
   In the IEEE   165
Public Safety, Health, and Welfare as Paramount   166
Environmentalism and Sustainability   167
A Participation Principle   168
2. Elaborating: Selective North American Cases and Issues   169
The Jakobsen, Payne, and ASCE Case (1930s)   169
The Hydrolevel Case (1960s)   171
The Bay Area Rapid Transit (BART) Case (1970s)   172
The *Challenger* Disaster (1980s)   173
Whistle-Blowing as a Duty to Public Disclosure   173
Concluding Non-Dialectic Postscript   175
3. Toward a Soft Dialectical Synthesis   176
A Duty *Plus Respicere* To Take More into Account   177
Practical Guidelines for Exercising a Duty *Plus Respicere*   178
Notes   179

10 The Concept of Sustainability: Origins and Ambivalences   183
1. A Historical and Philosophical Background for Sustainability   184
2. Immediate Origins of the Concept of Sustainable Development   186

3. Sustainable Development and Some Near Neighbors 191
4. Criticisms of Sustainability 196
5. Conclusion 199
Addenda: Economics, Philosophy, Engineering, and Ecomodernism 200
Notes 204

11 Engineering Ethics Education in the American Context: Retrospect and Prospect 207
A Brief History of Key Ideas in Engineering Ethics 208
Engineering Ethics: Some Quantitative Observations 210
Ethics into Engineering Education 211
Contemporary Possibilities: A Policy Turn? 214
Coda: Post-Engineering 216

12 Notes on Engineering Ethics in Global Perspective (*with Gary Lee Downey and Juan Lucena*) 221
Japan: Engineering and Profession as Household 222
Engineering Ethics as Institutional Protection in Hong Kong 227
French Engineers, Progress, and the Rational State 228
Germany: Engineering and *Bildung* 231
Engineering Ethics as Social REFORM in Sweden 234
The Dominican Republic: An Engineering Ethics Failure 235
Engineering Ethics as Alternative Development in Chile 236
Conclusion: Globalized Diversity 237
Acknowledgment 239
Addendum: A Further Note 240

13 Humanitarian Engineering (*with David Muñoz*) 241
1. Shifting Contexts and Constraints 241
2. Humanitarianism in History 243
   Humanitarianism, Humanism, and Human Rights 244
   Humanitarian Universalism 245
3. Five Phases in Modern Humanitarianism 246
   Phase I (1800s): Rise of the Humanitarian Movement Proper 246
   Phase II (Early 1900s): Humanitarianism beyond the Battlefield 247
   Phase III (1950s–1960s): Humanitarianism as Free World Ideology 247
   Phase IV (1970s–1990s): Alternative Humanitarianisms 248
   Phase V (2000–present): Humanitarianism Globalized and Questioned 248

4. Humanitarian Engineering   250
    The Fred Cuny Story   250
    Other Precursors and Influences   251
    Maurice Albertson and the U.S. Peace Corps   251
    Médecins sans Frontiers and Engineers without Borders   252
5. Challenges   253
    Practical Challenges   254
    Theoretical Challenges   255
Conclusion: Humanizing Technology   256

14 A Philosophical Inadequacy in Engineering   **259**
Engineering Defined   260
Historical Emergence   263
The Problem   265
Conclusion   271
Addendum: The Sociological Inadequacy of
   Engineering—Response to David Goldberg   273

15 The True Grand Challenge for Engineering: Self-Knowledge   **279**
An Axial Age   280
Two Cultures *Recidivus*   281
Why Humanities?   282
Re-Envisioning Engineering   284
Addendum: Simondon's Dream   287

16 From Engineering Ethics to Politics (*with Wang Nan*)   **291**
Pre-Philosophical Origins   292
Initiating Engineering-Philosophical Discussions: Germany   294
Initiating Engineering-Philosophical Discussions:
   United States   296
Globalization   300
From Ethics to Politics   304
Conclusion   307
Coda: Toward a Political Philosophy of Engineering   308

PART THREE   311

17 Engineering Policy   **313**
Conceptual Issue: What Is Policy?   313
Background: Classics in Science Policy   316
Science, Technology, and Engineering   318
Normative Arguments: Henry Petroski   319
Normative Arguments: Roger Pielke, Jr.   322

Normative Arguments: Natasha McCarthy   324
Conclusion   325

18 Energy Constraints (*with Jessica Smith*) — 327
Anthropologies of Energy   327
Philosophies of Energy   329
Type I versus Type II Energy Ethics   331

19 Can Philosophy Be Engineering? — 335
Learning from Trying   335
Toward an Engineering Epistemology and Metaphysics   338
The Question of Engineering   341
Questions of Engineering Ethics and Politics   343
Conclusion   344

20 In Conclusions — 347
Where To Begin?   347
Continuing   351
And More   355
Concluding Unsystematic Postscript: Toward a Techno-Human Condition or Clash of Anthropologies?   358

APPENDIX

On Engineering Use and Convenience — 365
The Charter of the Institution of Civil Engineers   365
Immediate Origins of the Charter   366
Description of a Civil Engineer: By Thomas Tredgold   368
Thomas Telford and the Institution in Cultural Context   369
Telford's Path: From Stone Mason to Engineer   370
Use and Convenience: Before and after Tredgold and Telford   375
The Distractions of Convenience   378
A Convenience Paradox   380
An Engineering Philosophy for Engineering   381
Objections and Qualifications   382

References   385

Index   423

About the Author   445

Preface

This *in media res* collection records a series of halting steps in a philosophical encounter with engineering. Although I have been thinking about engineering and technology for many years, I remain unsure of a final judgment and swing between trying to think with and think against engineering. Finality will be decided by history and the perilous, contingent trajectory that engineering, as the fount of modern technology, has introduced into human affairs as well as the nonhuman world.

Over the past three decades my "very ordinary brain" (much more ordinary than that of H. G. Wells, 1934) has stuttered, circling issues over and over, repeatedly referencing many of the same texts and arguments. What I am sure about is that engineering is an insufficiently thematized but deeply distinguishing feature of the present age, more powerfully definitive of modernity than science, technology, or democracy (three common alternative candidates). Modernity in any of its competing determinations either gives rise to or is animated by that unique form of world making and building that is engineering. Yet engineering seems to have been too hidden or banal a component of bourgeois capitalist civilization—less glamorous than the dramatic technologies it has created—to be accorded the philosophical attention it deserves.

Years ago I attempted to advance an understanding of technology through a movement from engineering to philosophy. Here my focus is simply connections between philosophy and engineering in order to delve deeper into the inner and outer dynamisms of engineering.

Some personal influences: My father was a registered professional mechanical engineer, an uncle was an agricultural engineer, and a son became an electrical computer engineer (then gave it up). Thinking with and against engineering has thus been sandwiched between generations, struggling to comprehend the affirmations and rejections they manifested. My occasional work as a vernacular house builder has further contributed a fraught relationship with the engineering way of making and thinking.

Still more influences have come from four decades of professional employment in engineering schools: Polytechnic University in Brooklyn, the College of Engineering at Pennsylvania State University, and the Colorado School of Mines. The enjoyments and frustrations of introducing the humanities into technical curricula cannot help but be reflected here. A late move to Renmin University of China, and a consequent shift from inviting engineering students to appreciate philosophy to encouraging philosophy students to attend to engineering and technology, in a cultural context other than North American, has been a rewarding opportunity. Finally, the community of scholars centered in the Society for the Philosophy of Technology (SPT) and the Forum for Philosophy, Engineering, and Technology (fPET) has continuously nourished and challenged the efforts reflected in these essays.

These essays have involved significant collaborations. Six have coauthors, although in each case they have been significantly revised or adapted. From the beginning of my scholarly research in the 1970s, I've worked with colleagues in pursuit of broader and deeper understandings of the nature and meaning of engineering and technology. As these collaborators are not responsible for the adapted versions, their names (with their permission) have been placed in parentheses. But none of these interlocutors should be presumed in any way contaminated with the arguments advanced in this new context.

Additionally, essays span a temporal period from the late 1990s to 2019. Because they come from across such an extended period, there are inevitable and considerable overlaps. Ideas and arguments are regularly revisited. This allows each essay to be read on its own while simultaneously serving both to emphasize central issues and in some cases to deepen them. In the center of the period covered, I coauthored a textbook (with R. Shannon Duval) on engineering ethics. The title, *Engineer's Toolkit: Engineering Ethics* (2000), suggested an interest in bridging theory and practice that is further emphasized in a move toward politics and policy. The same book led to another collaboration with Juan Bautista Bengoetxea in *Ética e ingeniaria* (2010).

At the same time, bridging theory and practice is not an end in itself. Simply to be more effective is not enough; it is a good to understand what is, even if we cannot change it. This same tension arises in a question that animates, often in a hidden manner, many of the discussions in these essays: Is the pursuit of a philosophy of engineering also a philosophy for engineers? The trajectory of many of the arguments suggests this is possible only with important qualifications—and perhaps some transformations in engineering itself.

There is an episodic character to these essays, with many dated references (but not so dated, I believe, as to vitiate basic arguments). My hope is that they hang together enough to stimulate further discussion, research, and teaching in the philosophy of engineering. I would be especially gratified if the book also helped deepen philosophical friendships with Chinese colleagues and their own reflective engagements with a world that is undergoing an engineered transformation more rapidly than any other in human history. As one younger colleague has argued, like much of Chinese culture, engineering too exists in tension between modernism and tradition. The same could be said for engineering in general and the many cultures being impacted by globalization. However, the tension that is worked out in China may well have implications for others.

Each set of essays is arranged in roughly chronological order. Part one is a scattershot of philosophical encounters; part two focuses on ethics and politics. Reprinted essays mostly repeat their original tones (spoken or written) and formats (i.e., using endnotes or in-line citations), although full citations have been pulled out of notes and integrated into a master set of references that offers a more unified indication of the scope of the literature engaged.

This book is a version of one published originally in Chinese in response to years of encouragement by Professor Wang Qian of Dalian University of Technology (DUT). In the summer of 2009, on the occasion of a short residency at DUT, Professor Wang proposed collecting a set of articles for Chinese translation. His encouragement eventually led to the publication of 工程与哲学：历史的、哲学的和批判的视角 *Gongcheng yu zhexue: Lishi de, zhexue de he pipan de shijiao* (Philosophy and Engineering: Historical-Philosophical and Critical Perspectives) (Beijing: Renmin Press, 2013). Another five years and encouragement by Isobel Cowper-Coles at Rowman and Littlefield International has yielded the present much revised publication. Librarians at the Colorado School of Mines provided excellent research support.

I want further to acknowledge stimulation and encouragement in diverse ways from the following colleagues: Adam Briggle at the University of North Texas, Huang Xiaowei at Tianjin University, Li Bocong and Wang Nan at University of Chinese Academy of Sciences, Liu Yongmou at Renmin University of China, Glen Miller at Texas A&M University, and Qin Zhu at Colorado School of Mines. Chen Yuqing worked to standardize references. The cover adapts a painting by my daughter Emilie Mitcham.

Denver/Beijing, April 2019

Acknowledgments

The original provenance along with, as appropriate, acknowledgment for permission to reprint and adapt these collected essays, all of which have been revised for the present volume, is as follows:

Essay 1: "Science, Technology, Engineering, and the Military" has not been published previously in English, although the argument was initiated in "The Spectrum of Ethical Issues Associated with the Military Support of Science and Technology," in Carl Mitcham and Philip Siekevitz, eds., *Ethical Issues Associated with Scientific and Technological Research for the Military* (proceedings of a conference, January 26–28, 1989), *Annals of the New York Academy of Sciences*, vol. 577 (1989), pp. 1–9. A Chinese version is in 工程与哲学 (2013), pp. 330–342.

Essay 2: "Ethics into Design" is reprinted from Richard Buchanan and Victor Margolin, eds., *Discovering Design: Explorations in Design Studies* (Chicago: University of Chicago Press, 1995), pp. 173–189. The argument was originally presented at a small international conference on "Discovering Design," University of Illinois at Chicago Circle, November 5–6, 1990. A Chinese version is in 工程与哲学 (2013), pp. 150–165.

Essay 3: "The Importance of Philosophy to Engineering" is reprinted from *Teorema*, vol. XVII/3 (1998), pp. 27–47, and remains available at www.unioviedo.es/Teorema. The argument was originally presented as a public lecture at the Technical University Delft, Netherlands, April 16, 1998, in conjunction with an international workshop on "The Empirical Turn in the Philosophy of Technology." A Chinese version is in 工程与哲学 (2013), pp. 11–32.

Essay 4: "From *Dasein* to Design: The Problematics of Turning Making into Thinking" conflates two previous articles: "*Dasein* versus Design: The

Problematics of Turning Making into Thinking," *International Journal of Technology and Design Education*, vol. 11 (2001), pp. 27–36; and (with J. Britt Holbrook) "Understanding Technological Design," in John R. Dakers, ed., *Defining Technological Literacy: Towards an Epistemological Framework* (New York: Palgrave Macmillan, 2006), pp. 105–120. A Chinese version of the former publication is in 工程与哲学 (2013), pp. 166–178.

Essay 5: "Professional Idealism among Scientists and Engineers: A Neglected Tradition in STS Studies" is reprinted with edits from *Technology in Society*, vol. 25 (2003), pp. 249–262. A Chinese version is in 工程与哲学 (2013), pp. 311–329.

Essay 6: "Can Engineering Be Philosophical?" has not been published in English. It is adapted from a plenary talk at an international research conference on "*Bildung* in Engineering," Aarhus University, Herning, Denmark, May 15, 2007. A Chinese version is in 工程与哲学 (2013), pp. 445–458.

Essay 7: "Convivial Software: An End-User Perspective on Free and Open Source Software" is reprinted with edits from *Ethics and Information Technology*, vol. 11, no. 4 (December 2009), pp. 299–310. It grew out of a presentation at the "Conferencia Internacional de Software Libre: Open Source International Conference," Málaga, Spain, February 18–20, 2004. An earlier, less-developed version appeared as "El software convivencial," *Argumentos de Razón Técnica*, issue no. 10 (2007), pp. 19–41, which benefited from comments and criticisms by Andoni Alonso and two anonymous reviewers. A Chinese version is in 工程与哲学 (2013), pp. 382–408.

Essay 8: "Comparing Approaches to Philosophy of Engineering: Including the Linguistic Philosophical Approach" (with Robert Mackey) is reprinted from Ibo van de Poel and David E. Goldberg, eds., *Philosophy and Engineering: An Emerging Agenda* (Dordrecht: Springer, 2010), pp. 49–59. A Chinese version is in 工程与哲学 (2013), pp. 33–45.

Essay 9: The core of "A Spectrum of Ideals in Engineering Ethics, Simplified" originated with "Engineering Design Research and Social Responsibility," in K. S. Shrader-Frechette, *Research Ethics* (Totowa, NJ: Rowman and Littlefield, 1994), pp. 153–168, which has been significantly expanded and rewritten. Related material also appears in "Ethics Is Not Enough: From Professionalism to the Political Philosophy of Engineering," in Satya Sundar Sethy, ed., *Contemporary Ethical Issues in Engineering* (Hershey, PA: IGI Global, 2015), pp. 48–80. A related Chinese version is in 工程与哲学 (2013), pp. 119–149.

Essay 10: "The Concept of Sustainability: Origins and Ambivalences" conflates two previous articles: "The Concept of Sustainable Development: Its Origins and Ambivalence," *Technology in Society*, vol. 17, no. 3 (1995), pp. 311–326; and "The Sustainability Question," in Roger S. Gottlieb, ed.,

The Ecological Community: Environmental Challenges for Philosophy, Politics, and Morality (New York: Routledge, 1997), pp. 359–379. The argument originated with a presentation for a conference of the *Consejo Latino Americano de las Ciencias Sociales* (CLACSO), Santiago, Chile, in late 1991. A Chinese version is in 工程与哲学 (2013), pp. 425–444.

Essay 11: "Engineering Ethics Education in the American Context: Retrospect and Prospect" is adapted from "A Historico-Ethical Perspective on Engineering Education: From Use and Convenience to Policy Engagement," in *Engineering Studies*, vol. 1, no. 1 (March 2009), pp. 35–53. A Chinese version is in 工程与哲学 (2013), pp. 214–241.

Essay 12: "Notes on Engineering Ethics in Global Perspective" includes some previously published text from the following: "Engineering Ethics in Asia," *Perspectives on the Professions*, vol. 12, no. 1 (August 1992), pp. 2–3; "Ethics Codes in Professional Ethics: Overview and Comparisons," *Encyclopedia of Science, Technology, and Ethics* (Detroit: Macmillan Reference, 2005), vol. 4, pp. 2176–2182; and Gary Lee Downey, Juan C. Lucena, and Carl Mitcham, "Engineering Ethics and Engineering Identities: Crossing National Borders," in Steen Hyldgaard Christensen, Christelle Didier, Andrew Jamison, Martin Meganck, Carl Mitcham, and Byron Newberry, eds., *Engineering Identities, Epistemologies, and Values: Engineering Education and Practice in Context*, vol. 2 (Dordrecht: Springer, 2015), pp. 81–98. The last text reprises some ideas from Gary Lee Downey, Juan C. Lucena, and Carl Mitcham, "Engineering Ethics and Identity: Emerging Initiatives in Comparative Perspective," in *Science and Engineering Ethics*, vol. 13 (2007), pp. 463–487, a Chinese version of which is in 工程与哲学 (2013), pp. 179–200.

Essay 13: "Humanitarian Engineering" appeared originally in a much shorter form as "The Humanitarian Context" (with David Muñoz) in Steen Hyldgaard Christensen, Martin Meganck, and Bernard Delahousse, eds., *Engineering in Context* (Copenhagen: Academica, 2009), pp. 183–195. An expanded version (again with Muñoz) was published as a pamphlet, *Humanitarian Engineering* (San Rafael, CA: Morgan and Claypool, 2010). The essay here cuts and adapts from each. A longer Chinese version is in 工程与哲学 (2013), pp. 247–307.

Essay 14: "A Philosophical Inadequacy in Engineering" is from *The Monist*, vol. 92, no. 3 (July 2009), pp. 339–356. A Chinese version is in 工程与哲学 (2013), pp. 46–63.

Essay 15: "The True Grand Challenge for Engineering: Self-Knowledge" is reprinted with edits from *Issues in Science and Technology*, vol. 31, no. 1 (Fall 2014), pp. 19–22.

Essay 16, "From Engineering Ethics to Politics" (coauthored with Wang Nan) is reprinted with edits from Steen Hyldgaard Christensen, Christelle

Didier, Andrew Jamison, Martin Meganck, Carl Mitcham, and Byron Newberry, eds., *Engineering Identities, Epistemologies, and Values: Engineering Education and Practice in Context*, vol. 2 (Dordrecht: Springer, 2015), pp. 307–324.

Essay 17: "Engineering Policy: Exploratory Reflections" is reprinted with edits from Carl Mitcham, Li Bocong, Byron Newberry, and Zhang Baichun, eds., *Philosophy of Engineering, East and West* (Dordrecht: Springer, 2018), pp. 247–259, which itself incorporated some material from two previous publications: "Petroski's Policy," *Technology and Culture*, vol. 52, no. 2 (April 2011), pp. 380–384 and "Ethics and Policy" (coauthored with Erik Fisher), in Ruth Chadwick, ed., *Encyclopedia of Applied Ethics*, 2nd ed. (San Diego, CA: Academic Press, 2012), vol. 2, pp. 165–172. A Chinese version of the analysis of the argument of Henry Petroski is in 工程与哲学 (2013), pp. 242–246.

Essay 18: "Energy Constraints" (with Jessica Smith) is reprinted with revisions from *Science and Engineering Ethics*, vol. 19 (2013), pp. 313–319.

Fragments in Search of an Introduction

Remarks on Engineering as a Theme in Philosophy

The philosophy of engineering is slowly becoming a defined discourse. In some respects it still functions as another name for the philosophy of technology, which can be parsed into a fourfold reflection on the ontology of artifacts, theories of technological knowledge, the structure of technological activity, and technological volition. As the engine that designs and produces with a historically distinct agency and uniquely modern artifacts while generating and utilizing technoscientific knowledge, engineering is at the dynamic core of technological creation. The philosophy of engineering thus provides privileged access to philosophical reflection on technology.

EMERGENCE

The philosophy of technology, insofar as it focuses on interactions with and among technologies (from medicine to computers and artificial intelligence) in lifeworld mediations, can marginalize the engineering agency that creates them. Henry Adams's dynamo was more than a symbolically powerful machine at the Paris World's Fair of 1900. It was the product of a new kind of dynamic agency.

Adams compared the symbolic implications of the Virgin Mary with those of the electric dynamo in a way that underappreciated deep differences between the way medieval artisans and modern engineers work. There are more than raw differences in manual and intellectual skill sets. In sculpting the Virgin Mary, a Christian artist would have been aware of the symbolic character of the artifact being created and worked to enhance its cultural power. The modern engineer normally has little to no conscious interest in the cultural salience of the electricity-generating machine but only in its functional utility. The Virgin is symbolic by design, the dynamo by accident—which is why Adams's (1918) experience was an insight into engineered culture.

Engineering agency is further overshadowed by the economic and political institutions that enroll it. Neither capitalism nor nationalism, nor internationalism (globalization), would be what they are without engineering. What is called the Anthropocene is not so much the work of humans qua humans or of capitalism (the Capitalocene) as of humans qua engineers (in affirmative captivity to capitalism), albeit as co-constructed through economics and politics. In the debate about whether the Anthropocene is dark or bright and the issues of social justice or cosmopolitics, another and better name for our new geological time frame might be the "Engineering Epoch"—to highlight the distinctive agency of an ongoing dynamic. It is not just a scene but a drama with a dominant but masked character: the phantom of the opera.

Historian Eric Schatzberg provides another take on the occlusion of engineering. For Schatzberg, a significant contributor to Henry Adams's "baffled amazement before the 'occult mechanism' of the electric dynamo" was a shift in the usage of key concepts. During the late nineteenth century, "art" was pulled away from "mechanical arts" and "industrial arts" to be aestheticized in the "fine arts," while "science" was increasingly identified with expert theoretical knowledge so that there arose what Leo Marx (1997 and 2010) termed a "semantic void."

> The net result of these conceptual changes was to remove human agency from the discourse of industrial modernity. . . . In the never-ending parade of speeches, books, and articles about the material progress of the age, writers granted agency to only a few . . . creative geniuses supposedly responsible for specific inventions. . . . [T]his limited scope for human agency became a key attribute of the instrumental discourse of technology, especially when *technology* was defined as the application of science.
> (Schatzberg, 2018, pp. 72–73)

The marginalization of engineering in the penumbra of both popular and intellectual discourse about technology is surely another factor, since it is not just human agency that is at stake but also a new form of agency. In the analysis of how German and American engineers used such terms as *Technik* and "technology," it is surprising how little "engineering" appears in *Technology: Critical History of a Concept*.

From the beginning, engineering lacked a prominent role in the philosophy of technology, both in its publications and in its professional institutionalization. For example, *Philosophy and Technology: Technology as a Philosophical*

Problem (Mitcham and Mackey, 1972) collected articles by 26 authors from the previous five decades (although 18 were post-1960), only 4 of which gave engineering any significant attention. The most notable was Mario Bunge, who argued for distinguishing the engineering sciences from natural science. Even the sole engineer-contributor wrote in terms of technology and invention, not engineering. Two years later, Friedrich Rapp's edited volume, *Contributions to a Philosophy of Technology: Studies in the Structure of Thinking in the Technological Sciences* (1974), gave Bunge pride of place and engineering a more prominent role but still seemed to privilege "technology" over "engineering" (even in "technological sciences" over "engineering sciences"). At the annual meetings of the Society for Philosophy and Technology (SPT), from 1981 to 2017, mostly in odd-numbered years, alternating between Europe and North America, presentations by engineers and on engineering have been fewer than might be expected.

The first English language monograph on the philosophy of technology was by Friedrich Rapp (1981, translated from 1974 German), which emphasized analytic issues and, referencing Bunge again (among others), devoted one chapter (out of five) to methodological issues in the engineering sciences. The first general textbook introduction by Frederick Ferré (1988) featured a discussion of the "bright vision" of eccentric utopian engineer Buckminster Fuller but minus serious engagement with engineering itself. Another introductory text by Don Ihde (1993) included only one brief historical reference to engineering. A later textbook by Val Dusek (2006) did only slightly better. Prior to the early twenty-first century, Rapp, Carl Mitcham (1994), and Joseph Pitt (2000) were the primary works to foreground engineering, and even they seemed to have reservations. Only Mitcham included "engineering" in the title.

Two primary journals in the field—*Techné: Research in Philosophy and Technology* (with roots going back to 1978) and *Philosophy and Technology* (2011–present)—have rarely included articles with "engineering" in their titles. In its digital incarnation, *Techné*, from 1996 through 2018, published a total of 495 articles (21.5/year) of which only 18 (<1/year) in total had titles with "engineering." In its initial eight years and 332 articles, *Philosophy and Technology* published six.

A similar situation existed in the science, technology, and society (STS) studies field. The lead journal *Science, Technology, & Human Values*, from 1976 through 2018, published few articles with "engineer*" in the title, and among those only a few handfuls exhibited significant philosophical (mostly ethical) content. Just as in philosophy, monographs in STS prior to the late 1990s seldom gave more than a walk-on role to engineers and engineering. Jacques Ellul's *The Technological Society* (1964, translated from 1954 French) does not mention engineers or engineering. One exception was Bruno

Latour's *Science in Action: How to Follow Scientists and Engineers through Society* (1987), which nevertheless conceptually wrapped them together as "scientists and engineers" in technoscience.

Two events initiated changes in the philosophy of technology that would begin to bring engineering out of the shadows:

- In 2000, philosophers at Delft University of Technology in the Netherlands initiated a research program on "The Dual Nature of Technical Artifacts" funded by the *Nederlandse Organisatie voor Wetenschappelijk Onderzoek* (NWO or Netherlands Organization for Scientific Research). The Dual Natures Program (as it became known) focused on the conceptual complexities of relating physical and functional descriptions of artifacts. This encouraged an "empirical turn" in philosophy of technology in which philosophers increased efforts to engage directly with engineers, creating a trajectory of research that subsequently helped establish in 2007 a Technology and Ethics Centre as a collaborative activity among Dutch technical universities.
- In October 2006, mechanical engineer and philosopher Taft Broome organized a workshop at MIT "to (1) establish whether an International Meeting on Philosophy of Engineering is feasible; and, if so (2) who will constitute the Planning Group beyond Oct. 20th; and answer such questions as (3) When? (4) Where? (5) Participants? (6) Funding?"

There were unsuccessful negotiations the next year at the SPT 2007 in July at Charleston, South Carolina, about integrating Broome's concerns into SPT. Instead, the Broome initiative led to creation of a Workshop on Philosophy and Engineering (WPE) later that year (October 2007) at TU Delft and then again the next year at the Royal Academy of Engineering in London (2008). The WPE group subsequently morphed into the Forum on Philosophy, Engineering, and Technology (fPET) and began to organize conferences on even-numbered years (beginning 2010), to complement the SPT biannual odd-year conferences.

The Royal Academy of Engineering followed up its hosting of WPE 2008 with two years of seminars on philosophy of engineering, inviting contributions from philosophers and engineers. The first series (RAE, 2010) focused on engineering knowledge, systems engineering, and artificial intelligence, with at least one philosopher and one engineer addressing each topic. The second series (RAE, 2011) dealt with metaphysics and ethics, again with presentations by both philosophers and engineers.

More expansively, the Dutch activities and WPE-fPET together contributed to establishment of a Springer book series in "Philosophy of Engineering and Technology" (2010–present) edited by Pieter Vermaas at TU Delft. However, even here in the first nine years of publication only 13 of 32 volumes have had "engineering" in their titles, and of these three focused on engineering education. Of the remaining 10, as might be expected, the most relevant are the 4 WPE-fPET proceedings volumes:

- 2010, vol. 2: Ibo van de Poel and David Goldberg, eds., *Philosophy and Engineering: An Emerging Agenda*;
- 2013, vol. 15: Diane Michelfelder, Natasha McCarthy, and David Goldberg, eds., *Philosophy and Engineering: Reflections on Practice, Principles and Process*;
- 2017, vol. 26: Diane Michelfelder, Byron Newberry, and Qin Zhu, eds., *Philosophy and Engineering: Exploring Boundaries, Expanding Connections*; and
- 2018, vol. 31: Albrecht Fritzsche and Sascha Julian Oks, eds., *The Future of Engineering: Philosophical Foundations, Ethical Problems and Application Cases*.

A fifth proceedings volume was published in the Boston Studies in the History and Philosophy of Science series in 2018 and edited by Mitcham, Li Bocong, Byron Newberry, and Zhang Baichun as *Philosophy of Engineering, East and West*.

The first volume in the WPE-fPET proceedings series is to its date the most concise, broad-spectrum introduction to philosophy and engineering interactions. A competitor might be *Philosophy of Technology and Engineering Sciences* (*Handbook of the Philosophy of Science*, vol. 9, 2009), edited by Anthonie Meijers in a project associated with the Dutch Dual Natures program. Its signal virtue was to grant engineering a substantial presence in the philosophy of technology, but at more than 41 chapters and almost 1,500 pages it does not serve well as an introduction; philosophy of engineering tends to get subsumed in an encompassing matrix. For an introduction, *Philosophy and Engineering* (*P&E*) does a better job. (Its lead editor, Ibo van de Poel, was another key member of the Dutch school in the philosophy of engineering and technology.)

In his introduction, van de Poel notes how although engineering is generally recognized as a technology-producing activity, there are substantial disagreements about its nature. Dutch philosophers, in a term coined by Hans Achterhuis (2001) at Twente University, have promoted an "empirical turn" in the philosophy of technology as an attempt to step away from definitional debates into the practical world of technological experience and engineering design. After his account of the already mentioned WPE-fPET series of gatherings, van de Poel identified four overlapping, nonexclusive topics in the new field: the nature of engineering; relations between science, technology, and engineering; a suite of conceptual, epistemological, methodological, metaphysical, and ethical issues; and interactions between philosophers and engineers.

Regarding definitions, the consensus in *P&E* was that design work is central to engineering. Yet disagreements remain between those such as Michael Davis, who argues for a historical-sociological definition of the profession, and others, such as Heinz Luegenbiehl, who pursue a more functional conceptualization. Further disagreements exist between philosophers who assume and question an Anglo-American social context as normative. Christelle Didier (1999 and 2000) maintains engineering is different in France, and Li Bocong argues for even greater differences in China. In both these national contexts, as well as others, engineering often functions differently than in the English-speaking West.

As for science, technology, and engineering relations, Euro-American philosophers quickly criticize any theory of engineering as applied science. Li Bocong, however, sets this issue aside in favor of what he terms a trichotomy or interactive triad of science, technology, and engineering. Science is discovery, technology is invention, and engineering is project construction that requires contributions from not only science and technology but also from economics, government, management, and more.

A number of other engineering-specific issues are distributed across the standard branches of philosophy. Conceptual issues include questions about the nature of engineering design, technical functions, invention, and creativity. Theories of technological knowledge ask, as with engineer Walter Vincenti (1990), what and how engineers know whatever it is that they know. Methodological questions work to distinguish engineering from scientific and other methods. Metaphysical or ontological issues arise when attempting to specify the precise reality of artifacts. Not without reason, ethical efforts to evaluate what engineers do and why they do it are the most prominent feature of philosophy and engineering discourse.

COMPLEMENTARY CONTEXTS: WEST AND EAST

In his contribution to *P&E*, Li Bocong (2010) began with a brief history of philosophical reflection on engineering, East and West. As he noted, the philosophy of engineering emerged simultaneously but independently in Chinese and Euro-American contexts.

In the Euro-American context, Li identified Paul Durbin's edited volume on *Critical Perspectives on Nonacademic Science and Engineering* (1991) as the first English volume in which a number of philosophers began to address engineering as such, while noting how Durbin himself avoided the phrase "philosophy of engineering." This took place only in a foreword by Steven Goldman and Stephen Cutcliffe:

> This collection of essays ranges very widely indeed, from a technical analysis of one facet of engineering reasons to the politics of design. Taken together, the essays begin to define the parameters of an as yet virtually nonexistent discipline, namely, philosophy of engineering. Their publication will, we hope, spur a continuing conversation that will make philosophy of engineering part of the ongoing study of technology.
>
> (p. 7)

Taft Broome's contribution to *Critical Perspectives* made a strong pitch for "philosophy of engineering" as something distinct from "philosophy and engineering" or "philosophy in engineering." "Philosophy *and* engineering" assumes a distinction between "learned disciplines," as is often the case in engineering ethics literature. Philosophy *in* engineering uses "the formal tools of philosophical inquiry to examine some aspect of engineering." With philosophy *of* engineering, "the problems subjected to examination by philosophy *in* engineering are synthesized into a coherent theory" (Broome, 1991, p. 256). Following a review of resistances to interdisciplinarity in the science, technology, and human values field among both philosophers and engineers, and of philosophical issues related to engineering, Broome calls for work by "philosopher/engineers."

In the following decade, two books illustrated Broome's typology, although not precisely as he conceived it: a monograph by MIT engineer Louis Bucciarelli on *Engineering Philosophy* (2003) and an edited collection by Danish humanities professor Steen Christensen and colleagues on *Philosophy in Engineering* (2007). The Christensen volume promoted bringing philosophy into engineering education; Bucciarelli found philosophy already present but hidden in engineering practice.

As an interdisciplinary engineer and STS faculty member, Bucciarelli both bridges and integrates the two disciplines. His 2003 volume is from a series of lectures given at TU Delft as a visiting professor and enlarges on arguments developed earlier in *Designing Engineers* (1994). Bucciarelli aimed to use philosophy to bring "to the surface . . . the essential, fundamental beliefs of what may be called an engineering mind set" (p. 4) by studying engineering narratives in the design process. This is important because of the ways both engineering and the world it designs are undergoing major changes. "In these times of change, if we are to claim some control over the future, we must allow that what needs to change includes more than the tools, the organization, the methods, the hardware and software, but more fundamentally, ways of perceiving and reading the world" (p. 7).

There is another distinction not mentioned by Broome: the philosophy *of engineering* versus philosophy *for engineering* or engineers. The philosophy for engineering is a rational explication and defense of the engineering way of life and seems closer to how Broome conceived philosophy of engineering. To some degree Bucciarelli was likewise concerned with philosophy for engineers: hence his term *engineering philosophy*. After all, his lectures were delivered at an engineering university and his tone was one of talking to and with engineers.

By contrast, philosophy *of* engineering in the philosophical sense suspends serving as the handmaid of a professional community. Philosophy of art or of religion may be only marginally for or of much use to artists or believers. At the same time it will include critical reflection on the inner life—both the cognitive and the ethical lives—of communities of practitioners, which may on occasion be of service to them. Regionalizations of philosophy may also help nonpractitioners develop some measure of appreciation of the practices on which they reflect.

For the Chinese context, Li Bocong, a professor of philosophy at the University of Chinese Academy of Sciences (UCAS), located the emergence of the philosophy of engineering in four key publications:

- Li Bocong, "我造物，故我在 *Wo zaowu, guwo zai* [I create therefore I am]," *Studies in Dialectics of Nature* (1993);
- Li Bocong, 工程哲学导论 *Gongcheng zhexue daolun* [Introduction to philosophy of engineering] (2002);
- Yin Ruiyu, Wang Liheng, Wang Yingluo, and Li Bocong, eds., 工程与哲学 *Gongcheng yu zhexue* [Engineering and philosophy] (2007); and
- Yin Ruiyu, Sun Yongfu, Wang Yingluo, Li Bocong, and Qiu Lianghui, eds., 工程与哲学 *Gongcheng yu zhexue* [Engineering and philosophy], vol. 2 (2018).

Equally significant, in 2003 UCAS established a Research Center for Engineering and Society (RCES). In collaboration with a few academicians in the Chinese Academy of Engineering (CAE), RCES organized a series of conferences and began publishing a new scholarly journal, 工程研究：跨学科视野中的工程 *Gongcheng yanjiu: Kuaxueke shiye zhong de gongcheng* [Engineering Studies: Interdisciplinary Perspectives on Engineering (hereafter referenced simply as *Engineering Studies: Interdisciplinary Perspectives*)]. In quick order there was also established a Chinese Society for Philosophy of Engineering (2004, with biannual meetings) as a section in the Society for the Dialectics of Nature (which also includes a Society for Philosophy and Technology, founded 2003 with biannual meetings). (Aside: The Chinese Society for the Dialectics of Nature is an umbrella unit in the Chinese Association for Science and Technology for research in the history and philosophy of science and technology. It was established early in the Reform and Opening to open up Marxist thinking.)

Following his historical note on the dual emergence of philosophy of engineering in the East and West, Li Bocong outlined his own philosophy of engineering. As van den Poel's introduction noted, Li's triad conceptions of science, technology, and engineering do not completely correspond to those common in Euro-American discourse. The Chinese term 工程 *gongcheng*, conventionally translated as "engineering," means something more like technological or engineering project.

> On the one hand, technology and engineering are closely related—there is no engineering without technology and technology can be applied to engineering. On the other hand, they are different—engineering always involves non-technological factors, and is the unity of both technological and non-technological factors. . . . Engineering activity contains economical factors, management factors, social factors, political factors, ethic factors and psychological factors, besides technological factors.
>
> (Li Bocong, 2010, p. 35)

In succeeding years, three more articles have appeared in translation advancing his perspective (Li Bocong, 2012, 2015, and 2018).

ENGINEERING STUDIES

Influenced by relationships with diverse Western scholars, STS studies began in China in the 1980s under the leadership of Yin Dengxiang at the Chinese Academy of Social Sciences (see Yin Dengxiang, 1997). Following a postdoctoral visit to Lehigh University and Pennsylvania State University, in 1992 Yin hosted Cutcliffe and Mitcham for a week-long series of lectures on the scope of American STS at that time. As in the West, however, engineering was not strongly thematized in this interdisciplinary field, which in the Cutcliffe-Mitcham interpretations included the six disciplines of history, sociology, and philosophy of science and technology (see, e.g., Spiegel-Rosing and De Solla Price, 1977; Durbin, 1980; and Cutcliffe, 2000). Bringing engineering to the forefront began to take place in parallel but independently in China and in the West. In China, however, the philosophy of engineering played a stronger role than in the West, where philosophy became marginalized in STS.

As mentioned, it was in conjunction with Li Bocong's research in philosophy of engineering that RCES began publishing the Chinese journal

Engineering Studies: Interdisciplinary Perspectives in 2005, first as an annual, then in 2009 as a quarterly, and from 2016 as a bimonthly publication. From the beginning, philosophy of engineering occupied a prominent position, not only because of Li's work but also because of the desire to place engineering in a Marxist theoretical framework.

In the West, the strong thematization of engineers and engineering was stimulated especially by the scholarly energy of Gary Lee Downey, an STS professor at Virginia Tech University. One version of the story is actually recorded in a lengthy interview conducted by Zhang Zhuhui, a former student of Li Bocong, and published in *Engineering Studies: Interdisciplinary Perspectives* (Downey and Zhang, 2015). Downey and Juan Lucena, at the time Downey's graduate student, first proposed the idea of engineering studies in a chapter of what became a field constructing publication, *The Handbook of Science and Technology Studies* (first edition, 1994; revised, 2001). That chapter remains a strongly informative analytic survey of research to that date in terms of four themes: engineering knowledge, engineering as technical work, engineering-related gender studies, and studies by engineers themselves; it is complemented by an update 20 years later organized along different lines (Downey, 2015).

The material referenced especially in the review of research on engineering knowledge calls attention to a wealth of historical and sociological work of philosophical interest. A significant percentage of this literature emerged from scholars associated with the Society for the History of Technology (SHOT) and was often published in its journal *Technology and Culture* (*T&C*, 1959–present). Scholars associated with the history of technology have made numerous contributions to philosophical discourse on engineering—the work of Edwin T. Layton, Jr., stands out—and the *T&C* reviews section is the single best place to get a sense of all book publications by historians, social scientists, and engineers that might be relevant to the philosophy of engineering. Yet even in *T&C*, articles with "technology" in their titles overshadow "engineering" articles.

Subsequent to his *Handbook* chapter, Downey (along with colleagues Maria Paula Diogo and Chyuan-Yuan Wu) in 2004 at a joint meeting of the Society for Social Studies of Science (4S) and European Association for the Study of Science and Technology (EASST) in Paris founded the International Network of Engineering Studies (INES). According to Downey, when a few years later INES created its own *Engineering Studies* journal, he "was aware of the name Li Bocong and his interest in engineering [so that he] invited him to join the journal's editorial board." But Downey "had no idea [Li] was already producing a publication using the [same] name" (Downey and Zhang, 2015, p. 21). Later Downey learned from Li Bocong that Li "had adopted the name [engineering studies] as a label for his own work after reading the [1994] *Handbook* chapter" and then, with colleagues, "went on to build a massive scholarly enterprise in Engineering Studies in China, by far the world's largest" (Downey and Zhang, 2015, p. 19).

NORTH AMERICAN INITIATIVES

Although in 1991 he did not use the phrase "philosophy of engineering," Durbin soon adopted it and published two articles summarizing a distinctive view of the field. One was "Multiple Facets of Philosophy and Engineering" (2010), which immediately followed Li Bocong's contribution to *P&E*. An earlier discussion served to supplement his *Critical Perspectives* volume and appeared as a chapter in his 2005 interpretative history of ideas and arguments in the SPT.

According to Durbin (2005), David Noble's *America by Design* (1979) is "the best history of the role of engineering, and the engineering professional societies" in the United States, especially as continued in *Forces of Production* (1984, p. 141). Against the background provided by Noble's neo-Marxist analysis, Durbin assessed the views of three contributors to his 1991 edited volume in three philosophical themes. In theme one, Steven Goldman is said to draw out the philosophical implications of Noble's analysis in a long quotation from an unpublished report on engineering education that he did for the U.S. Office of Technology Assessment in 1992. Quoting Goldman with approval:

> The rationality of engineering involves volition, and is necessarily uncertain, transient and nonunique, and is explicitly valuational and arbitrary. Engineering also poses a distinctive set of metaphysical problems. The judgment that engineering solutions "work" is a social judgment, so that sociological factors must be brought directly into engineering epistemology and ontology.
>
> (Durbin, 1991, p. 140; Goldman in Durbin, 2005, p. 142)

Theme two looked at "engineer's philosophy" in the thought of Billy Vaughn Koen (from 1991) and Samuel Florman. Durbin found both lacking in recognition of the political dimensions of engineering. These engineers—the former emphasizing method, the latter social context—were complemented by reference to a 1927 work by German engineer Friedrich Dessauer and his metaphysics of technical invention viewed from an eccentric conflation of Plato and Kant.

Theme three considered the theory of engineering as applied science—the position of Mario Bunge. Here the primary protagonists were the philosopher of science Ronald Layman and Spanish philosopher Ana Cueva Badallo. Using a sophisticated interpretation of "applied," Layman defended a version of engineering as applied science. Cueva Badallo, while sympathetic to Bunge, argued for a missing appreciation of how the engineering sciences are influenced by societal interests and imperatives.

In accord with the theme of his book—to identify the core controversies that have animated SPT, especially in regard to the potential social utility of philosophy—Durbin concluded with an awkward effort to categorize the philosophers discussed in terms of a left/right political spectrum. He continued this effort in three other engineering-focused chapters analyzing debates about computers, bioengineering, and biotechnology.

Durbin's later contribution to *P&E* revisited previous analyses to provide a cleaner picture of his views using what he called the four facets of a metaphorical diamond in the philosophy of engineering. These are (1) relations between engineering and science, (2) metaphysical criticisms of engineering, (3) pragmatist criticisms and attempts to reform engineering, and (4) more radical political criticisms of engineering. These four aspects are complemented with a brief criticism of the adequacy of social science and humanities components in engineering education curricula that references arguments by the German philosopher Günter Ropohl, another contributor to his 1991 collection.

In all three of his essays on the philosophy of engineering, Durbin repeatedly called attention to the thought of philosopher Steven Goldman, founding director of the STS Program at Lehigh University. Goldman's contribution to *Critical Perspectives*, "The Social Captivity of Engineering," was actually fifth in a series (1984, 1988, 1989, 1990). In conjunction with two publications that followed (2004 and 2018)—after a detour into business management theory and the history and philosophy of science—Goldman's work effectively constitutes a genuinely comprehensive philosophy of engineering.

With a broad awareness of engineering in its technical, educational, historical, sociological, popular culture, and management dimensions, Goldman locates engineering in Western culture as a unique form of knowledge and practice that is captive to society both intellectually and through its subordination to corporate management. Intellectual captivity is the result of a philosophical dominance of science that is conceived as a receptive knowing of reality in reliable (if not certain) terms, which uses engineering only for instrument construction and then makes the knowledge produced available for engineering use. With regard to management captivity:

> The definition of engineering problems, as well as what will count as acceptable solutions to them, explicitly depends on highly contingent value judgements that are external to the technical expertise engineers command. These value judgements derive from the projected economic, social and/or political consequences of the implementation of solutions to engineering problems. The assessment of these consequences in turn reflects the fact that engineering practice always takes place within highly specific, commercial and/or political action contexts. Thus engineers, in order to function as engineers, must have a boss, or at least a client.
>
> (Goldman, 2004, p. 166)

For Goldman, as for Li Bocong (although with other conceptual differences), engineering is necessarily in bed with all sorts of nonengineer actors.

Reaching into the heart of design, Goldman describes how this archetypical engineering action depends on what he calls the "Principle of Insufficient Reason." For any problem set before engineers by commercial or political power, there is never one right solution. The scientific knowledge that engineers exploit is never quite what is needed, either because it is too abstract or because there are multiple ways to exploit it. Absent sufficient reasons for making a design just so, engineers must practice a decisionist leap that, with luck, will satisfy a boss or client. Engineering is deeply volitional in its dependency on external will and in its complicity in willing.

One implication of Goldman's philosophy of engineering is to see engineering as a kind of acting. This in turn suggests the possibility of connections between engineering and the philosophy of action, which it seldom receives, and politics. Indeed, for Goldman, one upshot of his analysis is to reveal "a strong similarity between the engineering design process with its trade-off choices, and the process by which social institutions are created and public policies formulated" (Goldman, 2018, p. 16).

All authors in Durbin (1991) were nonexclusive contributors to the emerging philosophy of engineering discourse. It is significant that two of these were German philosophers (Hans Lenk and Günter Ropohl) who had long been involved with the Verein Deutscher Ingenieure (VDI or Association of German Engineers), which had since the 1950s hosted more serious collaborations with philosophers than any other professional engineering organization. Among Americans, however, there are two others who deserve more than passing mention: Edwin T. Layton, Jr., and Taft H. Broome, Jr.

Layton was a historian of technology of depth and philosophical acuity. In the early 1970s, a monograph on the engineering profession (Layton, 1971a) and two articles on engineering knowledge in *T&C* (Layton, 1971b and 1974) set the stage for what Eda Kranakis calls "an autonomous intellectual history of technology rooted in a sociological and historical understanding of engineering as a profession" (Kranakis, 2010, p. 549). More clearly than others before him, Layton called attention to the distinctive features of the engineering sciences. In a long memorial obituary in *T&C*, Kranakis went on to observe:

> His best work is incomparable in the degree to which profound ideas, powerful analytical frameworks, and impressive breadth of research and knowledge are united in concise yet engaging and readable prose. "Technology as Knowledge" is only ten pages, but it opened up a new world of thought about technology that still continues to shape research.
>
> (p. 557)

Further articles from this time period (1976 and 1977) should have been included in the 1986 second edition of *The Revolt of the Engineers*. His chapter on "A Historical Definition of Engineering" (in Durbin, 1991) was a further contribution to his project in the intellectual history of engineering thought and practice.

Broome, as already noted, has been a leading proponent of the need for a philosophy of engineering, and perhaps no one from the engineering side has done more reaching across the aisle to philosophy. As a professor of mechanical engineering, he began thinking of ethics at the university (1978) and the responsibilities of engineers for hazardous technologies (1985a and 1987), considered the problematics of professional commitment to public interests (1986 and 1989), and argued for more imaginative reflection in engineering ethics (1990) as well as for the practice of philosophy-engineering interdisciplinarity (in Durbin, 1991). Like Layton, he tackled the epistemology of engineering in relation to science (1985b) and the metaphysics of engineering (2010). Additionally, he has imaginatively drawn on the thought of Joseph Campbell to use the archetype of the hero's quest to describe the engineering life (1996 and 1997) and against such a backdrop has creatively and influentially reflected on his own experience (1999).

Two other Americans not included in Durbin (1991) deserve mention. One is philosopher Michael Davis. No one has done more serious philosophical work digging into the conceptual intricacies of engineering as a profession and engineering ethics (see, e.g., Davis, 1998, 2002, and 2005).

The other is aeronautical engineer Byron Newberry, who has since the early 2000s slowly published a significant body of carefully thought-out and crafted articles on ethical, epistemological, and metaphysical aspects of engineering (see, e.g., Newberry, 2004, 2007, 2013, and 2015). Newberry has also rendered valuable professional service to fPET.

ENGINEERING IN WORDS

Still another contributor to Durbin's seminal collection considered the origins and meaning of "engineer" and "engineering" in English (Mitcham, 1991; material subsequently integrated into Mitcham, 1994). Briefly, American usage assumes a rough distinction between engineer (more intellectual) and technician (more manual), white-collar versus blue-collar worker but can also subsume engineering into technology (as when institutes of technology teach engineering or technological projects require engineers to work under corporate owners, managers, politicians, and others). Engineers may have less cultural capital than scientists, but they possess more social capital and command higher salaries than technicians. After serving engineering time in corporations, engineers may transition to managers or even CEOs, technicians never. Indeed, engineering is increasingly seen as a pathway to becoming an innovative entrepreneur (as with Jeff Bezos, Sergey Brin, Tim Cook), and alliances are increasingly made between engineering and business schools in order to assist engineers in jumping on this bandwagon to success.

The Medieval Latin *ingeniator*, however, constructed and sometimes operated *ingenia* (catapults and other "war engines"). Samuel Johnson's *Dictionary of the English Language* (1755) defined engineers as those "who [direct] the artillery of an army" and Noah Webster's *American Dictionary of the English Language* (1828) described them as persons "skilled in mathematics and mechanics, who [form] plans of works for offense or defense, and [mark] out the ground for fortifications." Today military engineers, combining scientifically honed analytical skills and willful action, become generals.

The shift from Johnson to Webster identifies engineers less with manual operators (as in steam engine and locomotive engineers) and more with those who plan or think (in special but unspecified ways) before they act. This picks up on a supplementary Latin connotation by way of the Old French *engignier*, as one who contrives to make things work; the fifteenth-century English *yngynore* plots and lays snares. Puritan theologian Richard Sibbes (1577–1635) refers to Satan as a "great engineer." Satan is, of course, the archetypical image of willfulness.

Equating engineering with architecture (Latin *architectus*, Greek αρχιτέκτων, *archi-* "leading" plus *tekton* "artisan" or "builder") is questionable, if for no other reason than obvious contemporary tensions between the two professions. Architects are more concerned with the aesthetics of a building than architectural engineers who focus on decreasing construction inputs while making sure buildings will not collapse. (For more on the relationship, see Michael Davis in *P&E*.)

A British builder of public works (John Smeaton, 1724–1792) was the first in English to mark himself off as a "civil" (rather than military) engineer. (There are earlier cognates in Italian and French.) Civil engineers worked for governments to design and construct the physical infrastructure (roads, canals, water supply and sanitation systems, railroads) demanded by an emerging industrial capitalism and were rapidly enrolled into its intellectual infrastructure. When in 1828 Smeaton's heirs in the Institution of Civil Engineers (ICE) defined themselves for King George IV, they said they practiced "the art of directing the great sources of power in nature for the use and convenience of man."

Two hundred years later, emphasizing science over art and dropping "convenience," engineering was defined, in the words of *Webster's New International Dictionary* (1959) and the *McGraw-Hill Dictionary of Scientific and*

Technical Terms (first edition, 1974), as "the science by which properties of matter and the sources of energy [power (McGraw-Hill)] in nature are made useful to [humans] in structures, machines, and products." In what was from the late 1950s through the 1970s the most widely used American textbook of introduction to engineering, Ralph J. Smith (1956), retrieving "art," gave his version as "engineering is the art of applying science to the optimum conversion of natural resources to the benefit of man" (from the fourth edition, Smith et al., 1983, p. 9); additionally, "the conception and design of a structure, device, or system to meet specified conditions in an optimum manner is engineering" (p. 10). Further situating engineers in a spectrum of technical actors, Smith wrote (in a passage reflecting the gender bias of the time):

> The engineer is a man of ideas and . . . action [who] develops mental skills but seldom has the opportunity to develop manual skills. In concentrating on the application of science he can obtain only a limited knowledge of science itself. . . . The primary objective of the *scientist* is "to know," to discover new facts, develop new theories, and learn new truths about the *natural* world without concern for the practical application of new knowledge. . . . The engineer is concerned with the *man-made* world [and is responsible] for designing and planning research programs, development projects, industrial plants, production procedures, construction methods, sales programs, operation and maintenance procedures and structures, machines, circuits, and processes. . . . The *technician* usually specializes in one aspect of engineering, becoming a draftsman, a cost estimator, a time-study specialist, an equipment salesman, a trouble shooter on industrial controls, an inspector on technical apparatus, or an operator of complex test equipment. . . . [The] technician occupies a position intermediate between the engineer and the skilled *craftsman*. The craftsman, such as the electrician, machinist, welder, patternmaker, instrument-maker, and model-maker, uses his hands more than his head, tools more than instruments, and mathematics and science rarely.
>
> (Third edition, pp. 210–211)

The function of engineering and its multiple applications are developed in an educational curriculum that includes some natural science and mathematics, the so-called applied or engineering sciences (e.g., mechanics, strength of materials, thermodynamics, electronics), and design practice differentiated by what Newberry (2015) has described as "thing adjectives" (civil, mechanical, electrical) and "issue adjectives" (environmental, biomedical, sustainability). Although Smith's book, which was addressed to first-year engineering students, is rhetorically dated, sometimes earlier works can still be helpful in revealing the way things still are.

The motivation behind Smith has been taken up for the general audience by a number of popular apologists for engineers and engineering, of which two of the most prominent have been civil engineers Samuel Florman and

Henry Petroski. Florman (1976, 1981, 1987, and 1996) has been stronger in engagements with critics and philosophers of technology. Petroski (1985, 1994, 1995, 1996, 1997, 2003, 2004, 2006, 2010, and 2012) is better on the problematics and achievements of engineering design. This impressive and philosophically useful body of work is complemented by a number of other authors, such as physicist Sunny Auyang (*Engineering—An Endless Frontier*, 2004), civil engineer David Blockley (2012 and 2014, two volumes in the Oxford "Very Short Introductions" series), and biomedical engineer Guru Madhavan (*Applied Minds: How Engineers Think*, 2015; *Think Like an Engineer: Inside the Minds That Are Changing Our Lives*, 2016). Philosopher and head of policy at the Royal Society Natasha McCarthy's *Engineering: A Beginner's Guide* (2009) is another addition to this literature.

One paradoxical fact about the English words "engineer" and "engineering" is that since the period in which Smith wrote, the terms have declined in usage. A Google n-gram for "engineer" shows its peak usage to have been the early 1940s. For "engineering" the peak occurred in the 1980s. It is not yet clear whether American political efforts since the early 2000s to promote science, technology, engineering, and mathematics (STEM) as the primary educational fields for government support have halted or reversed this trend. Also of interest is that an n-gram on "philosophy of engineering" reveals an unusually high occurrence in the 1960s and a dramatic drop off to a fluctuating plateau from the 1980s into the early 2000s.

Words have consequences. This book is written in English, a colonizing language that is undergoing what Rosalind Williams has attributed to engineering: expansive disintegration. English is becoming at once a global language and differentiated by its geographically dispersed speakers. Any philosophy of English-speaking engineering must acknowledge its potentially parochial, imperialistic, and yet diversifying character. As almost any historian will qualify it (e.g., Friedel, 2007, especially chapter 10), engineering in England and America is not precisely the same as engineering in Italy, France, Germany—or in any other country that has taken it up.

Still, history reveals some commonalities. In a review of literature on the rise of engineering as a profession in Europe across the long eighteenth century, Hélène Vrin and Irina Gouzévitch (2011) proposed that European engineering can be differentially characterized using a common template relating to the state, to enterprise, and to systems of knowledge production and transmission. "The identity of the engineer was constituted within the force field created by these various authorities, specifically combined in different national contexts" (p. 159). States enrolled engineers by formal (French) or informal (British) means. Corporations conveyed and withheld capital in whatever ways were most effective in situ for profiting from engineering services. A diversity of guilds, societies, colleges, academies, or some mix of these and other institutions could serve as vehicles for engineering knowledge production and education.

English-speaking engineering should not be misunderstood as more universal than it is, and yet within its particularity there are surely more than particularities. A hypothesis here is that one of the universal particulars is a second-order expansive disintegration: the expansion of engineering as a dynamic disintegrating of premodern ways of making and building, of living and thinking that it is incumbent on philosophy to address.

PHILOSOPHY AND ENGINEERING

What *is* engineering? What is the *meaning* of engineering—that is, how is engineering related to other aspects of the world? These two basic philosophical questions raise issues that cut across the classic branches of philosophy: logic, epistemology, ontology, ethics, political philosophy, aesthetics. The essays here touch on some of these perspectives more than others. In fact, one way to distinguish different regionalizations of philosophy—such as the philosophy of engineering, science, or religion—is to observe differences in the relative weights of different branches. Ethics clearly weighs more heavily in this set of steps toward philosophy of engineering than it might in the philosophy of science or of religion. But there is no reason the weighing should remain fixed.

Indeed, ethics is a lighter presence in the 27 chapters of *P&E*, where more than half deal with epistemological and/or metaphysical issues. Key nonexclusive and overlapping topics include engineering methodology and knowledge (Vermaas, Marc de Vries, Mark Coeckelbergh, Taft Broome, Albrecht Fritzsche, and Billy Vaughn Koen), systems engineering (Maarten Ottens and Joel Moses), engineering modeling (Zachery Pirtle and Russ Abbott), science–engineering relationships (Pitt and De Vries), and the engineering worldview (David Goldberg, Natasha McCarthy, Broome, and Gene Moriarty). Engineer Goldberg argues the value of philosophy to help reform engineering education under conditions now transformed by supply chain capitalism. McCarthy sees Ludwig Wittgenstein's philosophical work as influenced by an engineering background (on this topic, see also Coeckelbergh et al. eds., 2018). But in a serious challenge to the philosophy–engineering relationship, Vermaas's report on the Dual Natures Program to analyze technical functions identifies a fundamental problematic:

> Attempts to analyze technical functions in collaboration with engineers . . . proved to be difficult. Precisely the engineering criteria of effectiveness and

efficiency . . . prevent mutual profitable collaboration: philosophical conceptual sophistication becomes for engineers quite quickly unproductive hair-splitting, and engineering pragmatism may become for philosophy conceptual shallowness.

(Vermaas, 2010, p. 62)

Yet as Vermaas would no doubt affirm, engagement with engineers is crucial to the philosophy of engineering. The question is how and in what ways.

In emphasizing ethics, philosophical reflection on engineering can focus on interactions between engineers (professional ethics), engineers and society (social responsibility), and design practice (value-sensitive design)—incorporating issues of individual and group action. But engineering spills out from the profession into culture at large, through the kinds of technological mediations explored in post-phenomenological analyses, but in other ways as well. Additionally, perhaps influenced by the philosophy of science, philosophy of engineering work has been growing on a range of theoretical questions, especially methodological ones.

One repeatedly contentious issue is whether critical reflection should focus on engineerings or engineering (and associated technologies or technology). Philosophical reflection on technology (as a whole) that was characteristic of the 1950s during the classic period of European philosophy of technology has, since the 1980s, been regularly stigmatized as essentialist, with arguments for an empirical turn to social construction and particulars. Interestingly, insofar as engineering is conceived in terms of agency and methodology there seems to be less demand for limiting attention to particulars. After all, electrical engineering creates thousands of technologies, from dynamos and power tools to electric cars. The basic metaphysical question of the relationship between the one and many nevertheless remains for any philosophical engagement with engineering (and technology).

The philosophy of technology, by sidelining the essence question, has in effect left it for others. For Brian Arthur, work as an electrical engineer and research on complexity in the economics of high technology left him wanting to understand "the 'technology-ness' of technology." On the question of "the nature of technology and of innovation" he found little help from philosophers (Arthur, 2009, pp. 1 and 3). (The two central chapters of his search turn to engineering.)

Andrew Feenberg (e.g., 1991, 1999, and 2017) is a philosopher of technology who tries to mark out a path between instrumental and substantive conceptions of technology in what he terms a critical theory of technology that synthesizes the limited truths of the other two approaches. Although he emphasizes the agency of technology users over engineering agency, his approach can help inform an understanding of creators and the complexities of engineering and the engineered world.

For engineers themselves, a particularly salient philosophical issue concerns engineering education. This is the focus, for instance, of the work of engineer David Goldberg (see Goldberg and Sommerville, 2014), who has been substantially involved with both SPT and fPET. In *P&E* Goldberg (2010) argued for bringing philosophy into the engineering curriculum as part of a broader revitalizing transformation of engineering. In fact, questions of the role for philosophy in engineering education are a regular feature in commentary on engineering education, both from engineers (e.g., Bulleit et al., 2015) and nonengineers (e.g., Kaag and Bhatia, 2014). For a counter idea of engineering itself as the basis for a liberal education, see the special theme double issue of *Engineering Studies* (Downey, 2015).

There are multiple vectors through which philosophy has been brought into engineering education, as can be seen in the *Cambridge Handbook of Engineering Education Research* (Johri and Olds, 2015). Broadly, as the 35 chapters in *Cambridge Handbook* witness, engineering education has become increasingly engaged with the philosophy of education. The seven chapters in part one, "Engineering Thinking and Knowing," touch on issues related to epistemology, as do some of the chapters in part two, "Engineering Learning Mechanisms and Approaches." Parts three and four, "Pathways into Diversity and Inclusiveness" and "Engineering Education and Institutional Practices," engage ethical and political philosophical questions, even though they seldom thematize them as such. Part five, "Research Methods and Assessment," reveals how much engineers and others are trying to apply engineering methods to the study and design of teaching and learning programs. Part six, "Cross-Cutting Issues and Perspectives," incorporates contributions by a number of leading scholars in the engineering studies field, such as Jon Leydens, Brent Jesiek, Joseph Herkert, Gary Downey, and Nancy Nersessian. The conclusion is by John Heywood, who has for years been a leading proponent for philosophy in engineering education.

Using the *P&E* (2010) volume as a framing device has caused these fragments of an introduction to ignore much relevant work by others, especially engineers. The other fPET proceedings volumes lack the attention they deserve, in lieu of which there is only this final, inadequate, fragmentary homage. Outside of the fPET constellation, as well, there are any number of other engineers and philosophers at work trying to think philosophically about engineering. The fPET influence crops up, for instance, in Turkish engineer Zekâi Şen's *Philosophical, Logical and Scientific Perspectives in*

Engineering (2014), with a positive acknowledgment of the 2007 gathering at TU Delft in his argument for philosophy as one of the three primary components of engineering indicated by his title.

ENGINEERING IS EVERYWHERE

Although engineering may not be applied science, maybe philosophy of engineering could function as applied philosophy of science? Martin Heidegger's idea of modern science as theoretical engineering provides the basis for such a hypothesis.

The philosophy of science focuses on methods of knowledge production but when broadly construed also considers societal and cultural influences of the (scientific, practically reliable) knowledge produced. This second aspect of the philosophy of science is not well integrated into standard, professional Anglo-American discourse. Discussions of the cultural impact of the Copernican, Darwinian, and Einsteinian revolutions tend to be left to social or cultural historians. Should this remain true for the Engineering Epoch? Because engineering does not so much produce knowledge as material structures, artifacts, and physical process that merge into the fabric of material culture and social order; its cultural influences tend to be more hidden, even while its physical impact is ultimately more pronounced.

There is a conflict of two cultures associated with engineering which mirrors C. P. Snow's between scientific and literary intellectuals: between two types of makes, engineers and artisans. This has been manifested since the Industrial Revolution in worker protests and craft criticisms of engineered production. The tension took a leap in significance with philosophical criticisms of the atomic bomb and other existential threats inserted into history through advanced engineering prowess. For Günther Anders, the engineered incineration of the Holocaust and Hiroshima are categorically different from hand-to-hand combat. The dread and resistance that was an initial postnuclear response by such philosophers as Anders and Hans Jonas has largely dissipated in acceptance of and even enthusiasm for engineered existence—along with a vision of designing our lives. Even in the face of the climate change outcomes of synergies between engineering and capitalism, the limelight is directed more toward the sophisticated redescriptions of Bruno Latour than the enlightened doomsaying of Jean-Pierre Dupuy (2009).

But the pervasive presence of engineering in what Louis Dumont (1986) calls the ideology of "artificialism" is spread through more than apocalypticism. John Dewey (1922) compared education to civil engineering bridge building and argued for applying similar methods to redesigning educational institutions. Edward Bernays (1947) conceptualized public relations as "the engineering of consent." Jürgen Habermas's analysis of the technical colonization of the lifeworld and Langdon Winner's observation about the political impact of the engineered infrastructure of New York City are further indicators. According to Feenberg,

> Modernity claims to be a rational form of social life, and it is in fact based on rationally designed technical artifacts and institutions informed by rational technical disciplines. This is unprecedented. Throughout human history rationality has been confined to specific tasks rather than organizing society as a whole. Once noticed, the strangeness of our modern way of life inspires reflection.
>
> (Feenberg, 2017, p. ix)

Engineering is everywhere but not everywhere recognized: another task for the philosophy of engineering.

ASPIRATIONS

In his contribution to *P&E*, Richard Bowen argued for an aspirational engineering ethics. Aspirations are not always realized but nevertheless useful to keep in mind. For Socrates, realizing the unrealizability of his aspiration was its realization.

The essays included here aspire to step in multiple directions. One is to recognize and acknowledge the many steps others have been taking toward a philosophy of engineering and in the philosophy of engineering, and thereby to suggest further philosophical reflection along existing pathways. At the same time, the goal is not a philosophy of engineering for engineers or even a philosophy of engineering for philosophers—but a philosophy of engineering for all of us caught in an engineering-engineered world. Engineering as a way of life is not just what engineers do. It is today characteristic of all of us who are living in the dynamism that Henry Adams sensed in Paris in 1900: much more than a machine that one can stand beside. As Latour, among others, has argued, the machines, the technologies, and indeed the world have been sucked up into a vortex in which the dynamo may have been an initiating member but in which it, like all of us, is now caught.

In a project called "An Enquiry into Modes of Existence" Latour asks, "How do we compose a common world?" His response:

> Not so long ago, the project that would have seen modernization spread over the whole planet came up against unexpected opposition from the planet itself. Should we give up, deny the problem, or grit our teeth and hope for a miracle?

Alternatively we could inquire into what this modern project has meant so as to find out how it can be begun again on a new footing.

(http://modesofexistence.org; accessed March 26, 2019)

A major part of what this modern project has meant is embedded in engineering as a distinctly modern form of agency. In his contribution to *P&E* nuclear engineer Billy Vaughn Koen asked, "Quo Vadis, Humans?" Thinking about and against such a question cannot be left to engineers or outside the philosophy of engineering.

For an *ex post facto* overview of the essays in this volume something that might have been expected in an introduction, see the last essay, "In Conclusions."

Part One

Definition

(after Lydia Davis)

By engineered I just mean a construct that's a little different from reality but real and really hard to understand.

Essay 1

Science, Technology, Engineering, and the Military

Philosophical and ethical issues associated with relationships among science, technology, engineering, and the military have been largely ignored in the contemporary professional philosophical community. The same is true in the engineering community. The philosophical community seems oblivious, perhaps reflecting a general tendency to shy away from practical engagements. The engineering community looks the other way, perhaps because so much of engineering employment is in one form or another related to the military.

OBSERVATIONS FROM HISTORY

Unlike philosophers, historians and social scientists have paid considerable attention to relationships between technology and warfare. These relationships make history. Any review of significant post–World War II work would have to include the following:

- William H. McNeill, *The Pursuit of Power: Technology, Armed Force, and Society since A.D. 1000* (1982);
- Merritt Roe Smith, ed., *Military Enterprise and Technological Change: Perspectives on the American Experience* (1985);
- Everett Mendelsohn, Merritt Roe Smith, and Peter Weingart, eds., *Science, Technology, and the Military* (1988);
- Martin van Creveld, *Technology and War: From 2000 BC to the Present* (1989); and
- Max Boot, *War Made New: Weapons, Warriors, and the Making of the Modern World* (2006).

Philosophical reflection can almost always benefit, at least at the beginning, by learning from what historians and social scientists have to say about a topic.

Something immediately obvious is that the term "engineering" is not in the title of any of these studies, nor is it common in their indices. Nevertheless, the activity is clearly assumed by the analyses. Why it should not be named as such is not clear and a point that invites linguistic philosophical consideration. Does absence of "engineering" simply reflect a general public failure to recognize the reality of our constructed world? Could it be an unconscious effort not to contaminate engineering, which since the 1700s has sought to separate from its military origins?

In this selective set of references, more attention is paid to the history and sociology of American military technology and engineering than of military engineering *tout court*. This is to be expected, given the post–World War II military dominance of the United States. However, as McNeill, Van Creveld, and Boot also illustrate, there do exist a number of general histories of technology and war, even though within them the United States plays the prominent role.

In all general histories, warfare is divided into different periods on the basis of contributions from engineering and technology; however, there are overlapping differences in the periods and technologies to which specific authors appeal. McNeill cites gunpowder and the engineered industrialization of warfare as turning points. Van Creveld distinguishes four basic ages of warfare: those of tools (from earliest times to 1500); of machines and therefore of mechanical engineering (1500–1830); of technological or engineered systems (1830–1945); and of engineering automation (1945 to the present). Boot distinguishes four revolutions in warfare: the gunpowder revolution, a first Industrial Revolution (powered by coal and steam and intimately involved with the development of engineering in the distinctly modern sense), a second Industrial Revolution (of oil and electricity and obviously dependent on the emergence of new forms of engineering), and an information revolution (again involved with the new engineering of computers). Although there are debates about the degree of technological determinism in military affairs, there is general consensus that engineered weapons are major influences on the character of warfare and that the influence is increasing as a result of advances in post–World War II engineering and technology (see Roland, 2009). Techno-engineered warfare is now an important determinant on the outcome of military conflicts, whether those conflicts are between states or state and non-state actors.

A preliminary effort to reengage philosophy in relevant critical reflection on engineering and technology in relation to military affairs would do well to proceed modestly. Philosophy may, for instance, contribute to existing discussions through conceptual clarification relevant to empirical questions. To a degree it may also provide frameworks for interdisciplinary reflection, promote the relating of relevant information and arguments, and offer criticism of selective theoretical assumptions. At some point, philosophy may even be able to advance substantive criticisms of its own—but it should do so with caution and only after due consideration. Philosophy seldom makes history.

What follows is no more than preparation for substantive philosophical engagement. It proceeds by sketching a general historico-philosophical background, then turns to a quick review of more immediate post–World War II discussions in the United States. Along the way it indicates some philosophical, often ethical, questions that deserve attention. Finally, it summarizes issues and questions that would be important especially to philosophical reflection on engineering.

HISTORICO-PHILOSOPHICAL BACKGROUND

As noted, warfare is commonly divided into historical periods on the basis of its distinctive technologies. One of the most well-identified periods in ancient military history is defined by use of the chariot (see Drews, 1993). Modern military history is often thought to have been ushered in with the use of gunpowder and the printing press (to disseminate nationalist propaganda); contemporary military history with the introduction of engineered weapons of mass destruction (WMDs), especially the atomic bomb, and the engineering of jet aircraft and rockets. Indicative of the importance of engineering in military affairs is the way WMDs undermined one of the supreme ethical-political achievements of the early modern period in Europe: the separation of military from civilian targets. WMDs turn nonmilitary populations into military targets in the manner of premodern military campaigns in which victors would slaughter or enslave whole populations of the vanquished. That civilian deaths are termed "collateral" damage or unintended is at best a weak qualification.

The engineering of the WMD known as a "neutron bomb" was designed to undermine the military/civilian distinction in an even more radical manner, by replacing the military/civilian distinction with one between living entities and physical infrastructure. The neutron bomb was designed to kill all persons (and many other living things) but leave infrastructure intact. The inventor, Samuel Cohen, argued this to be morally superior to killing people and destroying infrastructure, as survivors or victors could utilize the remaining material structures (see McFadden, 2010).

But the tables can be turned. The history of science (commonly periodized by the prominence of distinctive theories) and the history of technology (easily periodized by its materials and power resources) can also be distinguished into different eras on the basis of differential engagements with warfare. This is especially true in engineering or technology. Leo Strauss, for instance, divides modernity into three waves, initiated by three major political philosophers: Niccolò Machiavelli, Jean-Jacques Rousseau, and Friedrich Nietzsche (Strauss 1975). Although Strauss does not make the connection, each exhibits a distinctive philosophical attitude toward warfare.

In the first wave of modernity, modern science and modern engineering technology (as distinguished from premodern science and traditional technics) came to be regularly imagined in terms of violence and war. Machiavelli (the father of modern social science) defended his positivism with the metaphor of fortune as a woman who if unable to be seduced can nevertheless be ravished. According to Francis Bacon (one of the founders of modern technoscience), nature reveals her secrets more readily on the rack of experimentation than when left free and at peace and is subject to scientific conquest. According to René Descartes, human domination is to be asserted over nature through the new way of (technoscientific) thinking that he wishes to introduce. Descartes's contemporary, Thomas Hobbes, argued that the state of nature in human affairs is one of a "war of all against all." For more on this kind of imagery as a stimulus to technoscience, see Carolyn Merchant's feminist study *The Death of Nature* (1980).

Such images remain alive in contemporary references to scientific "wars" on polio, small pox, cancer, hunger, or HIV. This is true especially in the techno-engineering community most closely associated with the military, as has been analyzed by Carol Cohn (1990). Indeed, some scientists and science policy advisors have suggested that only fear of war can galvanize the public to contribute to science and engineering with sufficient generosity to create a world-class technoscientific establishment. Strong prima facie evidence exists that the most outstanding science is in fact associated with world-class military power and that attempts to fund science and technology through nonmilitary projects result in smaller percentages of the gross domestic product being devoted to research and development. As has been observed by Charles Boyle (1984), an engineering physicist at Trent Polytechnic in Nottingham,

> The military applications of science and technology form no mere fringe activity indulged in by a few atypical tinkerers and experimentalists. In financial and manpower terms, they are central to modern research and development.

Not just nuclear power but also drugs, pesticides, aircraft, radar, processed food, satellites, computers, transistors, lasers, and many other technologies have all been funded and developed initially by the military for military purposes—although often with collateral justifications that emphasize civilian spin-offs or applications.

Such arguments, however, call for conceptual clarifications and distinctions between research and development; between pure and applied or mission-oriented science; between science, engineering, and technology; and between military and civilian technoscience, in order to interpret and assess accurately a wealth of conflicting claims that can be marshaled from the available evidence. Such newly constructed terms as "technoscience" and "techno-engineering" deserve clarification as well. Could a better term be coined that more properly reflected the prominence of engineering over technology in "technoscience"? The epistemology and philosophy of science and technology might address the issue of the adequacy and accuracy of the science-as-war metaphor as well as possibilities for alternative forms of technoscience. The ethical legitimacy of using the threat or fear of war and the war metaphor to promote and fund technoscience, even for civilian peacetime benefit, is a further issue.

In the second wave of modernity, the Enlightenment elaboration of the science-as-war metaphor was restated as the theory that the warfare of human beings against each other could be replaced by a war of all (everyone) against nature. The basic social problem for Rousseau is how to bring peace out of the social situation that arises when individualistic humans gather in cities, where self-interested struggle and decadence predominate. Eschewing Rousseau's synthesizing general will, contemporary David Hume argued that human warfare is caused by material scarcity and a competition for resources. The best approach is for humans to unite through industry, using technoscience to remove scarcity and thus the cause of internecine conflict. Appeals to the need to discover "the moral equivalent of war" (from American pragmatist William James to U.S. president Jimmy Carter) are contemporary reformulations of such a theory.

By contrast, Plato's *Republic* and Lewis Mumford's *The Myth of the Machine* have advanced the thesis that warfare arose in association with the technical creation of "unnatural" or exceptional wealth and a resultant need to defend wealth against those who "naturally" desire to steal it. The first standing armies are associated with the transformation from hunting and gathering societies to domestic agriculture and emerging surpluses. Do such surpluses merely make possible a division of labor between agricultural workers and soldiers in a society already primed for war, or do they actually provoke the rise of warfare and the military class? What is the relationship between the ritualized fighting of preliterate tribes and so-called civilized warfare? Such questions point again toward philosophical issues, this time about human nature.

The third wave of modernity introduced another dimension to this discussion. Nietzsche criticizes the bourgeois domestication and commercialization of warfare, arguing there is something profound at work in struggle and conflict. This idea is further developed by Sigmund Freud, who postulated that human *Eros* or the drive for self-preservation is complemented by *Thanatos* or a death wish. Despair at the manifest aggressive utilizations of technology can occasionally be found in the reactions of scientists and engineers to the destructive applications of their research and development work. Rudolf Diesel, inventor of the diesel engine, committed suicide in 1913 in despair over the inability of his mechanical engineering to promote true human solidarity; his son, Eugene Diesel, under the experience of World War I, turned away from engineering and spent his life pointing out the social and cultural violence associated with technology.

A counter argument of Alfred Nobel, another engineer-inventor, and of military theorist Jan Block, is that increasingly advanced weapons make war so horrible it will become unthinkable and that modern weapons thus promote solidarity by a kind of technologically necessitated backlash against, or repression of, aggressive instincts. Although this theory was in some measure confirmed by the post–World War II nuclear peace, it could also be argued that a "return of the repressed" has been manifest in numerous proxy wars between smaller states and terrorism by nonstate actors—and perhaps even in the worldwide assault on the environment that has created acid rain, oceanic pollution, biodiversity loss, and climate change. Today disenchantment with the "side effects" or "unintended consequences" of the scientific and technological conquest of nature may be doing more to unite humanity than any positive vision of possibilities put forth by Enlightenment ideals.

Once again, philosophical examinations of relationships between human nature, aggression, and the motivations or intentions at the base of engineering and technology could add new depth to critical reflection on such phenomena. Some work in applied philosophy dealing with the conceptual and ethical dilemmas of nuclear weapons policies, in environmental ethics and philosophy and in emerging efforts in the philosophy of science policy, might make useful contributions as well.

AFTER WORLD WAR II IN THE UNITED STATES

Against such a general background, consider more immediate post–World War II science and technology policy debates in the United States. Before World War II, despite the existence of national arsenals and military research facilities, science and engineering were primarily funded by private enterprise (the DuPont chemical company and the work of Thomas Edison, for example), private foundations (mostly of a medical sort), or civilian agencies of the federal government (e.g., what are now the Department of Agriculture, the Department of Commerce, and the National Institutes of Health).

As a result of World War II mobilization of science and technology, the War Department—subsequently the Department of Defense—became a primary source of funding for technoscientific research.

To redress the balance, immediately after the war scientists and engineers themselves (especially in the person of Vannevar Bush, 1945) lobbied for the establishment of a federally funded, civilian-controlled foundation through which technoscientists, independent of the distortions introduced by military interests, might determine scientific funding priorities. The founding legislation for the National Science Foundation (NSF) nevertheless proclaims in its preface that it exists to promote not only "scientific progress" but also "national defense." To what extent has NSF been able to subordinate military interests to basic research? At the very least, NSF funding of science-technology has failed to approach the level of funding provided by the various agencies of the Department of Defense (through the Office of Naval Research and the Defense Advanced Research Projects Agency). Again, it would be useful to undertake conceptual analysis and clarification of relationships between engineering research and military interests in their political, sociological, epistemological, and ontological dimensions.

There are multiple factors that continue to tie technoscience and engineering to military interests. In one of the central contributions to a collaborative examination of *Technology in the Western Political Tradition* (Melzer, Weinberger, and Zinman, 1993), American philosopher Stanley Rosen argued simply that insofar as warfare is taken to be the defining feature of the human relationship with nature, "then it is impossible to restrict the development of *techne*." Referencing Thucydides's account of an exchange at Sparta between representatives of Corinth and Athens early in *The Peloponnesian Wars* (book I, chapter 71, section 3), Rosen notes how

> The Corinthians contrast the daring and innovative nature of the Athenians with the slow, cautious, and excessively traditional nature of the Lacedaimonians, and warn that the necessities of political change require technical innovation. It is necessary in politics "just as in *technē* always to master what comes next."
>
> (Rosen, 1993, p. 78)

In the modern period, military historian Max Boot argues that "most of the key inventions that changed the face of battle since the Middle Ages—the cannon, handgun, three-masted sailing ship, steam engine, machine gun, rifled breech-loader, telegraph, internal combustion engine, automobile, airplane, radio, microchip, laser, wireless telephone—were the products of individual inventors operating more or less on their own" (Boot, 2006, p. 457). Yet all these inventions and their quick military utilizations took place precisely in a cultural context in which the metaphor of a war between humans and nature holds sway and (following Hobbes) in which warfare is assumed to be the dominant interaction among humans, even as this interaction is argued to take pacified form in agonistic democracy.

The rationale for a high-tech bias in U.S. military policy is simply that a democracy is not capable of making the kinds of public sacrifices necessary to oppose in kind the large conventional military establishments of authoritarian regimes such as the former Soviet Union, which therefore had to be countered with more sophisticated weapons. A result of this policy was the Cold War technoscientific arms race and development of ever newer generations of ever more sophisticated weapons: generation after generation of chemical, nuclear, and biological weapons, missiles, detection and communication electronics, spy and interference electronics, lasers, precision-guided munitions or smart weapons, artificial intelligence, and more. The moral legitimacy of a policy that promotes and rationalizes such development, which has continued even after the end of the Cold War, in the name of continuing to defend democracy, should be open to ethical and political challenge. The readiness with which technoscience can be promoted and captured by such a policy may stimulate further questioning of science, engineering, and technology themselves. It would be informative as well to compare military spending on science and engineering in different countries.

The United States has in fact witnessed a number of debates among practicing technoscientists and engineers related to such questioning. Initially (1950s and 1960s), the focus was on the dubious moral justifications for chemical, nuclear, and biological WMDs. Subsequently (in the 1980s), debate arose regarding the Strategic Defense Initiative or "Star Wars." Themes that arose in such debates were not only the problematic nature of WMDs and associated deterrence strategies but also challenges to the political and economic distortions of a military-industrial complex and to demands for secrecy in technoscientific research. In one radical case, Norbert Wiener, the inventor of cybernetics and a pioneer of information control theory utilized by radar, came to a conclusion not unlike that of Archimedes (see Plutarch's "Life of Marcellus," xvii). He would no longer publish any scientific research that could be put to military use because it would inevitably be misused. For Wiener (1947), the ethics of technoscience required the withholding of information by technoscientists, not its free exchange. Since the early 2000s, however, even as technoscientific invention has continued apace and engineering innovation became the aim of all advanced economies, with continuing applications in the realms of warfare, this debate atrophied. It deserves to be revived.

MILITARY EMBEDS WITH PHILOSOPHY OF ENGINEERING

The modern profession of engineering originated in and then attempted to separate itself from the military. The previous preliminary review of relationships between science, engineering, technology, and the military suggests that the dis-embedding may have been less successful than is often assumed. It is thus important that any effort to promote a philosophy of engineering includes critical reflection on relationships with the military.

This critical reflection needs to include a spectrum of philosophical issues to be addressed by interdisciplinary collaboration not just among philosophers and engineers but also with historians, social scientists, public policy analysts, and others. To be considered would be

- conceptual distinctions between scientific, engineering, and technological research and practice in such forms as technoscience and techno-engineering;
- general analyses of the influence of the military on science-technology and the sociohistorical background of past and present debates on specific military-supported research projects and products;
- the kinds of arguments employed in the articulations of different positions; and
- the moral options open to persons on various sides of any issue.

In the course of exploring these four areas of reflection, the following more specific questions arise: Conceptually, what differences exist between mission-oriented (applied) and non-mission-oriented (pure) research and design? Between military and nonmilitary missions in research and engineering design? Between defensive and offensive military research? Between direct, indirect, and unintended consequences and problems? Does the scientist or engineer doing mission-oriented or military research and design have different political and moral responsibilities than one doing non-mission-oriented or nonmilitary research? Are there distinctions to be made among various kinds of scientific and technological research? What are the proper corporate and academic relationships to military research?

Historically and sociologically, what have been the imperatives associated with engineering research and design? To what extent have military needs always influenced the funding of technoscientific research and thus been associated with engineering advances? Are there differences between the military funding of university-based research, independently established government laboratories, and private-sector research institutions? How have scientists and engineers responded in specific situations? What kinds of general arguments have they made? How is the military engaged with science-technology policy in different countries? What have various moral philosophers, along with reflecting scientists and engineers, had to say about ethical issues in relation to military research and development? How does the political context influence different discussions of military technoscience and engineering? What are the various ways support for and protest against military research have been marshaled? How is psychological development related to moral analysis of the issues at hand?

Philosophically, what are the strengths and weaknesses of various ethical theories—utilitarian, deontological, natural law, virtue ethics, and more—in dealing with the moral issues of military research, development, and design? Are some ethical frameworks more often employed to justify and others to criticize military research? What is the relation of just war theory to military research? Are there ethical issues in this area that are not adequately addressed by traditional ethical theories? What is the proper role of secrecy in science? In technology? In engineering? What is the relationship between the moral arguments associated with public policy debates and those having to do with personal moral decisions?

In relation to practical action, does the professional ethics of engineers have anything special to say about military research, development, and design? Does the social status of the engineer alter moral responsibilities or obligations? What are the forms of action appropriate to different kinds of moral judgments of military research and design? How can individuals develop and promote their ethical understandings? Does ethical discussion really make any difference? Do military engineers acquire special legal responsibilities? What kinds of groups exist for the support of ethical responsibility among engineers doing research sponsored by the military? Does the political system within which decisions about research priorities are made—that is, in democratic versus totalitarian regimes—affect the moral responsibilities of the individual researcher?

CONCLUSION

During the early 2000s, a number of U.S. government officials called for focusing attention on knowledge and skills of science, technology, engineering, and mathematics as being crucial to economic competitiveness and public intelligence. The acronym STEM was coined to highlight their importance. Among those seeking work visas or to immigrate to the United States, the government sought to prioritize persons with STEM knowledge or skills. Representatives of the NSF, the National Academies, and the Department of Education all argued for the enhanced promotion of STEM education.

But insofar as STEM education is understood to focus on more than economic competitiveness—and to be valuable for public intelligence—the acronym might well be reinterpreted as referencing relationships between science, technology, engineering, and the military. At a more concrete level than mathematics, military interests and defense spending drive and influence the pursuit and practice of science, technology, and engineering. Increased public awareness of this fact and its implications are necessary to contribute to democratic intelligence and the philosophical understanding of engineering.

Such a need is only heightened by the development of new generations of engineered weapons, weapons that promise fundamental transformations in the character of warfare: cyber-weapons, remote-controlled weapons, robot weapons, and nano-weapons (CRRNs, for short). CRRNs are in the process of altering not just the practice of warfare but also civilian life. Cyber-warfare, remote-controlled drones, robot soldiers, and nano-arms will not just change the way humans engage in warfare; they will inevitably find civilian applications. CRRNs for military defense and civilian applications present new and special challenges related to the meaning of engineering in the contemporary world—and thus reiterate the requirement that philosophical reflection on relationships with the military be part of any effort to promote critical philosophical assessment of engineering.

ADDENDUM:
REENGINEERING WARFARE

A substantive contribution to the program proposed here is present in *Engineering and War: Militarism, Ethics, Institutions, Alternatives* (2014) by an interdisciplinary team of historian Ethan Blue, philosopher Michael Levine, and science and technology studies scholar Dean Nieusma. Written quite independently, it is another testimony to an emerging awareness of the important questions centering around engineering and the military.

With more empirical detail than the previous essay, *Engineering and War* begins by documenting "the close alignment of engineering and warfare" and a "surprising silence" in this regard in engineering ethics discourse (p. 12). It further provides a review of recent literature working to address this lacuna, especially Donna Riley (2008) and Richard Bowen (2009). The stage setting is followed by a review of just war theory ethics with an analysis of how it might implicate engineers and engineering integrity. The center of gravity of the book, however, is a social history of the military-industrial complex as it emerged in North America, providing a much richer account than the simple note "After World War II in the United States." Especially valuable is an account of the post–Cold War trajectory toward engineering "soft kill" weapons or weapons that disable without crossing the "death barrier."

> Older military technologies concentrated on exercising force through traumatic kinetic force: Bodies would be torn apart, buildings would be crushed.... The newer means of waging war . . . may have a different relationship to bodies. Weapons can focus on the molecular structure of bodies, the function of electricity in musculature, the senses of smell and of hearing and of taste. Drawing on a number of desired medical or physiological effects—such as pain, lethargy, or disorientation—contemporary "non-lethal" weapons are essentially "reverse engineered" to produce those disabling effects on individuals or groups.
>
> (p. 61)

A concluding chapter considers implications for the engineering community and how it has or might respond. It especially highlights how

> To the extent that soldiers' combat duties are being assumed by robotic, cybernetic, and other high-tech devices, it is engineers and engineering technicians who—without having to be physically located at or even near the site of the killing—are becoming the new frontline agents of warfare. It is engineers and engineering technicians who are creating the drones, instructing the robots, developing weapons guidance and delivery systems, and the like. Hence, in an all-cybernetic war, the opposing armies are not artificial intelligence networks, robots, drones, etc., though it is make to look like that in science-fiction films. Instead, it is the engineers and technicians who invent, perfect, maintain, and operate them. While engineers have always been behind the development of technologies of warfare, . . . newer technologies remove more and more of the "middlemen" in warfare. Engineers are increasingly becoming direct participants in acts of warfare.
>
> (p. 77)

Essay 2

Ethics into Design

Ethics constitutes an attempt to articulate and reflect on guidelines for human activity and conduct. Logic is the attempt to articulate and reflect on guidelines for human thought. Both ethics and logic further develop theories about the most general principles and foundations of their respective guidelines. But what is it that articulates and reflects on guidelines for that intermediary between thought and action called design?[1]

As an English word, "design" is a modern derivate of the Latin *designare*, to mark or point out, delineate, contrive, by way of the French *désigner*, to indicate or designate, and can be defined as planning for action or miniature action.[2] It is remarkable, however, that neither Greek nor Latin contains any word that exactly corresponds to the modern word "design." The closest Greek comes to a word for "design" in the modern sense is perhaps *hupographein*—to write out. Much more common are simply *ennoein* (*en*, in + *noein*, to think) and *dianoein* (*dia*, through + *noein*, to think).

For the Greeks, human conduct can be ordered toward the production of material artifacts or nonmaterial goods, through ποίησις (*poiesis*, making), activity with an extrinsic end, or it can be taken up with πρᾶξις (*praxis*, doing), activity with an intrinsic end. The pursuit of what is fitting in the domain of making is discovered through τέχνη (*techne*); in the domain of doing, through φρόνησῐς (*phronesis*). In a narrow sense *phronesis* is only one among many virtues; more broadly, it is the foundation of all virtue and thus coextensive with ethics.

Beyond the Greeks, planned making or doing—as distinct from simply intending to act, consideration of the ideals reflected or intended by different makings and doings, or the development of skills (*technai*) through practice—involves the systematic anticipatory analysis of human action. With regard to making, especially, such systematic anticipatory analysis entails miniature or modeled trial-and-error or experimental activity. In the modern context, this planning for making or rationally anticipatory miniature making, which was once severely restricted by both traditional frameworks and methodological limitations, has become the well-developed and dynamic activity of designing or design. The latter term can refer as well to the formal characteristics of the articulated plan or the static composition of the product brought forth by the scaled-up process that emerges from what has also been called "active contemplation."[3] An alternative might be "contemplative (theoretical) action."

The modern attempt to reflect on designing or design has engendered primarily studies of the social or aesthetic quality of designed products and analyses of the logic or methodology of design processes. The thesis here is that both aesthetic criticism and the logic of design must be complemented by the introduction of ethics into design studies, in order to contribute to the development of a genuinely comprehensive philosophy of design.

ON THE EXISTENCE OF DESIGN

But if it is so important, why does ethics not already exist in design? The simple answer is that ethics was not needed within design until quite recently because until quite recently the activity known as designing did not play a prominent role in human affairs.

The most fundamental question regarding design—an ontological question, as it were—is this: why is there design at all and not just nondesign? Certainly it is historically obvious that design has not always been and therefore need not necessarily be. In nature, for instance, the design process does not occur. According to modern science, nature brings forth by blind determination or random change. Hence there arise debates about whether human beings as designers are part of nature, and whether the science of nature is able to be unified with the human sciences and humanities, not to mention theology. (The idea that God created the world "by design" is a unique conflation of Greek rationalism and Judeo-Christian-Islamic revelation.) Even on an Aristotelean account, to be "by *phusis*, nature" and "by *nomos*, convention" (if not design) constitute two distinct ways of being.

To be "by design" in any possible (weak) premodern sense denotes no more than affinity with that unique human reality *nomos*, convention or custom, and *nous*, mind. Convention reified or in physical form is labeled artifice, that which has form not from within itself, like a rock or a tree, but from another, like a statue or a bed (see *Physics* II, 1). Prior to the development of design as rationalizing miniature making one could speak only of mental intention or static composition, thought or final material product, not any special or unique physical activity. The activity was simply making.

Vernacular human activity, especially vernacular making and building, insofar as it is restricted to traditional crafts, proceeds by intention but not necessarily by or through any systematic anticipatory analysis and modeling. Plato's shuttle maker looks to the form or idea of a shuttle and thereby does not have to design it (*Cratylus* 389a). Indeed, many central societal conventions and artifacts (e.g., traditional village customs and architectures) are, although human-made, not even the direct result of human intention.[4] (In the vernacular world, the "designing" actor is one who proceeds with schemes, deviously, improperly.) What is most characteristic of nonmodern making activities are trial-and-error full-scale fabrication or construction, intuition and apprenticeship, and techniques developed out of and guided by unarticulated or nondiscursive traditions and procedures. Reflection in relation to such making focuses more on the symbolic character of results than on the processes and methods of, say, efficiency in operation or production. To speak of design in crafts is to refer to something which is not yet, which occurs largely in unconscious or provisional forms—that is to say, design without design. Yves Deforge in one attempt to write about such "design before design" calls these phenomena "avatars of design."[5]

Design as a protoactivity is manifested originally in the arts in the form of sketches for paintings. The unfinished chambers of Egyptian tombs reveal that drawings sometimes preceded finished murals. But for Giorgio Vasari (1511–1574) and his contemporaries, *disegno* or drawing and preparatory sketches are the necessary foundation of painting. The need for arguments in defense of this position reveals its special historical character. And there are at least two observations that can be ventured about such anticipatory activity in the artistic realm. First, it exhibits a continuity with that to which it leads. The tomb drawings are even the same size as the final mural that will follow; the Renaissance sketches develop skills that are repeated on canvas or wall. Second, conspicuous by its absence is any quantitative or input-output analysis. At the time of the Renaissance, however, design also appears in a first distinctly modern form as the geometric construction of perspective, as a correlate of modern scientific naturalism, and as the precursor to engineering drawing.[6]

The distinctive feature of modern science as an activity rather than as a body of knowledge is experimental modeling. Through experimentation modern science constructs models of different natural processes, and by means of arguments based on a principle of proportionality uses them to reason from known or observed cause-effect relationships to unknown causes of known effects. Galileo Galilei (1564–1642) was the pioneer of such modeling in physics (falling bodies), which has since been extended to chemistry (atomic models), biology (models of DNA), and even human psychology (computer modeling of cognitive processes).[7]

Modern scientific experimentation constructs models of what (it thinks) already exists, to expand knowing. The activity of design constructs models of what (it thinks) might be, to extend making. For science, models take in or receive and simplify complex phenomena, thereby disclosing order. For modern technology, or scientifically refined making and using in all their diversity, models project complex possibilities in realistic form, thus determining or enabling the control of power. When this projective modeling exhibits a conceptual break with the final result toward which it is pointed, a break to be bridged by analogy, it takes on its distinctly modern character.

Design models in engineering can, for instance, be "true" models, although more commonly they are merely "adequate" or even "distorted" and "dissimilar." As one engineer has put it, "A distorted model is [one] in which some design condition is violated sufficiently to require correction of the prediction equation. Under certain conditions, particularly where flow of fluids is involved, it is impracticable, if not impossible, to satisfy all of the design conditions [under a common scale]."[8] Likewise, "dissimilar models are models which bear no apparent resemblance to the prototype but which, through suitable analogies, give accurate predictions of the behavior of the prototype."[9] (This should probably be "sufficiently accurate.") Another engineer distinguishes between models that are "totally direct," "totally indirect," "combination," "visual," and "competitive" with each being suited to test different aspects of a new idea.[10] All such models can be manifest in drawings, block diagrams, network schematics, mathematics, physical materials, and related systems of representation.[11]

Receptive, scientific modeling *embodies* knowledge; with regard to knowledge, embodiment necessarily entails simplifying *concepts*. Projective, technological modeling *disembodies* action; with regard to action, disembodiment that leaves *things* out, idealizes them. The former materializes, the latter dematerializes. The paradoxical aim of projective, dematerialized or idealized modeling is not so much explanation as practical leverage or effectiveness. The present and its desires are cast with great force and power into the future.

Because of the complexity of variables, theory alone cannot be used to deduce, for instance, the shape of an airfoil, or to determine the optimum spatial arrangements of elements within a given structure. Engineers have to "figure out" such things by simulation, often employing a variety of models. So they construct a miniature, model airfoil and test it in a wind tunnel (now in a computer program); by means of such activities they are testing not some illustrated theory but a represented artifact.[12] For structures, engineers create scaled-down floor plans or two-dimensional facades in order to play with alternative arrangements of shapes by means of sketched geometries or manipulated cutouts. In each case the model or mock-up constitutes a temporary reduction to be eventually scaled up in the production not of knowledge but of objects. Design uses created microscale cause-effect relations rendered in models to engineer known or creatable macroscale causes into the production of desirable or desired macroscale effects.[13]

ON THE SOCIAL DIMENSIONS OF MODERN DESIGN

As has been noted, for example by José Ortega y Gasset in his *Meditación de la técnica* (from lectures first delivered in 1933), traditional technics includes both the "invention of a plan of action"—which is not the same as a planning process—and the "execution of this plan."[14] Traditionally, both the formal-final and efficient causes remained within the mind and hand of an artisan. It is the modern separation of mental and manual, and the coordinate creation of inventor-engineer and worker, that grounds the original character of modern design. The two new categories of designing and working are not just thinking and making separated. Thinking and making are too inextricably conjoined in traditional craft for such a simple disjunction,[15] which is discerned only by critical abstraction. In the separation of intending and making are created instead an embodied, active form of intending (design) and a nonreflective but methodological form of making (labor).

This separation of formerly unified aspects of human active experience is further coordinated with the becoming autonomous of a whole range of elements in human culture. Religion and politics are to be independent, likewise with art and religion and politics and science and education; all, along with economics as a kind of paradigm, become what Karl Polanyi terms "disembedded" from social life as a whole.[16] This separating and becoming independent of previously interwoven dimensions of a way of life constitutes, for Jürgen Habermas, the essence of the modern project.[17]

The emergence of disembedded and autonomous design constitutes as well a movement from vernacular to professional design and has thus been variously defined by the two professions who claim it, engineers and artist-architects. The former emphasize the quantitative, analytic, but iterative character of a multiphase process that includes preparatory and evaluative moments. The latter presents design as embodied, poetic thinking. Louis Bucciarelli, from "an ethnographic perspective," has described engineering design as a social process,[18] whereas Richard Buchanan has argued for design as a kind of rhetoric. But what kind of social process? What form of rhetoric? What is to distinguish engineering and artistic design from the social process and rhetoric of politics? Whether engineering or architecture, accidentally reflecting social process or rhetoric, the defining activity is miniature making. For Bucciarelli this is found in a social process centering around distinct "object worlds"; for Buchanan it is a rhetoric of artifacts.

ON THE ETHICS OF DESIGNING

Possibility and contingency are the fundamental ground of ethics. On the one hand, in the absence of any recognition of alternative possibilities for some course of action, no ethical reflection is called for. On the other, if the course of action is strictly necessary, reflection can give rise only to theoretical explanation, not ethical judgment. One does not ask ethical questions of what cannot be or of what cannot be otherwise.

The historical discovery of design as systematic anticipatory analysis and modeling as a unique form of human action roughly contemporaneous with the rise of modern science and engineering uncovers a new way of being in the world. The most fundamental ethical question concerning design is this: to what extent is this new way of being in the world desirable or good?

It is now common to recognize that, as Langdon Winner has said, technologies are "forms of life,"[19] or as Buchanan has put it, "Design involves the vivid expression of competing ideas about social life."[20] But not only do different designs embody (implicitly or explicitly) distinct sociopolitical assumptions and visions of life, designing itself constitutes a new way of leading, or a leading into, different technological life worlds. Part of the unified newness of this way of leading into the techno-lifeworld, the activity or process of designing, can be indicated by noticing some difficulties or inadequacies of standard approaches to ethics in relation to it.

Consider, for example, what can be termed an ethics of correspondence, which judges action by the extent to which it is in harmony with or corresponds to what is already given by some preexisting order. Common forms of such an ethics of correspondence are found in appeals to tradition or to natural law. The attempt to judge the design act as lawful or unlawful in accord with the degree to which it harmonizes with and represents or opposes a tradition is contradicted by the core effort within design not to be guided by tradition, but to figure things out anew, to create new artifacts, to break with tradition, to innovate. Modern design becomes a new tradition precisely to the extent that it opposes tradition.

Perhaps, then, one should adopt a deontological approach and consider the intentions of the designer or the principles of the design act in terms of consistency and universalizability. Indeed, as something less than full-bodied action, designing might well be compared to having an intention. Although many of its particular maxims may be open to serious challenge, it is difficult to see how the design process as a whole should not be inherently universalizable. Criticisms of modern technological design often focus on the inherent consistency, the rightness and wrongness, of various design maxims. But without the design process as a whole, how could one possibly address the problems inherent in the designed techno-lifeworld?

Nevertheless, as making in miniature, design is something more than an intention. In however diminished a form, it is still physical activity. It is thus a busyness that, as such, does not encourage inner self-examination. Moreover, as physical activity, design is something that always has immediate physical consequences—even if they are, as it were, quite small, even minute. Its inner principle is the linking together of physical materials and energies in functional units to meet predetermined functional specifications, something to be worked out through models and testing. Design is inherently tipped toward action, is immanent activity, a proto-pragmatism.

Consider, then, an ethics of consequentialism, which would refer the moral character of action to the goodness or badness of its results. But the designing of an airfoil or a structure has no immediate socially significant consequences. How could one calculate costs and utilities except in the most indirect terms? Probably most such designing leads nowhere, since the majority of designs never serve as a basis for full-scale construction. Design is more like a self-contained game. Its full-scale consequences, whatever they may be, occur only at secondary or tertiary removes—once the design serves as a basis for construction. A consequentialist judgment of designing readily strikes any designer as an abstract, far-fetched focusing on remote results that an indefinite number of contingent variables may alter.

There are two further points that can be made about the difficulties of consequentialism. As Hannah Arendt has noted with regard to human action,[21] and as Hans Jonas has argued with regard to modern technology,[22] the remote consequences of activities are inherently difficult to predict. John Stuart Mill, anticipating such an argument, replies that the remote and unpredictable character of consequences can be mitigated by experience.[23] In more recent language, the difficulties of "act utilitarianism" can be met with "rule utilitarianism" grounded on common experience.[24] Human beings can learn that telling lies eventually has bad consequences most of the time. The problem with any appeal to experience in the case of modern design acts, however, is that insofar as designs are unique, their consequences are also continuously new. Principled change undermines the mitigating power of historical experience. (Could this account for the modern resistance to any reduction in the pace of technological change, and that continuously renewed optimism about design transformations that makes it so difficult to learn from failed experience and expectations?)

Yet the apparently diverse material products grounded in the new way of life defined by the principled pursuit of technological innovation through design do exhibit certain common features. Albert Borgmann has linked these together in what he terms the "device paradigm." Devices are to be contrasted with things. A thing, such as the fire-bearing hearth, entails bodily and social engagement. A device, such as a central heating unit, "procures mere warmth and disburdens us of all other elements." "Technological devices . . . have the function of procuring or making available a commodity such as warmth, transportation, or food . . . without burdening us in any way [by making them] commercially present, instantaneously, ubiquitously, safely, and easily."[25] The products of modern design are typically commodities that fit the device paradigm. Indeed, modern designing might even be described as "devising," the process of making present devices.

But devising and devices escape the reach of any full-bodied consequentialist criticism because of the apparently amorphous neutrality or ambiguity of commodities, of deontological restriction because of the apparently inherent morality of its intention merely to make available without presupposition, and of the ethics of correspondence because of their principled rejection of corresponding to anything. Devices are neutral commodities. How, in themselves, could they be considered lawful or unlawful, right or wrong, good or bad, since they are designed to be nothing but pure receptivity to any law, right, or good? The thermostat, the light switch, and the plastic bowl are simply available for use. These so-called neutral devices are through their neutrality the non-neutral harbingers of a new world.

But if neither traditional correspondence nor deontologism nor consequentialism has any immediate purchase on designing, how is one to address the problems manifest in the new techno-lifeworld?

TWO VERSIONS OF AN ETHICS IN DESIGN

Prescinding from any fundamental questioning of designing as a way of being in the world, it is still necessary to inquire about the presence of ethics in design. The modern systematic modeling of making—that is, design—has taken two distinct forms. One of these is technical, the other aesthetic. The former focuses on inner operational or functional relations within mechanical, chemical, electrical, and other artifacts and processes. The latter takes external appearance or composition as its concern. One evaluates its products in terms of an ideal of efficiency, striving with some minimal possible input of material and energy for a maximum (prespecified functional) output. The other seeks a formal concentration and depth of meaning.

To use less, engineers design increasingly complex but specialized objects devoid of decoration, although precisely because of their inner complexity the inner workings must be covered with some kind of ornamentation. To mean more, to become "charged and supercharged with meaning" (Ezra Pound), artists and architects render increasingly rich, ambiguous artifacts, textured and decorated in detail. In the modern capitalist context, the design of meaning almost necessarily implies the new profession of advertising.

Each design tradition also develops its own professional ethos, which constitutes an implicit ethics of design. In engineering there has been a stress upon subordination, if not obedience and sameness.[26] In the arts the commitment is to independence and difference. Each brings to the fore complementary aspects of the modern design experience: on the one hand, its authority and power: on the other, its revolution and independence. Extremes on both sides are reined in with appeals to responsibility.

The selective ethical responses to the problems summoned forth by the processes unleashed through modern design activity—from social disruption, dangerous machines, and oversold consumer products to crowded and polluted urban environments—further reflect these two traditions. One stresses the need for more efficiency and argues for pushing forward toward increasingly extensive and systematic expansions of design, from time-and-motion studies to operations research and human factors engineering. The other calls attention to anomie, alienation, over (or under) consumption, and cultural deterioration and calls for either a turn toward the arts and crafts, sexualized design, or the creation of postmodern bricolage. The problems of "bad design" are viewed as caused either by insufficient design or by too much and the wrong kind.[27]

One tradition thus promotes methodological and empirical studies of engineering design processes; the other develops broad interpretative studies of the aesthetic and cultural dimensions of artifacts.[28] Aesthetic sensitivity meets the engineering mentality in advertising, industrial design, and functionalism.[29] Engineering reaches out toward aesthetic criticisms with proposals for more socially conscious or holistic design programs.[30]

Both traditions depend on what may nevertheless be described as incomplete philosophical reflection. They uncritically seek either to export design methods across a whole spectrum of human activities or to import extraneous ideas into design. The proposal here is for the cultivated emergence of ethics within design as an effort to deepen the two traditions by moving from partial reflections and possible reforms to deeper understandings of the challenge of techno-lifeworld design and more comprehensive assessment of its problems.

NOTES TOWARD AN INNER ETHICS OF DESIGN

According to Aristotle, the study of ethics depends on the practice of ethics (*Nichomachean Ethics* I, 4; 1095b4–6). One cannot articulate and reflect on what one does not already have. Ethics cannot come from on high, as it were, to articulate guidelines for action. The attempt to cultivate ethics within design thus begins with the attempt to articulate and express the guidelines for that miniature action called designing such as they already exist. Only from here is it possible to move toward considerations of their adequacy, beginning perhaps with a recognition of special problems.

The fundamental ethical problem of design is created precisely by its principled separation from the inner and the outer worlds. It is not pure intention and part of an inner life, something that can be examined by means of self-reflection. Nor is it simply an overt action that readily calls for consequentialist evaluation. It is more like a game or play.

Indeed, in the premodern world, models functioned primarily as toys. Mayan toy carts and Alexandrian steam engines were never recast into the quotidian world as construction tools or industrial machines. With models one creates a provisionally self-contained or miniature world rather than thoughts that can be integrated into an inner life or actions that are part of everyday human affairs.

Models and their making thus easily take on a kind of independence, to constitute a phenomenon that demands evaluation on its own terms, whether technical or aesthetic. The inherent attractiveness of modern design activities lies not just in their potential utilitarian results but just as much in their technical beauties and beautiful techniques. Johan Huizinga, vulgarizing Friedrich Nietzsche and anticipating Jacques Derrida, speaks for the modern attempt to find new values in the midst of the destruction of the old when he describes play as segregated from all "the great categorical antitheses":

> Play lies outside the antithesis of wisdom and folly, and equally outside those of truth and falsehood, good and evil. . . . [I]t has no moral function. The valuations of vice and virtue do not apply here.[31]

The game, precisely because of what it is *qua* game, that is, a break from or setting aside of the world, asks not to be subject to the rules or judgments of the world. Children with dolls or with guns can behave in all sorts of ways that would not be acceptable were their toys real people or weapons. A game of cards has its own rules, which are all that must be obeyed in order to be a "moral" card player. Clay modeling needs only to keep the clay wet enough to manipulate but not so wet as to run; otherwise it is wholly without rules.

Precisely because of its independence from and potential opposition to traditional morality, ethical reflection from Plato to the Puritans has argued for circumscribing and delimiting the world of play. Play at work, for instance, limits production and causes accidents. Playful sex can degenerate into the promiscuous and pornographic.

Yet play need not be wholly rejected; it can also be delimited and preserved—perhaps in ways that maintain, even enhance, its very playfulness. Cut wholly free from any reference to the world, play can actually cease to be interesting. Pure play with words or numbers, as in *Finnegans Wake* or the higher reaches of mathematics, attracts fewer and fewer players and less and less of an audience. Under such situations it is appropriate to call for a revival of the relationship between play and life.

And insofar as play can be taken as a metaphor for design, this inner obligation that would preserve the activity from its own internal disintegration might be formulated as the following fundamental principle: "Remember the materials." "Return to real things." Do not let miniature making become so miniature that it ceases to reflect and engage the real world.

By way of attempting to elaborate on this suggestion, consider the following speculative observations:

1. The great temptation of any game is for it to become too self-contained, an activity of purely aesthetic pleasure or technical achievement. Insofar as all play becomes not a temporary separation from quotidian realities, but a pull away from life, it becomes subject to social criticism. The artist concerned only with form, the engineer concerned only with technical solutions—the pursuit of art for art's sake, engineering for the sake of engineering—can be challenged by more inclusive issues and social orders.
2. The human practice of designing simply as designing can be said to deepen the tendency inherent in all play by exhibiting a marked inclination to distance the designer from self-examination or social responsibility. Studies of the psychology and behavior of computer hackers dramatically confirm this point,[32] but it is hinted at as well by the ethos of each design tradition. The engineering tradition of obedience and the avant-garde tradition of independence in the arts are but two expressions of disjunctions, from self and community.
3. Designing, unlike more limited forms of play, constitutes a general pulling away from or bracketing of the world that can have immediate practical impact. The paradoxical strengths of the mathematization and modeling of modern design are that, more effectively than ever before, they separate from the world of experience and provide new levers for the technological manipulation of that world. Modern designing opens itself to being pulled

back into the world beyond anything that designers themselves might imagine, desire, or plan. Hence, again, there exists a fundamental obligation to remember the materials, return to real things, and not let miniature making become so miniature that it ceases to reflect and engage the world.

4. Perhaps nowhere is the challenge of remembering reality more important than in computer-aided design. Although tremendously powerful and attractive, computer-aided design is equally dangerous, precisely because even more than designing with pencil and paper against a background of practical experience with real-world artifacts, design with computers works in a rarefied medium with a facility that tends to deny the need for worldly experience. As Eugene Ferguson has argued,

To accomplish a design of any considerable complexity—a passenger elevator or a railroad locomotive or a large heat exchanger in an acid plant—requires a continuous stream of calculations, judgments, and compromises that should only be made by engineers experienced in the kind of system being designed. The "big" decisions obviously should be based on intimate, firsthand, internalized knowledge of elevators, locomotives, or heat exchangers.[33]

5. But just as obviously, in a society in which elevators, locomotives, and heat exchangers are increasingly run by computers, and children rather than playing with trains play with video game trains, it is difficult to cultivate an intimate, firsthand, internalized knowledge of material reality. Virtual experience is no substitute for physical experience. The problems of design are not isolated in design. They are part of, even at one with, the larger material world and culture as a whole. To return to real things is a challenge throughout the ways of life characteristic of postmodern society.

6. The real experience of struggling to return to real things taking ethics beyond fundamental principles into specific cases will be the basis for development of a *phronesis* of the techno-lifeworld.

The problems with design are not just technical or aesthetic but also ethical. Indeed, introducing ethics into design revels the deepest aspects of our difficulties. But the difficulties we face cannot begin to be addressed without clear-sightedness. To attempt to recognize them is itself to struggle for the right and the good.[34]

NOTES

1. For a different but related notion of the intermediary character of design, see Mills (1963), pp. 374–86.

2. Aspects of this definition are previously developed in Mitcham (1978), pp. 245–48; (1991), pp. 96ff.; and (1994a), pp. 220ff.

3. Buchanan (1989), pp. 98, 103.

4. See Hayek (1967). Hayek uses "design" in the weak sense as equivalent with intention.

5. Deforge (1990).

6. On the last point, see Booker (1963).

7. Key studies on the role of model construction in modern science can be found in Hesse (1966), Harre (1970), and Wallace (1984).

8. Murphy (1950), p. 61.

9. Murphy (1950), pp. 61–62.

10. Glegg (1981), pp. 44–45.

11. See Middendort (1986), pp. 156ff. (Note, in passing, that the positive connotations of "schematic representation" build on while transforming the traditional negative implications of a "scheme.")

12. For more on this point, see Vincenti (1990).

13. For an extended discussion of the dimensional problems engendered by such modeling, see Kline (1965).

14. Ortega y Gasset (1939), p. 365.

15. See Harrison (1978).

16. See K. Polanyi (1957).

17. See, for example, Habermas (1983).

18. Bucciarell (1988).

19. Winner (1986).

20. Buchanan (1989), p. 94.

21. See Arendt (1958), especially chapters 32–34.

22. See Jonas (1984), particularly chapter 1, "The Altered Nature of Human Action." For further analysis extending the ideas of both Arendt and Jonas, see Cooper (1991).

23. John Stuart Mill, *Utilitarianism* (1861), chapter 1, near the end.

24. See Frankena (1973), pp. 35ff.

25. Borgmann (1984), pp. 42, 77.

26. For more on this tradition, see Mitcham (1987).

27. For a good brief survey of the literature of these two traditions, see Margolin (1989).

28. See, for example, in the first instance, Vries et al. (1993); and in the second, Thackara (1988).

29. See, for example, Lindinger (1991).

30. See, for example, Papanek (1983 and 1984).

31. Huizinga (1955), p. 6.

32. See, for example, Turkle (1984), especially chapter 4.

33. Ferguson (1992), p. 37.

34. This essay owes some improvements, though still no doubt not enough, to critical comments from Tim Casey (University of Scranton).

Essay 3

The Importance of Philosophy to Engineering

My thesis is that, despite common presumptions to the contrary, philosophy is centrally important to engineering. When engineers and engineering students—not to mention those who make use of engineering services—dismiss philosophical analysis and reflection as marginal to the practice of engineering, they are mistaken on at least two counts: historical and professional.

It is also the case, I would argue, that engineering is important to philosophy, although philosophers have made woefully insufficient efforts to appreciate and assess the technical realities that they too often presume to criticize. Were philosophers to set their own discipline in order with respect to engineering, philosophy would no doubt be even more important to engineering than is presently the case.

Nevertheless, even granted the inadequate attention conferred on engineering by philosophy, philosophy is of critical and increasing significance to engineering. An argument in support of this thesis will, appropriately enough, rely in key respects on engineering experience. It will proceed by means of a historical review of engineering efforts to do philosophy in part as a self-defense against philosophical criticism. Then, in a central case study, it will summarize and reflect on efforts in the American professional engineering community to incorporate philosophy into engineering education curricula. Later sections will make a more reflective effort to speculate about the deepening relationships between engineering and philosophy in an increasingly engineered reality. Engineers, it will be suggested, are the unacknowledged philosophers of an emergent techno-lifeworld.

1. SELF-DEFENSE AND PHILOSOPHY

Let me begin, then, with the issue of self-defense. As preface to this issue, consider an engineering-like schematic presentation of the problem. The problem is that engineering and philosophy are typically conceived as two mutually exclusive domains, as can be crudely represented in Figure 3.1.

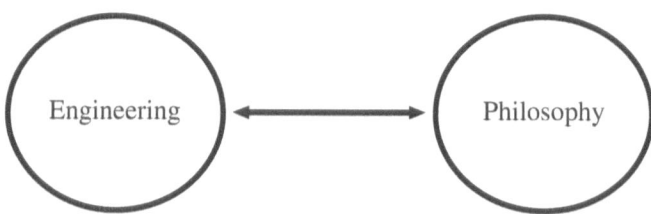

Figure 3.1

In the minds of most people, engineering and philosophy do not have much to do with each other. They are, as it were, giant islands separated by a large body of water.[1]

In fact, from the perspective of some members of the engineering community—not to mention those of the philosophy community—the situation is even worse. Engineering is customarily divided into a number of different branches: civil engineering, mechanical engineering, electrical engineering, chemical engineering, nuclear engineering, computer engineering, and more. Philosophy also includes different branches: logic, epistemology, metaphysics, ethics, aesthetics, and political philosophy. Representatives of some of these areas of the philosophy world, especially ethics and aesthetics, seem to have mounted cannons on their areas of the philosophy island in order to fire away at selected domains of the engineering world.

At least since the 1960s, members of the philosophical community or its fellow travelers have been accusing engineers of building nuclear weapons that could destroy civilization as we know it, manufacturing transportation systems that are a blight on urban culture, designing communication technologies that can enhance central or authoritarian controls by both governments and private corporations, and creating computers that depersonalize human life. Critics contend that engineers have been polluting the natural world with toxic chemicals and greenhouse gases while flooding the human world with ugly structures and useless consumer products.[2]

Martin Heidegger, one of the most prominent philosophers of the twentieth century, has even gone so far as to argue that all such ethical and aesthetic failures are grounded in a fundamental engineering attitude toward the world that reduces nature to resources in a dominating *Gestell* or "enframing."[3]

Heidegger is perhaps more subtle on this point than is always recognized. But on one common interpretation, Heidegger can be construed to say that Herbert Simon's "sciences of the artificial,"⁴ for example, promote a constrained and constraining ontology of mathematical reduction and an epistemology of virtual reality. Feminist critics have further associated engineering with patriarchal domination, the death of nature, and the loss of world-centering care.⁵

What such charges amount to is a major reactionary attack on the self-definition of engineering that goes back to the eighteenth-century formulation of Thomas Tredgold, and is reiterated in such standard reference works as the *Encyclopaedia Britannica* and *McGraw-Hill Encyclopedia of Science and Technology*. According to the classic and still standard definition that engineers give of their own profession, engineering is "the application of scientific principles to the optimal conversion of natural resources into structures, machines, products, systems, and processes for the benefit of humankind."⁶ The upshot of philosophical attacks would be to replace this traditional self-understanding with one that might read more like the following: "Engineering is the scientific art by which a particular group of human beings destroys nature and pollutes the world in ways that are useless or harmful to human life."⁷

Insofar as engineers have become aware of such attacks—and want to understand and to defend against them—philosophy is crucial to engineers. In the first instance, then, engineers have become involved with the study of philosophy in order to respond, to erect some fortifications against a philosophical onslaught. A whole school of engineer-philosophers has in fact taken up this challenge, but it is a school that is incompletely recognized even in engineering institutes and colleges—and certainly not in the liberal arts or humanities faculties in which most philosophy is taught. Allow me simply to mention in passing some representative contributors to this school or tradition.⁸

First is Ernst Kapp (1808–1896), a contemporary of Karl Marx. Although originally educated as a philosopher, Kapp emigrated from Germany to central Texas, where he became a pioneer and developed a view of technology as a complex extension or projection of human faculties and activities. In a subsequent articulation of this philosophical anthropology of technology, he actually coined the phrase "philosophy of technology" or "philosophy of engineering."⁹

Another representative is Peter Engelmeier (1855–1941), one of the founders of Russian professional engineering. A century ago Engelmeier, under the banner of the phrase "philosophy of technology," argued for a more than just technical education in the engineering profession. If engineers are to take their rightful place in world affairs, he argued, they must be educated not only in their technical fields but also in knowledge about the social impact and influence of technology.¹⁰

A third representative figure is Friedrich Dessauer (1881–1963), certainly a pivotal contributor to this tradition of engineering philosophy of technology. The inventor of deep-penetration X-ray therapy, a political opponent of Nazism, and a technical professional in dialogue with such philosophers as Karl Jaspers, José Ortega y Gasset, and Heidegger, among others, Dessauer put forth an interpretation of engineering invention as an experience that transcends the boundaries of Kantian phenomenal appearances and makes contact with noumenal things-in-themselves.[11]

Independent of Dessauer's interpretation, and as a final example of the engineering philosophy tradition, New York civil engineer Samuel Florman developed a related interpretation of "the existential pleasures of engineering" that both responds to many of its contemporary philosophical critics and defends engineering as in itself a fundamental human activity.[12] Engineering is not only instrumental to other human ends; it is in itself an existentially meaningful activity. Engineering possesses inherent or intrinsic as well as instrumental or extrinsic value.

In the first instance, then, philosophy is important to engineering because there are many who philosophically criticize engineering. In self-defense, if for no other reason, engineers should know something about philosophy in order to handle their critics. Moreover, some engineers have in fact taken up this challenge.

2. SELF-INTEREST AND PHILOSOPHY

Philosophy is also important, in a second instance, because engineers actually face problems internally or professionally that they admit cannot be resolved simply with engineering methods alone. I refer here primarily to professional-ethical issues.

There are times in the engineering world when engineers ask themselves questions about what they should be doing or how they should do it that cannot be solved by technical expertise alone. Although Clive Dym, an influential professor of engineering design, methodologically excludes aesthetics—and, by extension, ethics—from his analysis of design, in order to keep his discussion "bounded and manageable," he also grants that ethics often has a serious role to play in engineering design.[13] Questions of safety, risk, and environmental protection are only the more obvious manifestations of variables that call for ethical judgment in assessing their proper influence on design decisions. Philosophy (especially ethics) is an internal practical need of engineering—and is often so recognized by the professional engineering community.

To consider the point at issue here in a slightly fuller manner, let me compare the roles played by the sciences and the humanities and social sciences in engineering education. For this purpose, as an empirical case study consider the engineering education certification requirements in the United States. By proceeding in this manner, my aim is to let engineers, through their own professional community, speak for themselves about how they think philosophy is in the self-interest of engineers and to provide some complementary elaboration.

The organization that certifies U.S. engineering education programs is the Accreditation Board for Engineering and Technology, more commonly known by the acronym ABET. (ABET grew out of the Engineers' Council for Professional Development or ECPD, which was founded in 1932.)

According to present ABET accreditation criteria,[14] engineering programs require a minimum of

- one year of mathematics and the basic sciences;
- one half-year of humanities and social sciences; and
- one and a half years of engineering topics.

It is important to emphasize that these are minimal content requirements—and that the standard engineering degree in the United States requires four to five years of study.

These minimal content requirements exclude what are called "skills" courses focusing on the development of competence in written and oral communication, which are also required. If language communication skills course requirements are included with humanities and social sciences content course requirements—as they often are in traditional descriptions of the liberal arts—then ABET effectively requires engineering students to complete a year of *studia humanitatis*.

Consider now the justifications for the three primary components of engineering education provided by ABET.

The engineering topics criterion, of course, needs no justification, as it is engineering education that is at issue. Nevertheless, it is useful to note that engineering topics are explicitly said to include both the engineering sciences—as distinct from the basic sciences—and engineering design—as distinct from other types of design (IV.C.3.d.[3][a]).

As for the engineering sciences, these "have their roots in mathematics and basic sciences but carry knowledge further toward creative application" (IV.C.3.d.[3][b]). Such rootedness is what justifies course requirements in mathematics and the basic sciences. In the words of the ABET criteria, "The objective of the studies in basic sciences is to acquire fundamental knowledge about nature and its phenomena, including quantitative expression" (IV.C.3.d.[1][b]).

As for engineering design, this is defined as

> the process of devising a system, component, or process to meet desired needs. It is a decision-making process (often iterative), in which the basic sciences and mathematics and engineering sciences are applied to convert resources optimally to meet a stated objective.
>
> (IV.C.3.d.[3][c])

Such an understanding of engineering design obviously provides a second and supporting justification for mathematics and the basic sciences.

But what about the half-year of humanities and social science courses—or one year, if one includes studies of written and oral communications? What is the justification for including the humanities and social sciences as a major component of the curricular requirements for an engineering education?

Before citing the ABET criteria response to this question, note that the ABET criteria definition of engineering design silently drops one crucial aspect of the traditional definition of engineering. As mentioned earlier, Tredgold's and (until recently) the most commonly cited definition is that engineering is "the application of scientific principles to the optimal conversion of natural resources into structures, machines, products, systems, and processes for the benefit of humankind." ABET replaces the end or goal of being humanly useful and beneficial with simply meeting some "desired needs" or "stated objective." The normative aspect of the traditional definition is thus washed out in favor of a value-neutral or context-dependent process.

Therefore, at the point in the ABET criteria when the humanities and social sciences content requirements are described and justified, it is said,

> Studies in the humanities and social sciences serve not only to meet the objectives of a broad education but also to meet the objectives of the engineering profession. . . . In the interests of making engineers fully aware of their social responsibilities and better able to consider related factors in the decision-making process, institutions must require course work in the humanities and social sciences as an integral part of the engineering program. This philosophy cannot be overemphasized.
>
> (IV.C.3.d.[2][a])

In other words, once the goal of engineering design has been reduced from being humanly useful and beneficial to a context-dependent process, then the humanities and social sciences are presented as a means to understand and evaluate such contexts. Otherwise engineers would just be hired guns—and could serve the profession equally well as designers of concentration camps or of green (nonpolluting) chemical plants.

Thus, while mathematics and the basic sciences ground the engineering sciences, the liberal arts ground (in a different but related way) engineering design. Would it be too bold to conjecture that, just as the engineering sciences are thought to extend the basic sciences, by carrying "knowledge further toward creative application," so too engineering design may be described as creatively applying some modes of thought and ideals of the humanities and social sciences?

Consider briefly a contrast of two engineering experiences that may support, from quite different angles, just such a hypothesis. The first is imaginative, but real: that of Goethe's *Faust*. In *Faust II*, having abandoned first his liberal studies and then crude magic, Faust became a civil engineer erecting dams and draining marshes—while inadvertently killing innocent people.[15] The second is historical but imaginatively reconstructed: the case of Russian engineer Peter Palchinsky.[16] Executed by Stalin because he refused to separate technical knowledge and humanistic ideals, it is the ghost of the executed engineer Palchinsky that emerges triumphant in the *glasnost* that accompanied the demise of the Soviet Union.

This point is reiterated at the end of the ABET criteria content statement. After asserting that competence in communication "is essential for the engineering graduate" (IV.C.3.i), it is further affirmed that "an understanding of the ethical, social, economic, and safety considerations in engineering practice is essential for a successful engineering career" (IV.C.3.j).

ABET is currently in the process of revising and simplifying its criteria for accreditation. Its new criteria set, laid out in a document called "Engineering Criteria 2000," confirms the present argument by listing eleven "outcomes" upon which engineering programs will be assessed. Beginning in the year 2000, to be accredited by ABET, "engineering programs must demonstrate that their graduates have

a. an ability to apply knowledge of mathematics, science, and engineering;
b. an ability to design and conduct experiments as well as to analyze and interpret data;
c. an ability to design a system, component, or process to meet desired needs;
d. an ability to function on multidisciplinary teams;
e. an ability to identify, formulate, and solve engineering problems;
f. an understanding of professional and ethical responsibilities;
g. an ability to communicate effectively;
h. the broad education necessary to understand the impact of engineering solutions in a global and societal context;
i. a recognition of the need for and an ability to engage in life-long learning;
j. a knowledge of contemporary issues; and
k. an ability to use the techniques, skills, and modern engineering tools necessary for engineering practice.

Of these eleven outcomes, four—or over one-third—may readily be classified as engaged with the humanities and social sciences. Thus, again, in a four-to-five-year program, more than a year of course content can be expected to be *humanitas* focused.

"Such course work," appealing again to existing criteria,

> must meet the generally accepted definitions that humanities are the branches of knowledge concerned with man [sic] and his [sic] culture, while social sciences are the studies of individual relationships in and to society. Examples of traditional subjects in these areas are philosophy, religions, history, literature, fine arts, sociology, psychology, political science, anthropology, economics, and foreign languages. . . . Nontraditional subjects are exemplified by courses such as technology and human affairs, history of technology, and professional ethics and social responsibility.
>
> (IV.C.3.d.[2][b])

3. EXCURSUS: THREE QUESTIONS

This passage provokes at least three questions—questions that entail a brief excursus. The questions are as follows:

- One, what does it mean to invoke "generally accepted definitions" of the humanities and the social sciences? Are the humanities and the social sciences, including philosophy, historically or socially constructed?
- Two, exactly what is philosophy anyway? What is the relation between philosophy and the liberal arts? Is it perhaps the case that philosophy—having been named first—could be more important than or differentially significant from other humanities and social sciences?
- Three, in light of the generally accepted definition of philosophy as including ethics—together with statements here and previously regarding the importance of professional ethics to "a successful engineering career"—what is it, more concretely, that philosophy and ethics may do for engineering?

These are all serious questions. They are not to be answered either quickly or finally in this presentation. Indeed, they are the kind of open questions designed to provoke extended reflection more than to offer closure or straightforward solutions. It is nevertheless appropriate to begin to explore elements of what might be termed some boundary conditions on answers.

With regard to the first question, the passage is more insightful than many others in its cautionary reference to "generally accepted definitions" of the humanities and the social sciences. It is indeed the case that these definitions are historically, socially, societally, and culturally constructed. (For present purposes I use the terms "historical" and "social" as the primary qualifiers, but with recognition that in other contexts more careful distinctions would need to be drawn.) Such constructions as exist are also highly contested—in differentially constructed ways.

In the United States, this multilayered contest—with its contests about the contest—is known collectively if not affectionately as "the culture wars." One front in these wars is fought between protagonists of the "dead white men" (from Homer on) school of culture and the "politically correct" (we are the victims of discrimination) school—to use the warring parties aspersion-casting names for each other. In this sense the ABET criteria statement is at once cautious—and then anything but cautious, with its description of the humanities as "concerned with man and his culture."

Leaving aside this egregious gaffe, one may nonetheless note that early on engineers opened their own front in the culture wars. As John Staudenmaier has ably narrated in *Technology's Storytellers*, the establishment of the Society for the History of Technology in the late 1950s was done in part by engineers who found themselves left out of Western history just as much as women or various ethnic minorities.[17] History is technology as much as politics, the engineer historians argued. The humanities and social sciences have reflected the limited self-interests and ideological biases of nonengineers—not to say of those who use humanities and social sciences power/knowledge to discipline themselves and others.[18] Engineers have an interest in opening up the black boxes in history, to notice that political problems and their solutions often depend on engineering input, in order to include not so much another group of victims as unrecognized conquerors.[19]

The humanities and the social sciences, including philosophy, are thus historically and socially constructed. But it is also crucial to note that the same goes for engineering, though not so obviously. Both engineering and philosophy (emphasizing that element of the humanities and the social sciences most at issue here) have distinct historical origins and have not always been understood or practiced in the past as they are today.

Philosophy emerged as a recognized human way of life in the West in fifth century BCE Greece. According to Aristotle's account, philosophy originated when humans replaced speech about god or the gods with speech about *phusis* or nature.[20] Today, however, few members of that community which practices the discipline of philosophy—and discipline is not the same as a way of life—speak or write about *phusis* or nature. They are more likely to speak or write about phenomena and language.

Engineering also emerged as a recognized human activity at a particular point in history—namely the seventeenth and eighteenth centuries. The first engineers were members of the military who designed, constructed, operated, and maintained fortifications and engines of war such as battering rams, catapults, and cannons. The term "civil engineer" originally denoted the attempt to transfer the kind of activity and knowledge involved in such military concerns into nonmilitary contexts. The formulation of Tredgold's definition of engineering, as cited earlier, was part of the historical and social effort to bring about this displacement.

Indeed, both engineering and philosophy exhibit quite different characteristics across geographies as well as histories—even if one only compares cases from as closely related communities of discourse as those of Europe and North America.

It may be accepted, then, that both engineering and philosophy are historically and socially constructed. Such an admission would seem to grant to history and the social sciences priority in the liberal arts.

At the same time, history and society are not only about change; they are also about continuities. Historical and social construction is, after all, not *ex nihilo*. Indeed, it is perhaps better described not as construction but as reconstruction. Our efforts to name what is undergoing historical reconstruction—and thus what to some degree transcends history—are themselves subject to revision. At any one point in time, however, we must logically (if provisionally) accept our own sociohistorical constructions about how best to indicate such trans-sociohistorical—or perhaps better, multi-sociohistorical—features of our constructs.

With regard to the second question in this excursus, then, we inquire about what multi-sociohistorical features are exhibited by philosophy. What is it about philosophy that, since its origins, has enabled us to speak about the presence of this or related phenomena at other times and places? What is it that we now mean by philosophy?

Today the common or uniting elements in philosophy involve some mixture of the following:

- Conceptual analysis, to help clarify and correct both practical and theoretical uses of terms. This includes but is not limited to logic.
- Reflective examination of practice and thought, so as to deepen insight and understanding of, extend, or criticize both dimensions of experience. This includes the core areas of philosophy known as ethics, metaphysics, and epistemology, often with an emphasis on their rational methodologies.
- Thinking about aspects of experience that are more global than customarily dealt with by any one discipline. Here the emphasis is likely to be more substantive than methodological. Such thinking may also involve inter-, multi-, trans-, and antidisciplinary consideration of what is right and good (ethics), the structure of reality (metaphysics), and knowledge (epistemology).
- The practice of a distinctive way of life and thought, one taken to be good in itself, with its own unique knowledge of reality. Philosophy in this sense

may also be regionalized into the general guiding practices or principles of an individual or group, as when we refer to someone's personal philosophy or the philosophy of a firm.

In each of these manifestations, philosophy may be further described as engaged with nonempirical issues rather than empirical ones—though not without empirical or real-world reference. Each of the core areas of philosophy—ethics, epistemology, and metaphysics—exhibits both descriptive and normative dimensions. But it is the normative dimension that is at once crucial and most difficult to pursue without abandoning its conceptual and critical dimensions.

It may also be noted, historically again, that philosophy has functioned as a kind of seedbed from which many of the sciences and the humanities have sprung. Natural philosophy gave rise through mutation to natural science; it was philosophers such as Bacon and Descartes, together with natural philosophers such as Galileo and Newton, who constructed the physical sciences. It was social philosophers such as Comte, Marx, Durkheim, and Weber who constructed the social sciences. From philosophical reflection and conceptual analysis have also emerged economics, anthropology, psychology, religious studies, and other humanities and social science disciplines. The very idea of a discipline, defined either in terms of its object or in its method, is one that philosophy in its inter-, multi-, trans-, and antidisciplinary thinking both conceptually clarifies and reflectively criticizes.

In this way, particularly, philosophy does reasonably appear to be differentially significant from the other humanities and social sciences—to be, as it were, first among equals. Such significance provides reason to hypothesize that philosophy, more than the other humanities and social sciences, may matter to engineering in a special way.

Thus, with regard to the third question in this excursus—a question that returns again to the main theme—we may consider anew what it is that philosophy, especially philosophy in the form of ethics, contributes to professional engineering.

4. ENGINEERING AND ETHICS

It is certainly not the case that philosophy has sponsored engineering in anything like the way it has sponsored the sciences, the social sciences, and the humanities. Indeed, engineering has a strong tendency to distinguish itself from philosophy, not in a manner that would acknowledge philosophy as that from which it has emerged but as something in relation to which it is defiantly other.

As Louis Bucciarelli observes in his ethnographic studies of engineers, when students are working on engineering problems it is generally thought that they "ought not to get bogged down in useless 'philosophical' diversions."[21] As he notes on more than one occasion, in the realm of engineering, philosophy has strongly negative connotations. Yet at the conclusion of his study, Bucciarelli the engineer, having argued that engineering design is a social process, points out how this means there are alternatives. When there are alternatives, he says, then there can be better and worse. In such a situation, "the really important and interesting question becomes: what do we mean by a better design?"[22] This is an eminently philosophical question.

Only through conceptual analysis, rational reflection, and general modes of thought can such an issue adequately be addressed. Precisely because of numerous manifestations of this type of question—the question "What do we mean by a better design?"—engineers have constructed bridges between engineering and philosophy, especially to that branch of philosophy constituted by ethics, even though neither engineers nor philosophers may not always have recognized them as such. Summarized again by means of schematic diagram, the situation has been transformed from two mutually exclusive circles to something like that in Figure 3.2.

Figure 3.2

Four particularly well-known cases from the American context that have influenced this span construction are the Ford Pinto automobile gas tank that was subject to rear-end collision explosions,[23] the San Francisco Bay Area Rapid Transit (BART) Automatic Train Control system failure,[24] defective DC-10 cargo bay door and engine mounts,[25] and the field joints on the solid rocket booster of the space shuttle *Challenger*[26]—examples representing in turn the fields of automotive, computer, aeronautical, and mechanical engineering. Scrutiny of the ethics-related design and operational dilemmas in these and related cases have been advanced by engineers themselves such as Stephen Unger,[27] Roland Schinzinger,[28] Charles Harris and Michael Rabins,[29] Aarne Vesilind and Alastair Gunn,[30] and others. Their work has

- undertaken conceptual analyses of right and wrong, good and bad, in engineering practice;
- sought a reflective deepening to insight and understanding into the ethical dimensions of engineering experience; and
- pursued interdisciplinary, cooperative research into professional ethics codes, disciplinary procedures, moral educational strategies, and more.

Beyond the efforts of these engineering ethicists to analyze professional codes of conduct, reflectively enhance the ethical dimensions of engineering practice, reconstruct professional organizations to better support appropriate engineering autonomy, and engage in interdisciplinary pedagogical efforts, one can discern right in the core of the engineering analysis of design a fundamentally ethical impulse. For want of a better phrase, let me call this the imperative to remain connected.[31]

A failure to remain connected to social bonds and the limitations of the human condition is, for instance, one way to define the problem of Faust as engineer. A determination to remain connected to what is pragmatically known about the world is what has cost many engineers such as Palchinsky their jobs if not their lives.

One of the drivers behind Clive Dym's computer modeling of design representation, for instance, is to promote communication between design engineers and construction personnel that would avoid the kind of disaster precipitated, as in the Kansas City Hyatt Regency atrium walkway failure, by a fabricator failure to grasp the significance of a crucial design specification.[32] The Hyatt Regency contractor error was, in turn, set up by a design engineering failure to recognize the construction problem entailed by the crucial design specification at issue.

Hanger rods long enough to transmit a second floor walkway load through the fourth floor walkway, directly to the roof trusses above, were not available. The contractor, not understanding the load transfer dynamics involved, substituted two rods instead, in effect hanging the second floor walkway from the fourth floor walkway. The identified need for better communication—that is, better connection—between design intention and constructive reification is a moral as well as a technical imperative.

It may well be the case that, as engineer Henry Petroski argues, design failures are inherent in the fallible practice of engineering and the learning curve that constitutes technical progress.[33] But conceptual analysis and reflective examination reveal that not all failures are equal. Moreover, philosophical analysis and reflection are part of the very process by which engineers learn from design failures. Again, Dym's work on the languages of representation in design is a case in point.

It is central to the argument at this point to note that disciplines ought not to be conceived as barriers to all trespassers but rather as selective niches for the promotion of differential growth. We are all to some extent engineers, insofar as we design, construct, and operate in the microworlds of our lives. Something as simple as packing a box is a quotidian mini design problem: how can we best fit all the items we want to store or ship into a predefined space? Likewise, we are all to some degree students of philosophy, insofar as we undertake to conceptually analyze, reflect on, and generalize about aspects of our lives and works: what items belong and what type of box packers do we want to be?

Because this is the case—as in most others—we are selectively enhanced persons, it makes sense for us to reach out and call to another differentially enhanced individual or community of practitioners for assistance. Because engineers already practice philosophy to some extent, it makes sense for them to build bridges to philosophers (who also already to some degree practice engineering) and invite their participation to a collective endeavor. This is precisely what engineers such as Unger, Schinzinger, and Rabins have done—to which philosophers such as Tom Rogers, Mike Martin, and Michael Pritchard have responded.[34] In each case we have more than simple bridge building between engineering and ethics. What we now see is the actual partial merging or overlapping of the engineering and philosophy worlds, graphically represented in Figure 3.3.

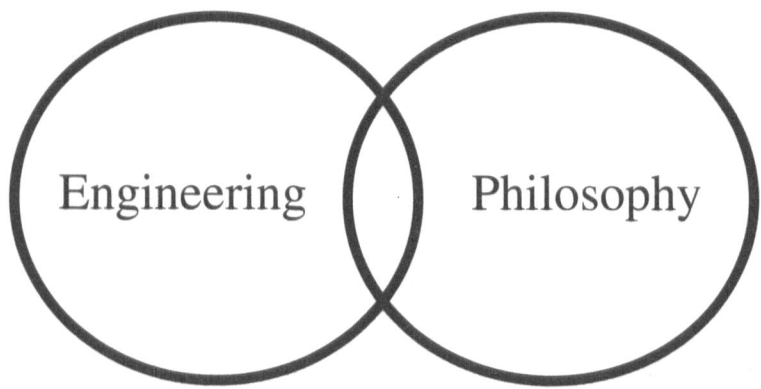

Figure 3.3

5. BEYOND APPLIED ETHICS: SELF-KNOWLEDGE AND PHILOSOPHY

Engineering in the past may have been historically and socially constructed so as to alienate philosophy. Philosophy in the past may also have sought to keep engineering at arm's length. But times and the world change. Engineering has changed. It has, I would even venture to suggest, become much more philosophical. Indeed, engineering is ripe not just with philosophical problems but with a philosophically significant way of life. Philosophy, for its part, is becoming much more open to engineering thought and practice—though not as fully or as fast as some think appropriate.

Why is philosophy important to engineering? The first reason, I have argued, is self-defense against philosophical critics. The second reason is self-interest, to help deal with issues of social context and ethics within engineering practice. But there is also a third reason why philosophy is important to engineering: Engineering is modeling a new philosophy of life. In this instance there is some tectonic plate movement. Not just bridges are being constructed between islands, but continents actually begin to overlap and geologically alter each other.

But just as tectonic plate movement is imperceptibly slow and thus difficult to recognize, so too is this third interaction of engineering and philosophy. It is also an interaction that is both grounded in and calls for increased self-knowledge on the part of all participants.

What might conceptual analysis, reflective insight, and interdisciplinary thinking have to contribute to engineering? To pose the question in this way is virtually to answer it. Is not engineering, too, characterized by conceptual analysis, reflective insight, and interdisciplinary thinking?

As we increasingly reconstruct the world, we increasingly recognize the world as constructed. As human beings have moved from a natural to a carpentered and then engineered world, surely it is no accident that natures and essences have been called into question, that process has replaced substance, that knowledge is increasingly framed by economics and politics as much as cognitive methodology, and that ethical issues have moved to the forefront in public as well as technical discussions across a broad spectrum of human activities, from medicine to computers.

The applied philosophical discourses of bioethics, environmental ethics, computer ethics, and engineering ethics are nevertheless no more than the tip of an iceberg breaking apart in a sea of metaphysical speculations (from scientific cosmologies to the new existentialisms of risk projection, electronic networking, and virtual reality), epistemological explosions (transhuman and remote sensation and perception, automated instrumental data gathering and analysis, research articles as advertisements and promotional campaigns for the next round of funding grants), and aesthetic constructions (graphic media presentations and probability analyses, hypertext communications, macro- to micro-engineering projects, interactive Internet websites). Food, housing, transportation, communications, economics, art, literature, music, sex are all being transformed by technological makings. These remakings are themselves the continuous subjects of exoteric and esoteric theoretical discussions, philosophical debates, and ideological disputation.

Our world may be shot through with technology, but our technology is in turn interpenetrated with philosophical dialogue. Indeed, it is precisely such lifeworld transformations that postmodern philosophy has made the primary subject of its discourses, even as engineers create the very transformations that philosophers talk about. But the engineers have remained silent. Because of their silence, they have in a paradoxical manner marginalized their powers—failed to recognize themselves and their practices as central to the cultural superstructure they engender, which in turn engenders them.

Consider one case in point: the Xerox Palo Alto Research Center (PARC). This engineering research center, perhaps even more than Bell Labs, is one of the truly great innovation hubs of history. In the late 1960s and early 1970s it invented virtually all the major elements of what became the personal computer revolution: the graphic interface, the mouse, and more. But its corporate sponsor failed to capitalize on its pioneering technical innovations.[35] Xerox PARC creativity was stimulated in part by its philosophical interactions with and sensitivity to cultural developments. At the same time, on one reasonable interpretation it failed to be able to promote these innovations because of its passive receptivity with regard to the philosophical stimuli of the culture.

By contrast, Mark Weiser, the current chief technologist at Xerox PARC, has been influenced by the philosophical reflections of Herbert Simon, Michael Polanyi, Hans Georg Gadamer, and Martin Heidegger to imaginatively project design beyond mainframes and personal computers toward a third wave of what he terms "ubiquitous computing" or "ubicomp" for short.[36] With ubicomp Weiser and other engineers at Xerox PARC are working to let computers merge into the background of our lives, to blend in with the environment. Similar radical engineering innovation centers such as the Media Lab at MIT[37] nevertheless exhibit a strong tendency only to absorb postmodern philosophical influences, even while they exhibit or live them out.

The engineering design process embodies and exhibits precisely the kind of contingent, decentered, boundary crossing, and emergent ordering processes that postmodernity analyzes, explores, and celebrates. Engineers live but do not speak postmodernism.

Engineers are the unacknowledged philosophers of the postmodern world. What is distinctive about the material base of postmodernity is that it is an engineered materiality. Robert Venturi's playful postmodern architecture is the playfulness of a skilled engineer.[38] François Lyotard's postmodern condition of self-reference mimics the self-referential iterative practices and processes of engineering design.[39] Donna Haraway's border-crossing coyote-cyborg could not exist without biomedical technology.[40]

For literally thousands of years human making and using relied on what was given in nature. Under such conditions, artifice remained unalterably limited in both quantity and substantiality. Indeed, its lack of quantity was reflected in a hand-and-mind-crafted particularity, the evident beauty of which was never more than skin deep. "If a bed were to sprout," wrote Aristotle, "not a bed would come up but an oak tree."[41]

The engineering extraction from nature of both hidden materials and energies, together with the design of minded machines, made possible the quantitative proliferation of artifice and its coordinate standardization. Standardization appeared to deprive the world of crafted beauty as a necessary trade-off for affluence. The standardization that engineers constructed, not just with their machines and industrial processes but behind the scenes through the negotiation of technical codes, nonetheless foreshadowed a fabricated substantiality at the base of a new ecology of artifice.

With the extension of engineering processes into the micro, nano, genetic, molecular, atomic, and even subatomic levels, our new artifacts, when they sprout, sprout not their old matters deprived of form but in newly informed structures.

No one has lived and lives more deeply in this world of living artifice than engineers. Engineers are only beginning to share their design lives with the larger world by means of conceptual analysis and critical reflection. This is an analysis and reflection from which the philosophical world would profit, and to which it might contribute, if it would but make the effort.

Why is philosophy important to engineering? Ultimately and most profoundly it is because engineering is philosophy—and through philosophy engineering can be enabled to become more itself.

Engineers of the world philosophize! You have nothing to lose but your silence!

NOTES

1. The classic presentation of this view is, of course, is Snow (1959).
2. Well-known examples include Ellul (1954) and Mumford (1967 and 1970).

3. Heidegger (1954), pp. 13–44.

4. Herbert Simon, *The Sciences of the Artificial* (Cambridge: MIT Press, 1969; 2nd ed., 1981; 3rd ed., 1996).

5. See, for example, Caroline Merchant, *The Death of Nature: Women, Ecology, and the Scientific Revolution* (San Francisco, CA: Harper and Row, 1980).

6. *New Encyclopaedia Britannica*, 15th ed. (Chicago, IL: Encyclopaedia Britannica, 1995), *Micropaedia*, vol. 4, p. 496. The *McGraw-Hill Encyclopedia of Science and Technology*, 8th ed. (New York: McGraw-Hill, 1997), vol. 6, p. 435, modestly truncates then expands on this definition when it describes engineering as "most simply, the art of directing the great sources of power in nature for the use and the convenience of humans. In its modern form [it] involves people, money, materials, machines, and energy." Thomas Tredgold's original wording was, "Engineering is the art of directing the great sources of power in nature for the use and convenience of man" (from 1828).

7. Cf. Lewis (1947), p. 35: "What we call man's power of nature turns out to be a power exercised by some men over other men with nature as its instrument."

8. For more extended narratives concerning the engineering philosophies of technology cited below, and related ideas, see Mitcham (1994a), pp. 19–38.

9. Kapp (1877).

10. The best study of Engelmeier is Gorakhov (1997).

11. See, for example, Dessauer (1956), which is a completely rewritten and much expanded version of Dessauer (1927).

12. Florman (1976). See also Florman (1981, 1987, and 1996).

13. See Dym (1994), p. 15. (In personal conversation, Delft, Netherlands, April 17, 1998, Dym acknowledged the importance of ethics.)

14. All quotations from ABET materials were taken from documents available on the Internet at http://www.abet.org (retrieved April 1998).

15. Goethe, *Faust II* (c. 1830), act V. See also Schillinger (1984).

16. See Graham (1993).

17. Staudenmaier (1985), especially chapter 1, pp. 1–8.

18. The allusion, of course, is to Foucault. See, for example, Foucault (1980).

19. Although "black box" opening has become identified as a program of (sometimes historically oriented) sociologists of technology such as Bruno Latour and Wiebe Bijker, the idea was also proposed early by historian Edwin Layton (1977), p. 198, and developed independently by Rosenberg (1982) prior to its appearance in Bijker et al. (1987).

20. Aristotle, *Metaphysics* XII, 6; 1071b27.

21. Bucciarelli (1994), pp. 105–106.

22. Bucciarelli (1994), p. 197.

23. Birsch and Fielder (1994).

24. Anderson et al. (1980).

25. See Curd and May (1984); and Fielder and Birsch (1992).

26. Boisjoly (1991) and Vaughan (1996).

27. Unger (1994).

28. Martin and Schinzinger (1996).

29. Harris et al. (1995).
30. Vesilind and Gunn (1998).
31. For a different but related explication of this engineering ethical imperative, see Mitcham (1994b), pp. 153–96 and 221–23.
32. R. D. Marshall et al. (1982). For Dym's analysis, see Dym (1998).
33. Petroski (1985).
34. Philosopher C. Thomas Rogers participated with Unger in engineering ethics research work, and is cited in Unger (1994), pp. 115, 135. Philosopher Mike W. Martin coauthored with engineer Roland Schinzinger, *Ethics in Engineering*. Philosophers Michael S. Pritchard and Charles Harris, Jr., have worked extensively with Michael Rabins, a collaboration reflected not only in their book *Engineering Ethics: Concepts and Cases* but also a collection of more than 30 case study scenarios available at http://ethics.tamu.edu (accessed April 1998).
35. Smith and Alexander (1988).
36. Weiser (1991). Further information is available at http://sandbox.xerox.com/hypertext/weiser/UbiHome.html (accessed April 1998). Cf. also the philosophical receptivity evidenced in engineers Winograd and Flores (1987).
37. See Brand (1987).
38. Venturi (1977).
39. Lyotard (1979).
40. Haraway (1991).
41. Aristotle, *Physics* II, 1; 193b10.

Essay 4

From *Dasein* to Design

The Problematics of Turning Making into Thinking

(with J. Britt Holbrook)

> The proper study of mankind *is* the science of design.
>
> —Herbert A. Simon (1968, p. 83)

A representative American myth film from the last year of the twentieth century, *October Sky*, tells the story of Homer Hickam, Jr. (a.k.a. Sonny), growing up in rural Coalwood, West Virginia. Life there had been the same for generations; most young men had little expectation other than acceptance of the fate of working in the mines. That was certainly what Homer Hickam, Sr., himself a miner, expected for his son. But the Soviet launch of *Sputnik* on October 4, 1957, shook the world even in West Virginia. It inspired Homer, Jr., and five fellow high school students to start the "Big Creek Missile Agency," which, after numerous failures, became progressively more successful at designing rockets. In 1960 they won gold and silver medals in the National Science Fair and earned college scholarship tickets out of Coalwood. Sonny Hickam went on to become a NASA engineer, working on projects such as the International Space Station. The movie is based on his 1994 memoir in the *Smithsonian Air and Space* magazine, which was later expanded into an autobiography, *Rocket Boys* (1998).

Sonny Hickam's story is emblematic of more than the Horatio Alger myth; it pictures a transition in our history and in our existence that, appropriating and repurposing a term from Martin Heidegger, may be described as a movement from *Dasein* to design. For most of history human existence has centered on growing into a preexisting way of life: the there-ness of learning a language; accepting and sharing the subsistence joys and sufferings of a particular family, tribe, or village; celebrating the cycles of the days and years of that place and time; and learning to be there as deeply and as fully as possible. In *Sein und Zeit* (1927), Heidegger takes the common German word *Dasein*, meaning literally "there-being" but often translated as "existence," and invests it with new philosophical name for "that entity which in its Being has this very Being as an issue, comports itself toward its Being as its ownmost possibility" (*Being and Time* I, 1 H 42, trans. Macquarrie and Robinson). In contrast to a life of unquestioning acceptance of what is given, *Dasein* in Heidegger comes to designate that entity for which its own being is a question. Heideggerian *Dasein* is fated to choose if not design the life it will lead.

October Sky (an anagram of the original book title) is told as a story of an achievement in this new, nontraditional *Dasein* through engineering success, escaping from Coalwood by engineering rockets to escape the earth. Although there is some nostalgia, it is for youth rather than the richness and depth of a stabilized place and life. There is little if any sense of what Milan Kundera, for instance, describes as *The Unbearable Lightness of Being* (1984), which may be read from the point of view of traditional *Dasein* as a narrative of the often thin and insubstantial character of the designed life. Paradoxically, after his famous turning, Heidegger too undertook a radical questioning of the kind of designed life that his 1927 interpretation of *Dasein* might be read to imply.

Two years prior to Heidegger, Alfred North Whitehead drafted his own celebration of the designing and designed life. In the first of three central chapters of *Science and the Modern World* (1925), Whitehead presented engineering as the differentia of the nineteenth century and "the invention of invention" as its greatest contribution to human civilization. Invention ceased to be an accidental or rare event and became a methodologically pursued and economically promoted process. In *American Genesis* (1989), historian Thomas P. Hughes even went so far as to compare the achievements of inventors such as Thomas Edison and Henry Ford to those of Renaissance artists. During the twentieth century something similar happened with technological design, a human activity much more directly related to the Renaissance. According to the lead story in an early twenty-first-century issue of *Business Week*,

> When people talked about innovation in the '90s, they invariably meant technology. When people speak about innovation today, it is more than likely they mean design. Consumers, who are choking on choice, look at design as the new differentiator. In a sea of look-alike products and services, design creates the "Wow!" factor. Managers, facing fierce global competition, look to design for the kind of innovation that generates organic growth, new revenues, and wider profit margins.
>
> (Nussbaum, 2005, p. 62)

Beyond its possible economic ramifications, we now speak of inventing and designing things as diverse as paintings and engineering products, processes, and systems; suburban residential communities, cities, and the environment; the layouts of magazines, newspapers, and computer web pages; and political, military, and advertising campaigns. Jeans, drugs—and even babies—have all been (or conceivably could be) designed.

THE ETYMOLOGY OF "DESIGN"

There is no word in classical Hebrew or Greek that translates our term "design." This etymological fact immediately reinforces the idea that there is something peculiarly modern about design, that design may be indicative of a way of being-in-the-world that is distinctive of modernity.

The term "design" has roots in the Latin *designare*, meaning to mark out, trace, denote, or devise, with cognates in Italian (*disegno*, drawing) and French (*dessein*, plan or purpose; and *dessin*, drawing or sketch). The English "design" can function as either verb (meaning to mark out, nominate, appoint; to plan, propose, intend; and to sketch, draw) or noun (mental plan and artistic shape). In these senses it first appeared in the mid-to-late sixteenth and early seventeenth centuries.

Attentive philosophers and theologians might object, contending that the notion of design has been central to an argument for the existence of God since the Middle Ages; some may even wish to go back as far as Cicero. (See *De natura deorum* ii, 34, for Cicero's description of providence in nature by means of a contrast between art and chance.) But the so-called argument from design for the existence of God does not actually use the word "design" until the eighteenth century. In Thomas Aquinas's famous fifth way (*Summa theologiae* I, q.2, a.3) the argument is from the "governance of the world" to a governor, more a political than an engineering image. The fifth way is more properly described as an argument toward teleology than an argument from design. Even Isaac Newton, undoubtedly the most influential advocate of scientific order in nature and a proponent of the idea that such order can only be accounted for by the idea of God, refers to the "counsel and dominion" of God rather than any design capabilities (*Principia Mathematica*, Book III, "General Scholium").

It was Bernard Nieuwentyt and, most famously, William Paley, who during the early Industrial Revolution turned teleology into design by describing many things in nature if not the world as a whole as like a watch (a machine) that must have had a watchmaker (a designer). This new argument from design coincided with the rise of mechanical philosophy. It is this idea of design in nature—design that presupposes a (divine) designer—against which David Hume argued in his *Dialogues Concerning Natural Religion*, and it is this idea of design for which Charles Darwin sought to offer an alternative account, design by natural selection or unintentional design—a view often used by classical economists as well to explain the source of order in the market.

Since the late scientific and early industrial revolutions, the verb "design" has developed progressive associations with engineering and technology. As denoting "the act of conceiving and planning the structure and parameter values of a system, device, process, or work of art" (*McGraw-Hill Dictionary of Scientific and Technical Terms*) or any such conception and plan, it has also come to be applied to an increasingly diverse range of human experiences. For the artist and architect, design is an activity honoring tradition, civic life, and aesthetic principles, and thus not something readily transferred to the world at large. For engineers, design is simply conscious if not rationalized construction, something much more generalizable. Engineers design structures, devices, processes, and systems. Adopting this sense, even artists can be described in a new way as designers of paintings and statues. Industrial designers likewise design consumer products, graphic designers advertisements and displays, and entrepreneurs business strategies. Musicians design their compositions, poets and novelists their works of literature. Scientists design hypotheses and how to test them. Physicians design treatments for their patients. Teachers design new curricula, parents how to raise their children. In the early 2000s, the United States and the United Kingdom attempted to design a new, democratic Iraq (a designer country) to take its proper place in a redesigned Middle East.

Unlike aesthetic design, engineering design has been extended from mechanical structure across economic and political affairs into psychology and biology. There are designer drugs and designer organisms (genes)—all mass-produced (thus annulling particularity) and planned out beforehand (in an attempt to escape uncertainty if not worry and risk). In the arena of sports, athletes design strategies for improving their games: some undertake specific drug regimens designed both to enhance performance and to avoid detection; some rely on improved equipment design (e.g., in golf, where design improvements have in turn made it necessary to redesign golf courses). For many, not just for professional athletes, nutritionists design meal plans, and personal trainers design workouts.

People today design their lives—that is, they try to think out in advance how to go about doing something and in the process to bring to bear as much scientific information and technical competence as possible. For group behaviors, designs take the form of policies. In what are now the most characteristic design- and policy-constructing activities, people use computers to assemble and organize this information, to render it in graphic form, and thus to test out alternative courses of action before actually undertaking them. Testing is no longer by lived experience but by modeling and simulation. People construct in miniature, on the screen, in virtual form, their actions—in order, they say, to live chosen lives more effectively, more fully, more responsibly.

In essence, design thus turns making (not to say living) into thinking—a thinking beforehand about how to make (or how to live). Is this simply an unqualified good? Or is it as problematic in new ways as that more here-and-now making (and living) that it progressively replaces?

TECHNOLOGICAL DESIGN HISTORY

Technological design has a history, and a specifically modern history, which may be briefly narrated as follows. Prior to the rise of modern engineering, design was hidden or embedded in the craft of making as a characteristic activity of human beings. Artisans in their particularities of body, place, and history were at one and the same time those who conceived or imagined artifacts and then worked to fabricate them. Aboriginally, they were also the users. Artisans in wooded geographies worked with wood—wood growing there (not elsewhere); artisans living in rock-rich landscapes worked with stone—stone quarried there (not elsewhere). Only kings transported wood and stone for great distances to construct temples and palaces. Artisans worked also with the strength and skills of their own bodies and within the traditions of their peoples or cultures; they used and lived with what they made. Artifacts so fabricated had their unique bodies, places, and histories, which guided and circumscribed human engagements. There were no generalized things or devices. Tools were task specific and gendered (see, for example, Illich, 1983). Even when particularities were exchanged in trade they remained specific in fact. All things had a natural particularity that is today at once highly prized and difficult to reconstruct.

The historical impetus behind the rise of engineering design as a new and specialized activity was the industrial mode of mass production. Prior to mechanical industrialization, the artisan was at once designer and worker—yet neither. Artisans, as it were, designed artifacts in the course of constructing them so that making seldom involved anything like a separate moment of thinking out or planning beforehand but proceeded instead as intuitive cut-and-try fabrication, letting oneself be guided by materials and tradition, and even by personal relationships in the community. Testing took place continuously right in the making and then again in the using.

The design process was dis-embedded from this rich, intimate context, by the demands of mechanization and its increasing divisions of labor. Coordinate with the replacement of human power with coal- and steamdriven prime movers, and the gearing of power into repetitive motions in order to mass-produce identical products, two things happened. Design historian Penny

Sparke summarizes these in *An Introduction to Design and Culture in the Twentieth Century* (1986):

> First came the need for a designer as pattern maker for artifacts that could be mass produced; second came the need for designer as form giver for artifacts that could be mass marketed. Mass marketing in turn may be read as a turning of users into consumers in much the same way as mass production turns workers into laborers, repetitive shopping serving as an echo of the repetitive mechanical motions required of industrial laborers. Now niche mass marketing asks designers to try to re-create the feel of particularities—something that nevertheless often feels, when designed, artificial or false.

In Sparke's words, "design is characterized by a dual alliance with both mass production and mass consumption":

> Like Janus, design looks in two directions at the same time: as a silent quality of all mass-produced goods it plays a generally unacknowledged but vital role in all our lives; as named concept within the mass media it is, however, much more visible and generally recognized. . . . [But] design as an adjunct of marketing has grown out of design as an aspect of mass production.
>
> (Sparke, 1986, pp. xix–xx)

It is difficult for us not to read back into craft work some primitive form of design. But this is as great a mistake as thinking of the hand-copied book (literally, a manuscript) as a primitive form of the book that rolled off Gutenberg's press. Instead, engineering design is properly seen as both a response to and a promotion of industrial production. To quote Sparke again, "Changes in production and in design . . . advanced in leaps and bounds and made otherwise complex tasks simple." Textile production provides an oft-cited exemplar. The spinning jenny and the Jacquard loom fundamentally transformed

> the way textiles were both conceived and made and intensified the changes in the process of design that the division of labor had already instigated. No longer could the [artisan] make spontaneous decisions about the appearance of the final product during its manufacture but . . . the desired pattern had to be fully planned and broken down into its component parts before manufacture began. This method of designing prior to production was echoed in fabric printing with the use of the mechanical roller, and in ceramics production where molds were used increasingly in the mass-production sector. The effects of these organizational changes in production methods were felt both in the appearance of the final products . . . and in the structure of labor patterns within the factory, with the emergence of a new breed of "art-workers" who translated the ideas of fine artists into mass production.
>
> (Sparke, 1986, p. 4)

The effects were felt as well in the new breed of "engineering-workers" who thought out the internal structure of products at the mechanical, chemical, and eventually electronic, molecular, and even genetic levels. This structural design was complemented by design of the mechanical, electrical, chemical, and other processes by which such products could be produced—and, eventually, by the psychological manipulative "engineering of consent" (Bernays, 1947) through propaganda.

ENGINEERING DESIGN AS THE TURNING OF MAKING INTO THINKING

Technological design connotes consciousness, intention in making, using, or acting. But making remains fundamental. The consciousness at issue in making can, as indicated, be of (a) the surface, the shape, the look or feel of the object (industrial and/or graphic design), and/or of (b) that which is beneath or behind the surface (engineering design). These two senses of technological design have engendered two quite different literatures. With regard to industrial- or graphic-design literature, the focus is on aesthetics and communication, especially in the context of competitive markets (advertising). (A leading effort to bridge this hiatus in the worlds of design is the work of Margolin and Buchanan, 1995.) With regard to engineering-design literature, the focus is on making-building and making-building processes, especially the methodologies for thinking or conceiving such processes with greater efficiency (see, for example, Cross, 1989).

Engineering design thus constitutes a distinctive way of turning making into thinking, engendering not only a special kind of making but also a unique way of thinking. The attempt comprehensively to explicate engineering-design thinking and making was first undertaken by the Dutch engineer and philosopher Hendrick van Riessen in his book *Filosofie en techniek* (1949).

Taking off from van Riessen, one may describe modern technology as a whole as distinguished from premodern or traditional technology insofar as artifacts are designed and produced from what he termed "neutral" elements. "Neutral" here means isolated, decontextualized, dis-embedded, standardized, and interchangeable or replaceable. Such neutral elements (processes as well as material parts), precisely because of their dis-embedded character, are largely deprived of qualitative value.

Reduction in contextual or qualitative value means in turn that, in the course of being integrated into a design, these neutral elements may be subjected to quantitative assessment—that is, numbered, measured, weighed in relation to eventual production and use. The design process itself is thus commonly described as composed of two basic moments: analysis (breaking down, identifying, and assessing the parts) and synthesis (a new bringing together or integrating the parts).

Take, for example, the designing of some machine. The machine is first defined in broad terms, commonly with regard to inputs, outputs, and operational specifications. The design engineer, thinking in frameworks provided by the engineering sciences (mechanics, thermodynamics, strength of materials, etc.), decides on the general or global character of the design (whether the machine is to incorporate a heat engine, electric motor, etc.). Within such parameters a dialogue then begins between the analysis of more detailed functions and parts, available technological building blocks (as summarized in engineering handbooks on parts, materials, and processes), and their synthesis into a functioning unit to meet the original specifications.

The engineering design is a synthesis composed of neutral modular elements, elements that mean nothing until they are integrated into some product, process, or system. The same elements may be integrated into many different machines, and their integration is always contingent or with little reference to larger inherited contexts. Furthermore, the design process itself has been progressively modularized. What was in the nineteenth century originally manifest in calculation and mechanical (i.e., modularized) drawing is increasingly computerized, the computer process being composed of numerous modular subroutines and bits of information or objects.

Engineering design is to traditional making as composing complex computer graphics with clip art is to freehand drawing. The difference between engineering design and clip-art composition is that in engineering design the aesthetic eye tends to be overshadowed by technical standards and calculations founded again in the engineering sciences: formulas for determining the bearing load of a beam, the reduction of friction in a mechanical assembly, energy consumption efficiency, and more. It is as if various options in clip-art composition were to be tested by calculations of minimal area covered with maximum number of pieces used or some other such mathematical determination.

As a consequence of this new state of affairs, design engineers, although not deprived of creativity, are called upon to exercise their creativity in what may be termed calculative ingenuity. The same engineer is also increasingly alienated or separated from actual fabrication. Making is preceded by a calculative thinking out in advance. Early on in the emergence of engineering design, there remained numerous opportunities for feedback from the actual construction or making that would take place in the shop or in the factory. Insofar as this calculative thinking out in advance increasingly and effectively analyzes or modularizes parts and processes, there is less and less opportunity or need for such feedback. Indeed, making may even be described as itself reduced to a thinking out, insofar as engineers are able fully to prepare in their minds and on computer screens artifacts and production processes. This trajectory in engineering design is illustrated in Karl Sabbagh's *21st-Century Jet: The Making and Marketing of the Boeing 777* (1996), a narrative account of the first wide-body passenger jet designed completely with computers, for which no prototype or test model was ever constructed.

THE PROBLEMATICS OF ENGINEERING DESIGN

In an essay stimulated by van Riessen, the Dutch aerospace engineer Ad Vlot reflected on the problematics of this new way of thinking before making. Vlot writes:

> In the traditional form of technology the same craftsperson is the designer and the maker, and the parts of a product are unique for that specific product. Standardized bolts, rivets, switches, transistors, filters, etc. are typical of modern technology. In this way the technological design method that isolates the different disciplines corresponding to the various neutral elements becomes a powerful tool used by the engineer in combination with automation to control the designed artifact before it becomes operational. [But] this systematic design method does not correspond with reality, in which everything is unique, variable, and embedded in relationships. Therefore the art of engineering involves the "feeling of the engineer" for dealing with frictions between the result of theoretical optimization and reality such as variations, wear, environmental influences and, above all, human freedom that may disturb the process or functioning of the artifacts.
>
> (Vlot, 2000, p. 210)

For Vlot there thus exists the possibility that thought-out products may not be sufficiently thought out. In conjunction with the professional engineering obligation "to hold paramount the safety, health, and welfare of the public" (as it is termed in the U.S. National Society of Professional Engineers 1993 Code of Ethics, Fundamental Canons, 1), this implies a practical responsibility to think harder and more broadly than might otherwise be the case. Design engineers have what may called a duty *plus respicere*, to take more into account: that is, a professional obligation to expand design thinking in order to consider more aspects of reality.

Yet as Vlot acknowledges, the issue of human freedom—in contrast to the issues of variations, wear, and environmental influences—presents special challenges for any duty *plus respicere*. Human freedom includes the freedom to use improperly (to fail to follow directions for use), to misuse (to employ as means for some unintended or socially disapproved end), or to abuse (to treat in such a manner as to degrade or destroy more rapidly than planned).

From one perspective—that of Michel de Certeau (1984)—misuse may actually manifest human freedom and creativity in a way that should not be abridged. Technicalfunction theorists, especially those influenced by the concept of proper functions in biology, have a tendency to read the use plan of a designer as a moral prescription (Franssen, 2005). But what is wrong with using a screwdriver as a chisel?

Bart Kemper (2004), however, argues that at least some artifacts ought to be designed taking into account possibilities for "evil intent." Kemper focuses in particular on otherwise benign artifacts that might be turned into WMD: vans and fertilizer (e.g., in the case of the Oklahoma City car bomb attack carried out by Timothy McVeigh) or airplanes and skyscrapers (e.g., in the case of the 9/11 attacks on the World Trade Center and Pentagon). Vans and fertilizers and planes and skyscrapers can all be improperly used or misused in various relatively benign ways (as when a van is turned into living quarters or so much fertilizer is put on plants that they die). Although such designer-unintended uses might be worth modest attention in the design process, the stakes are raised dramatically when evil intent is able to employ such products in ways that can kill hundreds or thousands of people.

In such cases, Kemper argues, engineers have an obligation to try to design them so as to make misuse difficult. The principle would be a simple extension of that manifested in designs such as the safety shields to keep workers from harming themselves with large punch-press machines or passive restraints in automobiles to strap in the driver and passengers whether they like it or not. Nitrogen fertilizers could be so diluted as to make them more difficult to use in powerful explosives; airplanes can be equipped with stronger cockpit doors and tall buildings with structures that better withstand planes crashing into them or provide better means of egress for building occupants.

A DUTY *PLUS RESPICERE* AND ITS DISCONTENTS

It may be objected that an engineering duty *plus respicere* to take such aspects of human behavior and reality into account—especially the potential evil intent of terrorists—is unreasonable on two counts: (a) such considerations are outside the scope of engineering knowledge and practice and (b) no one, especially not an engineer, can predict all aspects of the future.

First, what is the character of engineering such that engineers might be called on to consider apparently nonengineering issues like the intentions of terrorists? In the popular mind engineering is commonly taken to be focused on making matter or energy useful and convenient for people in the most efficient manner. Engineering students certainly often enroll in their programs of study expecting to be able to rely on their mathematical, scientific, and computer aptitudes and to avoid developing interpersonal or communication skills. Engineering is reduced to determinate (often calculated in terms of efficiency within specified boundary conditions) problem solving—and it seems unfair to ask persons with such technical abilities to consider the many possibilities of misuse.

Engineers do not, however, simply aim at some abstract ideal of efficiency; they aim at specific instances of efficiency as determined by particular specifications. Sometimes what is important in designing a bridge is that it will be durable and last a long time; other times what is important is that it can be constructed as quickly as possible. The value of efficiency is formal and contextual, given substance by other values external to the design process. Engineering as the pursuit of efficiency is subordinate to whatever external specifications define the relevant inputs and outputs. At the same time, external specifications are almost always negotiated with the engineers to whom they are given. Engineering clients seldom know for sure how to specify what they want; indeed, they often have to learn from the engineers they hire that some of their imagined specifications are at odds—that there are trade-offs between cost and quality, for instance—whereas others are impossible. No engineer will take a job that makes unrealistic demands on a budget while requiring previously unattainable levels of safety and reliability or that specifies a system with a perpetual motion machine as its prime mover.

Thus the understanding of engineering as design often calls for some engineering-client negotiation. To some extent design rests on natural science (physics, chemistry, and biology) and on engineering science (mechanics, thermodynamics, electronics), but the design of products, processes, and systems also requires appreciation of human needs and how the social world works. Engineered artifacts must ultimately find a home in human practice and make their way in the marketplaces of money and ideas or find themselves relegated to what one author has described as "inventions necessity is not the mother of" (Jones, 1973). Engineering-design failures can be as much social as technical. The most effective engineers are those who either have a sense of economic need or interest or know how to enter into dialogue with their employers or the public to assess in advance how designs might be used. It is thus not unreasonable to ask them to consider as well how their designs might be misused—perhaps by unparticipating interlocutors and unanticipated users.

Moreover, the extensive technical work that engineers do to guard against technical failure provides preparation for parallel analyses to avoid social misuse. Robert Charles Metzger's *Debugging by Thinking* (2003) argues for a multidisciplinary deployment of logic, mathematics, psychology, safety analysis, computer science, and engineering in order to avoid and correct software errors. This is virtually an operationalization of the duty *plus respicere*. From

Metzger's perspective there is something professionally irresponsible about the too common release of software programs that have been insufficiently tested for multiple weaknesses. But such interdisciplinary skills exercised at the technical level readily blend into the exercise of related analyses at the social level. Indeed, just as earlier complaints about the difficulties of taking into account multiple possibilities for unintended consequences in the environmental area have led to the development of methods for environmental impact assessment, so it seems reasonable to postulate the formulation of methods for misuse-and-evil-intent possibility assessments.

This last point also suggests an initial response to the second objection: that no engineer can be expected to be able to predict all aspects of the future. The future, and the future consequences of designs, can be studied. Futurology is the most general effort to undertake such a study and has generally not been directed toward assessing the results of particular design projects. Often it seeks to predict or to call for the emergence of various designs. But the proliferation of scientific, sociological, and other efforts to develop technology assessment methods and to analyze and empirically examine unintended consequences, uncertainty, and risk all point toward a rich variety of possible theoretical and practical responses to the general challenge set forth in the second objection. (In this regard see, for example, Averill, 2005.) To accept the objection at face value is simply to make an excuse or to refuse to acknowledge the work being done to advance knowledge of unintended consequences. The unintended consequences of design may be difficult to predict or not fully predictable, but they are certainly able to be assessed more fully than is commonly done.

THE METAPHYSICS OF ENGINEERING DESIGN

There is nevertheless a deeper sense in which the future may be described as unpredictable—a sense that deserves consideration and exhibits, as it were, an ontological tension with the very idea of technological design. Design and designing constitute an attempt to prefigure the future that sees the future as open to indefinite manipulation. The design stance (to reinterpret what is for Daniel Dennett, 1987, an epistemological term) involves a conception of the human as both a subject and object of design, planning, thinking out, and the further activities of making, remaking, and transforming. As such it opposes a traditional understanding of the human as fundamentally rooted or particularized, although in ways that are open to the more than human.

The traditional or premodern view of the world is that it constitutes a reality that ultimately calls for acceptance. The world may be tinkered with around the margins, but the human condition is one that humans do not have the option of escaping. The metaphysical basis of engineering design was, for one philosophical observer, dramatically disclosed in a common reaction to the 1957 launch of *Sputnik*, as simply the first step toward an escape from human imprisonment to the earth. According to Hannah Arendt,

> The banality [of such a view] should not make us overlook how extraordinary in fact it was; for although Christians have spoken of the earth as a vale of tears and philosophers have looked upon their body as a prison of mind or soul, nobody in the history of mankind has ever conceived of the earth as a prison for men's bodies or shown such eagerness to go literally from here to the moon.
> (Arendt, 1958, p. 2)

The engineered achievement of space flight represents the apogee of a historico-philosophical trajectory that began with early modern arguments for a "conquest of nature" (Francis Bacon) that would enable humans to become its "masters and possessors" (René Descartes). At its deepest level, technological design aims to enlarge human freedom to world-transformational levels.

The paradox of the designed world, which refuses to accept the givenness of the human condition, is that it nevertheless creates a world to which even the designer is forced to submit. In the first instance this does not appear to be submission at all. The designed world is, after all, a world of human creation. Yet all designed products, processes, and systems have some, however marginal or residual, undesigned consequences. This is as true of engineering as of other types of design (including artistic or graphic design). Just as philosophers of science have argued that all scientific theories are underdetermined by the evidence, so are designs underdetermined by the intentions and analyses of their designers. Just as the world is susceptible to design, it is also in its contingencies and complexities always resistant to design. As design takes hold and its consequences (intended and unintended) proliferate, the need or demand for more design multiplies. More and more thinking and planning and designing seems to be required just to live with what has been designed. What other solution to design shortcomings is there than simply more design? There appears to be less and less space, even in a radically pluralist culture, for anyone who would live the undesigned life.

Nowhere is the demand for more and more design (and its limitations) more apparent than in relation to the issue of sustainability. Despite the claims of those who deploy the argument from design to postulate from the intricacies of nature the reality of a creator (or designer) God, the real reason the multiple parts of nature function together as they do both in organisms and across ecosystems is that they have not been consciously designed. Nature works as well as it does because it has taken form slowly, adjusting parts to each other and the whole over periods of time that are quite beyond human scale, not by conscious design but by trial-and-error selection. As humans introduce new changes into nature of a speed and character that is non-natural, these changes interfere with existing relationships and introduce perturbations that design must work increasingly hard to comprehend and control. The essence of engineering design is to try to thwart a slow process by extracting itself from embeddedness in a complex interactive system and to insert top down into it an innovation that does not emerge from the bottom up. Environmental sustainability and sustainable development thus demand progressively more research and conscious planning. The disciplines of environmental science and engineering point to projects for earth-systems engineering and dreams for managing the planet (U.S. National Academy of Engineering, 2002). (Extended, such dreams take on prospects for redesigning Mars and other planets through terraforming.)

Yet even among engineers there are recognitions of the fact that even the most consciously thought-out dreams can repeatedly bump up against the complexities of nature—and the consequent difficulties of design. The very fact that design seems to demand more and more design points to an inherent resistance to design. This resistance to design is a limitation that designers (and users) ultimately have an obligation to acknowledge in both theory and practice.

AUTHENTICITY IN ENGINEERING DESIGN

If the world is at once susceptible to and resistant to design, how ought people respond? One extreme would be to emphasize susceptibility and ignore resistance. Maybe resistance can be overcome by better or deeper or more thoroughgoing design (nanotechnological design or genetic engineering?)—can be overwhelmed by more and more design. Another extreme would be to emphasize resistance and argue that susceptibility to brute intelligent design (using AI?) is at best illusory and at worst harmful. Unable to design completely, humans ought not to design at all.

Precisely because they seek to ignore one aspect of the complex ambivalence that is the reality of the experience of engineering design, both extremes exhibit a kind of inauthenticity. Adapting yet redesigning a notion of authenticity found in Søren Kierkegaard, perhaps it would be possible to confront the dialectic of susceptibility and resistance to design by acknowledging a dialectic. The human world is a complex of both finitude and infinitude, necessity and possibility, determination and freedom. Authenticity consists in accepting in proper proportions both my freedom and my unfreedom; inauthenticity consists in failing to accept some aspect of either or both. In the world of traditional making it may be suggested that the most common temptation to inauthenticity was to deny all freedom (what Kierkegaard called the "despair of necessity").

In the world of design, the most common temptation to inauthenticity may be instead to deny all determination (Kierkegaard's "despair of possibility"). Certainly Kierkegaard's words have some resonance in the world of design, when he writes that

> if possibility outruns necessity so that the self runs away from itself in possibility, it has no necessity to which it is to return; this is possibility's despair. This self becomes an abstract possibility, it flounders around in possibility until it is exhausted.
> (Kierkegaard, 1980 [1949], pp. 35–36)

On the one hand, to reject all opportunities for design (susceptibility to design) is to fall inauthentically into "the despair of necessity." On the other, to design while denying necessity (resistance to design) is to fall inauthentically into "the despair of possibility." Authenticity entails recognizing an obligation to design with an openness to the limits of design—that is, recognizing in engineering design a duty *plus respicere*, a professional obligation to expand design thinking in order to try to take more, including now the limitations of design, into account.

Patrick Feng (2000), in a related argument, raises the issue of a reaction to technology that he suggests is a "barrier to ethical design." According to Feng,

> For many people, the idea that technology is moving faster than we can adapt seems commonsensical. When a 1995 opinion poll asked American consumers whether they agreed with the statement "technology has almost gotten out of control," an astounding 63 percent of respondents said yes. Media reports reinforce this belief: with few exceptions, the press in America tends to talk about technology as if it were an external force beyond human control.
> (Feng, 2000, p. 209)

Such a fatalism with regard to technology, along with its flip side, a blind faith in technological progress, is inauthentic. A more authentic attitude toward technology would be characterized by Feng's claim that *technology both shapes and is shaped by its social context* (p. 212, author's emphasis). Such an authentic attitude "opens the door to including ethical discussions in the actual *design* of artifacts" (p. 212, author's emphasis). At the root of a possible ethics of design there will thus rest a recognition that no matter how good or sophisticated the technological design, to some degree results can always transcend intentions—and the recurring presence of such results are phenomena that human designers, even in their attempts to design around them, must accept. Further recognition that such acceptance may imply a limitation not just *in* design but *on* design may be suggested as the beginning of wisdom for design.

Essay 5

Professional Idealism among Scientists and Engineers

A Neglected Tradition in STS Studies

"It was the best of times, it was the worst of times." With a sentence that has become an often-evoked statement of the ambivalence of modernity, Charles Dickens opens *A Tale of Two Cities*, first published in 1859, the same year as Charles Darwin's *The Origin of the Species*. Dickens's narrative of complex and intertwined destinies in Paris and in London during the French Revolution—not wholly unlike that of St. Augustine's extended reflection on the presence of a heavenly City of God in the very midst of the earthly city of temporal power—is one that could apply equally well to what C.P. Snow, a hundred years later, in 1959, termed the "two cultures."

Snow's view of the two cultures, which distinguishes scientific from literary intellectuals, presents scientists and engineers as progressivist revolutionaries hounded by the rearguard critics of an old-order literary establishment. Scientific intellectuals are moral leaders "impatient to see if something can be done" (Snow, 1959, p. 7), optimistic about the future, and the only hope for dealing with such "menaces" as nuclear war, overpopulation, and the gap between the rich and the poor. Literary intellectuals, by contrast, are "natural Luddites," whose irrational reactions against both the scientific and industrial revolutions helped pave the way for the horrors of Auschwitz. In Snow's portrait, scientists and engineers are members of a progressive cadre who are able to quote Shakespeare and more but are fundamentally opposed by antitechnologists who fail to appreciate science sufficiently even to be able to state the Second Law of Thermodynamics.

During the year in which Snow delivered his lecture at Cambridge University, however, the biologist Rachel Carson was beginning work on a book to be serialized in a literary journal, *The New Yorker*, that would be as ardently critical of technoscientific developments as any articulated by such natural

Luddites as T. S. Eliot or William Butler Yeats, two of Snow's whipping boys. Indeed, it was as a scientist that Carson put on the intellectual map a menace not yet recognized by Snow—that of environmental pollution. When the U.S. Environmental Protection Agency was established in 1970, less than a decade after the publication of *Silent Spring* (1962), it was considered by many to constitute political homage to Carson's strenuous efforts to educate not just the public but also her fellow scientists.

This important but bureaucratic legacy must not, however, obscure the radical bifurcation she introduced into science itself. "We stand now where two roads diverge," she wrote at the end of her book.

> But unlike the roads in Robert Frost's familiar poem, they are not equally fair. The road we have long been traveling is deceptively easy, a smooth superhighway on which we progress with great speed, but at its end lies disaster. The other fork of the road—the one "less traveled by"—offers our last, our only chance to reach a destination that assures the preservation of our earth.
> (Carson, 1962, p. 277)

The scientific project to conquer nature through chemical pesticides was, she argued, "conceived in arrogance . . . when it was supposed that nature exists for the convenience of [humans]" (p. 297). Her alternative is a science that, based on an awareness of the richness and complexity of life, would attempt to share "our earth with other creatures" (p. 296).

What Snow failed to notice is that the line between his two cultures— between the endorsement and criticism of technoscience—like that between the two cities, does not so much divide scientific from literary intellectuals as it runs through the midst of both communities. Indeed, the divide between optimistic promotion and critical assessment of technoscience is one that runs through all of us, often separating our own hearts and minds. "That which I would do, I do not," wrote St. Paul, and "that which I would not do, that I do" (Romans 7:15). The life that we as a modern society would construct, our scientific knowledge and engineering prowess often fall short of being able to realize; the material culture we would not construct, the unintended consequences of our scientific-engineering actions too many times construct for us.

What is remarkable is the extent to which scientists and engineers have known this as much if not more so than their humanities colleagues. At the 1945 explosion of the first atomic bomb near Alamogordo, New Mexico, physicist J. Robert Oppenheimer quoted to himself words from the *Bhagavad Gita*: "I am become Death, the destroyer of worlds" (Giovannitti and Freed, 1965, p. 197). In a lecture at MIT shortly thereafter he observed, "In some sort of crude sense which no vulgarity, no humor, no over-statement can quite extinguish, the physicists have known sin" (Oppenheimer, 1947, p. 6).

For Albert Einstein, "The bomb . . . and other discoveries present us with . . . a problem not of physics but of ethics" (Einstein, 1960, pp. 384 and 385). During the 1950s, according to French mathematician and philosopher Michel Serres,

> For the first time since its creation, perhaps since Galileo, science—which had always been on the side of good, on the side of technology and cures, continuously rescuing, stimulating work and health, reason and its enlightenments—begins to create real problems on the other side of the ethical universe.
> (Serres, 1995, p. 17)

Among others, Raphael Sassower, in his felicitously titled (although not as felicitously argued) book, *Technoscientific Angst* (1997), has called attention to this intrascientific tradition of the criticism of science—a tradition that Snow himself attributed solely to literary intellectuals. What I propose is simply to extend the recognition of such scientific intellectual criticism of technoscience by calling further attention to this insufficiently appreciated phenomenon and its literature,

Let me note in passing that this tradition of what I will call "professional technoscientific idealism" has had to struggle for recognition not only among scientists, engineers, and the general public but even within that interdisciplinary scholarly field known as science-technology-society (or STS) studies, which should have paid most attention to it. In what follows I thus simply review some historical moments in the practice of this tradition, which might equally well be termed "scientific-engineering idealism," calling brief attention to the Federation of American Scientists (FAS), the *Bulletin of the Atomic Scientists*, the Pugwash movement, and the Union of Concerned Scientists (UCS). Next I provide a slightly more detailed report on the Committee for Scientific Freedom and Responsibility (CSFR) of the American Association for the Advancement of Science (AAAS). I conclude by reiterating a speculative interpretation of the importance of the historical and sociological material under review.

1. FAS, THE *BULLETIN*, PUGWASH, AND UCS

Any historical narrative of the making of the atomic bomb has to mention at some point Albert Einstein's initiating letter. In 1939, Einstein signed a letter (drafted by Leo Szilard) to President Franklin D. Roosevelt promoting the need for a program to develop nuclear weapons. Two decades later, in 1955, Einstein also issued a public manifesto (authored this time by Bertrand Russell) to his fellow scientists calling on them to work to restrict and control nuclear weapons. Comparison of these two documents is instructive.

The 1939 letter is a carefully worded, spare six paragraphs pointing out to the president the reality of atomic energy and the need for "quick action" on the part of the U.S. government, in light of German actions pointed toward exploiting the possible use of such energy to construct "extremely powerful bombs." (The full text may be found in Grodzins and Rabinowitch, eds., 1963, pp. 11–12.) The letter in fact had the desired effect of stimulating establishment of the Manhattan Project—which led first to Alamogordo and then to Hiroshima and Nagasaki.

What is remarkable, however, is how completely atypical this letter is of Einstein's thought. Much more common during this period of his life were public pronouncements in support of socialism and war resistance—although he was also pointedly critical of fascism. But when speaking to scientists and engineers, as at the California Institute of Technology in 1931, he was equally likely to criticize technoscience itself. "It is not enough that you should understand about applied science in order that your work may increase [human] blessings," he told the Cal Tech students and faculty.

> Concern for man himself and his fate must always form the chief interest of all technical endeavors; concern for the great unsolved problems of the organization of labor and the distribution of goods in order that the creations of our mind shall be a blessing and not a curse to mankind. Never forget this in the midst of all your diagrams and equations.
>
> (Einstein, 1960, p. 122)

For Einstein it was important that scientists and engineers not overestimate the importance of their work and give themselves up to the perennial attractions present in the excitements of science and engineering in ways that would distract them from being truly human.

The 1955 public manifesto, which he endorsed just a few days before his death, is much more in harmony with the predominant spirit of Einstein's views. Unlike the private letter to President Roosevelt, the public manifesto opens with dramatic recognition of the difficult situation created by Hiroshima and Nagasaki, and the subsequent continuing development of nuclear weapons. "In the tragic situation which confronts humanity, we feel that scientists should assemble in conference to appraise the perils that have arisen as a result of the development of weapons of mass destruction." (For the full text of this manifesto, see Rotblat, ed., 1972, pp. 137–40.) It goes on in 15 strongly worded paragraphs to argue that scientists must educate the public about the underappreciated dangers of nuclear weapons and radioactive fallout, then proposes a resolution to promote:

> In view of the fact that in any future world war nuclear weapons will certainly be employed, and that such weapons threaten the continued existence of

mankind, we urge the Governments of the world to realize, and to acknowledge publically, that their purpose cannot be furthered by a world war, and we urge them, consequently, to find peaceful means for the settlement of all matters of dispute between them.

The original manifesto was signed not only by Einstein but also by eleven other Nobel laureates and senior scientist-engineers.

One need not overlook the naiveté of the world government and socialist sentiments that lie behind some of the rhetoric of this manifesto. Indeed, there may well be a sense in which the signers of the document, although they were able to quote Shakespeare, did not really understand his dramatic teaching at any depth. Nevertheless, there is present in this document not only a criticism of governments but also a direct criticism of scientists and engineers and of the fact that they have been insufficiently involved in public affairs. This is an ethical-political criticism of science—or at least of technoscience as socially constructed—that is already clearly present in Einstein's Cal Tech talk of 1931 and that resurfaces not just in the Einstein–Russell manifesto of 1955 but earlier, in more institutional form, with the establishment of the Federation of American Scientists and the publication of the *Bulletin of the Atomic Scientists*.

The Federation of American Scientists

The Federation of American Scientists (or FAS)—originally named the Federation of Atomic Scientists—was founded in 1945 by members of the Manhattan Project in order to respond to the implications and dangers of the nuclear age. Today with 5,000 members, including more than 50 Nobel Prize winners (i.e., over half of the living U.S. science laureates), and its headquarters in Washington, DC, FAS is the senior organization of professional scientists dedicated to nuclear arms control and disarmament, often providing expert testimony before the U.S. Congress. FAS is a nonprofit organization licensed to lobby in the public interest. According to its very active website (fas.org), the FAS "conducts analysis and advocacy on science, technology and public policy, including national security, nuclear weapons, arms sales, biological hazards, secrecy, and space policy."

During its early years, the FAS lobbied in support of the creation of a civilian-controlled national science foundation and criticized attempts to impose national security restrictions on scientists. Alice Kimball Smith (1965) and Donald Strickland (1968) provide useful historical studies of its origins. Then after working publicly and behind the scenes for 15 years to bring to a halt the momentum of the atmospheric testing of nuclear weapons, achievement of the Limited Nuclear Test Ban Treaty in 1963 led to a period of quiescence, perhaps exhaustion.

Jeremy Stone, a Stanford mathematician, became the new FAS director in 1970 and began a concerted effort to revitalize the organization (for his personal account, see Stone, 1999). In response to the Nixon-Brezhnev period of detente, for instance, FAS promoted public and political action to defend the human rights of scientists such Andrei Sakharov, supported both the SALT and START treaty negotiations, and lobbied in favor of the nuclear freeze movement. Later the FAS criticized the Strategic Defense Initiative and published a number of important studies such as *First Use of Nuclear Weapons: Under the Constitution, Who Decides?* (1987) and the *International Handbook on Chemical Weapons Proliferation* (1991). Today its periodical, *FAS Public Interest Report*, remains a leading source of information on these and related issues.

Bulletin of the Atomic Scientists

Closely related to FAS is the *Bulletin of the Atomic Scientists*. Both originated at the same time and out of the same sense of urgency about responding to the implications and dangers of the nuclear age. In a fundraising appeal for the *Bulletin*, for instance, Einstein made his famous remark:

> The unleashed power of the atom has changed everything save our modes of thinking, and thus we drift toward unparalleled catastrophe. . . . [A] new type of thinking is essential if mankind is to survive and move toward higher levels.
>
> (Einstein, 1960, p. 376)

What Einstein referred to as a "new type of thinking" is subject to diverse interpretations, not just for society and politics but also for scientists and engineers. Nevertheless, as early as the summer of 1944, a few scientists gathered around Nobel laureate James Franck at the University of Chicago Metallurgical Laboratory (or Met Lab, eventually the Argonne National Laboratory), where the first controlled nuclear chain reaction had been achieved in 1941, opened discussion about the long-range consequences of their work. The received wisdom was that atomic science could be used for good or evil—for unbelievable advances in biomedicine, metallurgy, and perhaps even electrical energy production, or for a worldwide nuclear arms race and wars of unimaginable destruction. Yet given the enormity of the potential evil, especially in comparison with the potential good, it became increasingly questionable to some whether this characterization of the problem was not just another instance of an old type of thinking.

Struggles to think this new situation were first articulated in June 1945 in what became known as the Franck report to U.S. Secretary of War Henry L. Stimson. The report, taking a "more active stand" than in the past, when "scientists could disclaim direct responsibility for the use to which mankind . . . put their disinterested discoveries," called for the international control of atomic energy (Grodzins and Rabinowitch, eds., 1963, pp. 19 and 20, with the complete Franck report available on pp. 19–27). It further argued that the unannounced dropping of atomic bombs on Japan would so compromise the moral leadership of the United States as to make internationalization impossible to achieve. The probable outcome would be a postwar nuclear arms race.

There is no evidence that Stimson ever read the Franck report, and it is certainly the case that it had little if any influence on war (or postwar) strategy. But immediately after the war, scientists associated with the Franck report founded the Atomic Scientists of Chicago (ASC), which published the first issue of the *Bulletin* in December 1945. "The American people," said the *Bulletin*'s opening editorial, must work "unceasingly for the establishment of international control of atomic weapons, as a first step toward permanent peace." The *Bulletin* remains today a major representative of professional scientific idealism and a continuing source of internalist criticism of technoscience.

The Pugwash Movement

The FAS and *Bulletin* emerged almost simultaneously in the second half of the 1940s in response to the atomic bomb. Ten years later, the detonation of the first thermonuclear device, the hydrogen bomb, gave rise to a second wave of scientific engagement with world affairs.

The first wave appeared impotent; international control had failed, and the nuclear arms race was in full stride. The logic of nuclear weapons development seemed to follow a path that was, in Oppenheimer's apt phrase, "technically so sweet that you could not argue" against it (U.S. Atomic Energy Commission, 1954, p. 251). In this situation the already mentioned Einstein-Russell manifesto was published in the *Bulletin* in 1955, and gave rise two years later, in the small fishing village of Pugwash, Nova Scotia, where the U.S. left-wing industrialist Cyrus Eaton had a vacation house that he offered for their use, to the first truly international meeting of scientists opposed to the nuclear arms race.

Beginning with the first meeting in 1957, Pugwash Conferences on Science and World Affairs have been held at least once every year. They are always international and rotate around the globe. The record of these conferences is available in Joseph Rotblat's *Scientists in the Quest for Peace* (1972) and subsequent publications. And as Rotblat notes in a later Pugwash symposium proceedings, it was a Pugwash proposal for "black-box" unmanned automatic seismic detectors that made possible the 1963 Limited Nuclear Test Ban Treaty (Rotblat, ed., 1982, p. 139).

In the mid-1970s, it may also be noted, university students in the United States took the initiative to form Student Pugwash, which also began holding a series of educational conferences. Although not so much public policy oriented as public policy education oriented, Student Pugwash has served as an important bridge between older and younger generations of scientists. (The proceedings of the first Student Pugwash USA conference are available in Lakoff, ed., 1980.)

In 1995 Pugwash and Rotblat, sole survivor of the original 27-member group that met in Nova Scotia forty years earlier, were awarded the Nobel Peace Prize. Since then, especially, Rotblat has argued tirelessly for the adoption of some form of Hippocratic Oath for scientists. Only this, he argues, will make it possible for many individuals to step out of that technoscientific momentum that so often overwhelms personal judgment (see, e.g., Rotblat, 1999 and 2000). (It should also be noted that Rotblat was the only nuclear scientist to depart Los Alamos once it became clear that the Nazis' bomb project had been abandoned, thus depriving the Manhattan Project of its original ethical and political rationale.)

The Union of Concerned Scientists

The UCS, born out of an opposition to the Vietnam War, represents what may be termed a third wave of technoscientific activism. Once again it arose out of a profound sense that technoscience had, however socially constructed, taken on a life of its own that was becoming a threat to humanity. It also expanded from issues of nuclear arms to questions of conventional military technologies and concern for the environment.

In December 1968 at MIT—at the research heart of what President Dwight Eisenhower in his farewell address of scarcely eight years earlier had termed the "military-industrial complex"—50 senior faculty issued a statement that began by declaring how the "misuse of scientific and technical knowledge presents a major threat to the existence of mankind." Because previous responses of the scientific community have been "hopelessly fragmented,"

> We therefore call on scientists and engineers at MIT, and throughout the country, to unite for concerted action and leadership: Action against dangers already unleashed and leadership toward a more responsible exploitation of scientific knowledge.
> (UCS website at ucsusa.org.)

The immediate impact of this call to action was adaptation of the university "teach in"—a form of Vietnam War protest developed on campuses throughout the United States—to the exploration of science and public policy issues. On March 4, 1969, in place of their regular classes, MIT faculty and students initiated a series of critical reflections on the nature and use of technoscience. The results were wide ranging, inspiring not only a classic STS essay by Paul Goodman (1969) entitled "Can Technology Be Humane?" but also the formation of the UCS, which is today an independent nonprofit alliance of 70,000 committed citizens and leading scientists across the United States with a staff of about 50. "We combine rigorous scientific research with public education and citizen advocacy to help build a clean, healthy environment and a safer world," states the UCS website (ucsusa.org).

One UCS goal has been to provide journalists and the public with authoritative sources of technoscientific information other than those provided by the government. For instance, during the 1970s two UCS members, Henry Kendall and Dan Ford, used the Freedom of Information Act to supply reporters with technical information about weaknesses in the regulation of the nuclear power industry. Then after the Three Mile Island nuclear accident, the UCS undertook independent monitoring of nuclear power plants. In the 1980s the UCS, like the FAS and *Bulletin of the Atomic Scientists*, became heavily involved in criticizing President Ronald Reagan's Strategic Defense Initiative.

In a memoir that builds on his years of experience, former chair of the UCS board Kendall makes a distinction between two kinds of harms from technoscientific change. One is those that are immediate, affect broad segments of the population, and thus "generate political forces that have the strength to bring the problems under control." Another is those "whose damaging consequences may not be apparent until well into the future" and/or "do not induce society to correct them" (Kendall, 2000, p. 3). In this second context technoscience tends to exhibit a kind of autonomy that is difficult to address. "All we who can gauge [such] threats can do," Kendall concludes, "is soldier on, exploiting what tools we have, gaining as much ground as time permits" (p. 304).

The FAS, *Bulletin*, Pugwash, and the UCS all provide concrete examples of scientists serving as moral critics of technoscience and its social constructions. They have both witnessed and in many cases contributed to the technological construction of society—especially with nuclear weapons—and then struggled to oppose it. In so doing they have attempted to practice the social reconstruction of technology, although not with any simple or unequivocal success. Given the exigencies of techno-economics, not to mention those of human nature, there seem to be tendencies or trajectories of use embodied, if not in the technoscience alone then certainly in the technoscience-society interface, that it is often extremely difficult to sidestep them.

2. COMMITTEE FOR SCIENTIFIC FREEDOM AND RESPONSIBILITY

During the 1970s, as another instance of the third wave of post–World War II technoscientific criticism of technoscience, the American Association for the Advancement of Science (AAAS)—the largest interdisciplinary scientific society in the world—established a special Committee on Scientific Freedom and Responsibility (CSFR). In this committee it is possible to see even more vividly than with the FAS, *Bulletin*, Pugwash, or the UCS, the incorporation of technoscientific criticism into mainstream technoscientific institutions—and the broadening of the critical agenda, from nuclear weapons to environmental issues and questions concerning professional codes of ethics, human rights, and technoscientific practices.

The AAAS, founded in 1848, is the oldest and most prestigious general professional organization of scientists in the United States. Its original three goals were somewhat self-serving:

- to promote interaction among scientists;
- to strengthen the systematic pursuit of science; and
- to secure more (political and financial) support for science.

But issues of scientific ethics and responsibility began to be voiced early, and during the 1930s such discussion achieved a significant flowering. (For one broader though inadequate history of this pre–World War II period, see Kuznick, 1987.)

The rising emphasis on questions concerning relations between science and society were promoted especially by James McKeen Cattell, editor of *Science*, the AAAS journal. Cattell was a socialist who in 1936 succeeded in getting his candidate, Forest Ray Moulton, elected AAAS permanent secretary. The following year Moulton announced that the AAAS annual meeting would begin a series of conferences on "Science and Society." The presidential address that year was even titled "Science and Ethics" and reflected a strongly increasing trend to break out of strictly technical papers and to criticize the social contexts within which science operated.

This trend was necessarily stymied by World War II, during which critical questioning was subordinated to the pursuit of technoscientific weapons. Then immediately after the war, independent organizations such as those already mentioned siphoned off some of the potential for radical reflection within AAAS. But by the mid-1960s such action began again to find serious institutional purchase, which led to the creation of a standing Committee on Scientific Freedom and Responsibility (CSFR).

CSFR had its origins in the appointment, in 1970, of a temporary Committee on Scientific Freedom and Responsibility composed of seven leading scientists and political figures. This committee was charged

- to study the general boundary conditions for the exercise of scientific freedom and responsibility;
- to develop criteria and procedures for dealing with freedom and responsibility; and
- to recommend how the AAAS might act to defend scientific freedom and promote scientific responsibility.

Its official report, which was issued in 1975, became known as the Edsall report, after John Edsall, a committee member and Harvard professor of biochemistry, who served as its primary author.

The Edsall report is an important historical document. It begins by noting the two-edged character of science: science has increased knowledge and cured disease; it has also made possible nuclear warfare and chemical pollution of the environment. It recognizes the "tense and often bitter controversy" among scientists about these issues, especially in regard to what society owes to science and what science owes to society (Edsall, 1975, p. 2).

In its own response to these disputes, the report argues, with reference to its first and most general charge, that rights are subordinate to responsibilities. "Scientific freedom," it asserts, "is an acquired right, generally accepted by society as necessary for the advancement of knowledge from which society may benefit" (p. 5). Scientific freedom is thus not a fundamental human right such as freedom of speech but a right that is to be justified on consequentialist grounds. From this perspective the report considers various proposals for abridging scientific freedom and makes, with regard to basic science, the following three arguments:

a. No area of research should be proscribed, though some methods may be. Proscribed methods are fundamentally those that violate human dignity by failing to meet standards of informed consent, especially with regard to risks.
b. Because of its potential benefits, fetal research should not be restricted.
c. Despite national security interests, it must be recognized that secrecy obstructs science.

Turning to applied science, the report admits that "the results of innovation are always more complex than the innovators intended, and usually more complex than they could even imagine" (p. 23). This suggests the need for technology assessment. The report explicitly rejects any idea that "there is something inherently evil about technology," as technology of some sort is necessary for human life (p. 26). At the same time, the report also rejects "the notion of the so-called 'technological imperative'" (p. 26). It is not only possible to reject the development of certain technological possibilities, rejection has in fact been done.

With reference to its second charge, the development of criteria and procedures for dealing with freedom and responsibility issues, the report begins with three case studies of whistle-blowing: one having to do with radiation exposure, another with engineering safety, and the last with chemical exposure. In each case, the establishment of criteria for judging the situation depend on many variables. The report only feels competent to note some of the factors that must be taken into account: effects on human health and safety, quality of life, economic impact, and more. As for procedures, the report sides strongly with the need to protect whistle-blowers and argues that professional scientific and engineering societies should serve as protectors of the public interest.

Finally, with reference to its third charge—that is, ongoing AAAS activity—the committee recommends the creation of a permanent Committee on Scientific Freedom and Responsibility. Accepting this recommendation, the AAAS instituted a standing CSFR.

CSFR is composed of twelve members, each divided into three groups of four serving staggered three-year terms. The committee meets two to three times a year to support policy development and provide advice to AAAS officers, staff, and associates. A charter defines the duties of CSFR, which may conveniently be described as closely paralleling the first two original committee charges. That is, the standing CSFR is involved in ongoing studies of the boundary conditions for the exercise of scientific freedom and responsibility and with the development of criteria and procedures for dealing with freedom and responsibility issues.

Perhaps the single most significant evolutionary shift within CSFR over the last three decades has been increased emphasis on human rights work both in defense of scientists deprived of their civil rights—not simply their rights to scientific freedom—and the use of forensic science to advance the civil rights of all, scientists and nonscientists alike. Moreover, there has clearly been a turn toward a deontological rather than a consequentialist justification of scientific freedom conjoined with public activities. A revision of the CSFR charter undertaken in the mid-1990s, for instance, "affirms at the outset that scientific freedom is grounded in basic human rights and implies special responsibilities to extend and disseminate knowledge for the good of humanity" (see aaas.org/csfr).

Specific CSFR activities run the gamut from organizing sessions on freedom and responsibility topics at AAAS annual meetings and undertaking related special research projects to investigating individual complaints of rights violations and assisting with the investigation and protection of civil rights, especially in countries with poor human rights records. Sample instances of such activities include the following:

- developing a series of teaching videos dealing with scientific integrity;
- conducting a research conference on the problem of anonymity on the Internet;
- sending teams to help human rights groups in developing countries create secure computer database systems that the government cannot tap into;
- creating an international registry of scientists who have been deprived of their rights;
- organizing an independent public conference on secrecy in science; and
- giving an annual award for scientific freedom and responsibility.

IMPLICATIONS

What are the implications of these truncated narratives? The response has at least two parts. In part my goal has been to offer a brief interpretation that highlights a neglected aspect of technoscientific experience, thereby deepening our understanding of the human–technoscience experience. In part, however, the aim is also a philosophical assessment of this experience, thereby constituting an attempt to understand science-technology-society relationships—and through this understanding, to contribute to influencing them. Let me conclude, then, with an abbreviated overview of what I take to be the most salient features of these actions in which engineers and scientists, through the practice of a professional technoscientific idealism, have played roles as moral critics of science and technology.

First, it belies the suggestion that is sometimes made or assumed that technoscientists are uniform proponents of technoscience. There is more than one type of scientific intellectual.

Second, it points up, however indirectly, the new integration of science, engineering, and technology. Science, engineering, and technology have merged into technoscience—thus, paradoxically, adding new justifications for needs to reflect on their proper ethical and political assessment. On the one hand, scientific theory is influenced by developments in technological instrumentation and pursued with a concentration made possible only with funds derived from the extraordinary surpluses of industrial technology; on the other, modern scientific knowledge is precisely of such a character as to open up the natural world to a level of human exploitation and transformation unprecedented in history.

This technoscientific power was at once dramatically demonstrated and raised to new heights by the discoveries of nuclear reactions and the pursuit of these discoveries in the creation and development of nuclear weapons. For the first time in human history it became an issue for many of those who recognized the radical character of their technoscientific work that precisely as scientists and engineers they had acquired responsibilities beyond any they had anticipated and that they had obligations to help educate politicians and the public about the new realities that had, through their actions, been introduced into human affairs.

Third, although at first sight the technoscientific criticisms of technoscience may appear to focus merely on possible external misuses, this is not the whole story. The argument that technoscience is good in itself but subject to deformation by certain social contexts is an inadequate appreciation of the real-world character of technoscience. There are clearly cases in which engineers and scientists have also been what C. P. Snow called "natural Luddites," criticizing science itself. When the Edsall report, for instance, argued that in some cases it is simply not possible to know all the potential negative consequences of a technology, this would appear to have implications for technoscience itself. When Einstein argued that nuclear weapons require new types of thinking; when scientists and engineers associated with the FAS, the *Bulletin of Atomic Scientists*, Pugwash, and the UCS contended, to use the words of UCS leader Kendall, that "nuclear power is intrinsically dangerous" (Hively, 1988, p. 19)—such views clearly implied that at least one technology exhibits a character that resists any subordination as a neutral means, that is, that it will have an influence over any end for which it is used.

Moreover, the experience of Einstein and others that there is a temptation within science to forget human beings and their fate is also surely a criticism of technoscience itself. Indeed, the struggle against such a temptation to forget is, it may be suggested, a kind of ethical responsibility the acceptance of which constitutes a peculiarly germane form of moral leadership. It is a leadership that is especially needed in our advanced and advancing technoscientific time.

Finally, this tradition of technoscientific criticism of technoscience is not merely a thing of the past, even in our turbocapitalist charged technological present. Among phenomena that could be cited to back up this claim, the World Conference on Science held in Budapest in 1999 is one of the largest. The scientists and engineers who participated in this international meeting, in which Joseph Rotblat gave a keynote address calling again for a Hippocratic Oath for scientists, and who signed the two final documents—"Declaration on Science and the Use of Scientific Knowledge" and "Science Agenda-Framework for Action" (www.unesco.org/science/wcs)—were exhibiting professional technoscientific idealism. The manifesto signed by 108 Nobel laureates in science, calling for "co-operative international action" especially by the highly developed nations "to counter both global warming and a weaponized world" and released in December 2001 on the 100th anniversary of the first Nobel Prizes, provides yet another contemporary exemplification of this tradition.

A more individual or personal example of the tradition can be found in the recent work of Bill Joy, a computer scientist and cofounder of Sun Microsystems and its current director of research. In "Why the Future Doesn't Need Us," Joy (2000) makes explicit reference directly to the tradition of the technoscientific criticism of technoscience manifested precisely in the Federation of American Scientists, *Bulletin of the Atomic Scientists*, Pugwash, and the UCS in order to argue that this tradition needs to be revived and extended. The age of nuclear, biological, and chemical weapons of mass destruction is being superseded by a new age of genetic engineering, nanotechnology, and robotics that constitutes as great a leap in power and danger as nuclear weapons did over the weapons created by the chemical industry. According to Joy, if scientists and engineers are to exercise their true responsibilities to themselves and to their fellow citizens, they must imitate their post–World War II predecessors and carry forward the torch of ethical reflection and criticism.

Finally, lest it be objected that only the leading scientists and engineers have the luxury of practicing professional idealism, one can only affirm the existence of numerous less heralded cases. A few that have bubbled up into print are an argument by Carl Safina (1998), a lecturer at the Yale School of Forestry and Environmental Studies, for scientists to get involved in policy debates and the profiles of a number of "Ecologists on a Mission to Save the World" (Brown 2000) published in *Science*.

The bottom line is that professional idealism is a major part of the technoscientific enterprise—one that is often overlooked in STS studies and marginalized in our science and engineering education curricula, even (amazingly enough) in courses devoted to the teaching of ethics in science (see Mitcham et al., 2001) and engineering. The traditions of modern science and engineering deserve better.

Essay 6

Can Engineering Be Philosophical?

This conference is dedicated to "*Bildung* in Engineering" as "a dream in the heads of humanists"—and the possibility that this may be more than a dream, perhaps even a requirement, for any engineering curriculum able to address the future that engineering itself is creating. As our world becomes progressively defined, from molecular to planetary levels, as intentionally designed and managed, can we imagine the humanities as of more than another consumer product or something with entertainment value? As one contribution to such imagining, let us ask, "Can engineering be philosophical?"

Can engineering be philosophical? The question is complicated. In different contexts, I have argued both negative and positive responses. On the present occasion let me revisit the question and simply try to disclose some of its complexities—along with a spectrum of responses.

But first it may be good to ask, What difference does it make whether or not engineering can be philosophical? Is there a sense in which engineering ought to be engaged with philosophy? Insofar as the answer is positive, then we would be required to explore the possibility. That is, our question becomes one concerning whether philosophy is important to engineering in a way that would make it worthwhile to construct a bridge between the two. Insofar as the answer is negative, we can assume no bridge is necessary and we can dedicate ourselves to constructing real rather than conceptual bridges. Either way, the question and its answers presume some kind of opposition between engineering and philosophy—a presumption on which it is appropriate to reflect, before considering more directly the obligations and options for a philosophical engineering.

OPPOSITIONS

Let us thus begin with oppositions. One way to appreciate the assumption of an opposition between engineering and philosophy is to outline a history of ideas, taking departure from a reflection on the meaning of *Bildung*.

The notion of *Bildung* as education that is something more than technical training—that is, as culture or the cultivation and perfective formation of human nature—draws strengths from the Renaissance *studia humanitatis*, the medieval *artes liberales*, Ciceronian *humanitas*, and Platonic *paideia*. The core idea embedded in these variegated but overlapping traditions is that human nature is at root a potency that, when properly nurtured or cared for, is able to grow forth or manifest itself in special ways of thinking and acting—in a manner analogous to that by which the seed sprouts into a tree or blossoms into flower and fruit. In classical education this cultivation of human nature was found most fully in friendship and philosophy with that wisdom that provides guidance in the modulating of worldly affairs.

As an aside relevant to our globalizing philosophical community, one might also note possible relationships to the Chinese intellectual and educational tradition. Although *Bildung* is commonly translated into Chinese as 教育 (*jiao yu*), it should also be associated with the Confucian notion of 仁 (*ren*) or the perfecting virtue of benevolence. *Bildung* is not just any *jiao yu* but *jiao yu* that cultivates *ren*.

In the Christian mystical tradition, however, as represented for example by Meister Eckhart (drawing on a tradition than can be traced back to Evagrius Ponticus), there emerged a notion of *Bildung* as spiritual formation to bring about a release of the soul from worldly attachments and a corresponding subsumption into the divine. In mysticism wisdom is transformed into a *gnosis* that transcends the world; philosophy is replaced by theology. (Here again there are perhaps ways in which the teachings of Daoism could be compared with mystical *Bildung*.)

The modern period constitutes a challenge to both classical pagan philosophy and Christian theology. The idea that philosophy in the classic sense might constitute the flowering or flourishing of human nature as well as the faith that theology might be a means for transcending humanity were both opposed by the moderns. The moderns argued there was something weak and impotent about the ancient traditions of thought and reflection not to mention mysticism. Niccolò Machiavelli went so far as to praise Muhammad for his use of the sword, at the expense of Jesus who declined to do so.

According to Machiavelli's younger contemporary, Francis Bacon, in his preface to *The Great Instauration* (1620), the "wisdom . . . derived principally from the Greeks is but like the boyhood of knowledge, and has the characteristic property of boys: it can talk but it cannot generate; for it is fruitful of controversies but barren of works." Moreover, for Bacon, the fundamental issue Greek philosophers were prone to pursue in a culture of discourse, a discourse that absorbed energies that might otherwise have been dedicated to the production of effects—that is, questions concerning the good—had been settled by revelation. According to Bacon's interpretation of Scripture, it was the attempt "to judge of good and evil" rather than simply accepting the command to act with charity, that led to the Fall. Thus are human beings admonished to pursue a science that will be "for the benefit and use of life" as well as governed "in charity." Bacon predicted engineering two centuries before its full arrival on the historical scene.

On the basis of such a critical assessment, Bacon and his followers sought to create a new kind of philosophy: one that would pursue "the conquest of nature for the relief of man's estate" or, to quote from his younger contemporary, René Descartes, would render human beings "masters and possessors of nature." The pursuit of this new philosophy led to the rise of modern technoscience and the progressive analysis of nature into elements interacting according to laws that could serve as means for a human manipulation of the world. Nature itself ceased to be envisioned as manifesting a growth from potency to actuality, which anyone who lived within it was inherently oriented toward respecting and cultivating; it became, instead, a machine of possibilities to be worked and deployed for any purposes the competent operator might desire.

For Bacon human desires nevertheless remained governed by the commands of revelation. Two centuries later, Karl Marx restated the Baconian commitment when he declared in the eleventh *Theses on Feurerbach* (1845) that the purpose of philosophy is "not to interpret the world in various ways [but] to change it." Although no longer a believer after the manner of Bacon or Descartes, Marx believed that the human good continued to exhibit an unproblematic character conceived as freedom that should be pursued by a science not just of nature but also of society. By means of revolutionary criticism Marx simply sought to replace the atheistic, authoritarian politics of power adumbrated by Machiavelli and promoted by Thomas Hobbes with a newly empowered atheistic, democratic science of political economy. The levers of power that natural science and engineering had placed in human hands were illegitimately being used by only a few humans to control nature for their restricted benefit—relegating other humans to the status of instrumental means. Once this was generally recognized—if necessary, by force—then all humans would be able to share power and cease to be oppressed by either nature or other humans—and be freed even more fully to do whatever they might chose.

In this new context, that which may be referred to variously as *Bildung*, the humanities, or philosophy has assumed characteristics quite different from those it originally exhibited. Speaking generally, there have emerged three not necessarily mutually exclusive possibilities for the humanities. The first possibility is for the humanities to become subordinate to the advancement of science, engineering, technology, and other forms of revolutionary this-worldly transformation. Literary culture can employ rhetoric to praise, to marshal public support for, and to popularize modern science. Philosophy, for its part, can contribute by means of conceptual or epistemological analysis, as in the philosophy of science.

A second possibility is for the humanities and philosophy to imitate the sciences or revolutionary activity. Dedicated efforts in systematizing textual analysis and advancements in logic or conceptual analysis exemplify the former; the literature of social criticism and consciousness-raising drama or poetry illustrate the latter. Philosophy and the humanities can also become creative, after the manner of engineering and technology. "Make it new," proclaimed the American poet Ezra Pound.

The cult of creativity blends into a third possibility, one in which the humanities and philosophy inhabit a more or less independent or autonomous world—art for art's sake or postmodernist, self-referential playfulness. Such practices can be found in both literature and philosophy.

In all three instances, philosophy becomes at best a shadow of its premodern self. In relation to human nature, for instance, the idea of a potency to be realized by means of disciplined cultivation is replaced by the pursuit of possibilities in creative expression: expressive praise of the value-free knowledge and power of modern science and technology, the austere rhetoric of scientific and technological arguments themselves, or self-expression as alternatives or oppositions to science and engineering.

The modern *Bildungsroman* as it emerged in the late eighteenth and early nineteenth centuries illustrates the transformation. In Wolfgang Goethe's novel, *Wilhelm Meister's Apprenticeship* (1775) the apprenticeship in question is less to some master than to life—life defined more in terms of becoming than of being. It is an apprenticeship through which the protagonist learns not subordination to an existing standard of achievement but a self-expression that exceeds all predeterminations. In place of potency, with its deep inner orientation toward a mature order that it struggles through discipline to achieve and then preserve, there is possibility and a plethora of options for virtually unlimited self-realizations. It is instructive to compare more classical coming-of-age or maturation narratives such as those found in the biography of Siddhartha Gautama or Moses or Plutarch's *Lives*, all of which lead to the replacement of a false with a true self through a process of becoming what one already is and thereby achieving a kind of perfection or rest. Wilhelm

Meister, however, marks the arrival of a new sense of biography that will become characteristic of the modern form, the becoming of a new and ever greater, self-transforming self: transhumanism before the time.

Like Wilhelm Meister, who comes to affirm change rather than rest as the end, the *Bildung* theory put forth by Wilhelm von Humboldt as the foundation for a new university curriculum sees education as promoting, in harmony with the reality of a world of change as revealed by science and technology, continuous formation—beyond certainties and devoid of definitive results. As echoed in the ethics of the American pragmatist John Dewey, all ends are replaced by provisional ends in view. Any efforts to defend something more determinate or substantive are interpreted as rearguard if not reactionary conservatism.

Yet was it not precisely this *Bildung* as process of will development that made it possible for the most *Bildung*-infused nation in history to give itself over in the middle of the twentieth century to the catastrophe of National Socialism? Is it not also possible that engineering *Bildung* in some form has been complicit in many more minor disasters, such as those often associated with the teaching of engineering ethics? Do such possibilities not invite us to be cautious in reflection on the role of *Bildung* in engineering? Against such a background, what are the possibilities for a philosophical, *Bildung*-embodying engineering?

OBLIGATIONS

There are two basic ways for engineering to be philosophical. One is to become philosophical from within, the other from without. The idea that there might be a human obligation to transform the world, which in turn might constitute engineering as a human way of being in the world, is the foundation for becoming philosophical from within.

Engineering as a distinctly modern profession emerged in the 1700s, from military origins in the designing and operating of engines of war and Dutch efforts to control water in the lowlands—and in the process became infused with the methods as well as the substantive content of science. As such, engineering inherited a distinctly modern antipathy toward the *Bildung* of premodern philosophy—a philosophy in which nature was interpreted as a good to be approached through suffering but with gratitude and grace and which, in the name of the limitations of human knowledge, counseled restraint in human action, especially human action as manifested in making and using.

That the distinctly modern rejection of such counsel was not unknown to the ancients is witnessed by more than one classical text. Consider, for instance, the argument of Callicles, in Plato's *Gorgias*, that philosophy is a kind of playful activity appropriate to children but commonly replaced among adults by affairs of state and power, an argument that Socrates treats with inquisitive irony. The classic definition of modern or scientific engineering (which can be traced back to the British civil engineer Thomas Tredgold) as the systematic skill "of directing the great sources of power in nature for the use and convenience of [humans]," sides with Callicles—both in its pursuit of power and its dedication to this-worldly achievement, with little if any questioning of common beliefs about the priority of human convenience over competing insights into the good.

Indeed, what Marx, in harmony with the moderns, saw as the unproblematic goal of revolutionary economics—that is, the consolidation and democratic distribution of human power—his contemporary Ernst Kapp saw as dependent on and more fully realized through *Technik*. In his *Grundlinien einer Philosophie der Technik* (1877), Kapp proposed a philosophy of engineering technology as the true means for the humanization of self and world—that is, as the overcoming of an unjustified constraint on human possibilities for freedom. This view, anticipated in G. W. F. Hegel's description of the master-slave dialectic in *Der Phänomenologie des Geistes* (1807), founded a tradition of what has been called engineering philosophy of technology. This is a tradition in which the Russian engineer Peter Engelmeier would imagine politics being replaced by technocracy; in which the German engineer Friedrich Dessauer would interpret engineered invention as breaching the Kantian barrier between phenomena and noumena; and in which the American engineer Samuel Florman would find an existential pleasure and creativity that rivals all alternative aesthetic achievements. From within the engineering profession there has thus emerged what may be called an engineering philosophy of technology as *Bildung* of world mastery and self-expression—one that would reinterpret all ostensibly nonengineering aspect of human experience as engineering *manqué*.

As evidence for the continuing vitality of this tradition there is no better argument than that of nuclear engineer Billy Vaughn Koen. According to Koen's *Discussion of the Method: Conducting the Engineer's Approach to Problem Solving* (2003), "The engineering method captures all of the characteristics of a universal method that you will need to create the world you desire" (p. xi).

> If you desire change, if this change is to be the best available, if the situation is complex and poorly understood, and if the solution is constrained by limited resources, then you too are in the presence of an engineering problem. What human has not been in this situation? If you cause this change by using the heuristics that you think . . . are the best available, then you too are an engineer. . . . *To be human is to be an engineer.*
>
> (p. 58, emphasis in the original)

But this is not just a description; it is an obligation—the fundamental obligation of engineering. From within such a framework engineering itself becomes philosophical and properly incorporates into itself any features of the humanities that would further its advance.

Such a brief for a philosophy or *Bildung* of engineering has been made—although without using the terms as such—by the U.S. National Academy of Engineering in a report on *The Engineer of 2020: Visions of Engineering in the New Century* (2004), with its follow-up recommendations for application in *Educating the Engineering of 2020: Adapting Engineering Education to the New Century* (2005). The basic argument of these reports is at one with proposals for going beyond engineering science and design skills because of the need to attract more engineers and to give engineers a greater ability to act in a more complex world. From all such perspectives there emerges what appears to be a rhetorical question: Is there any option to human self-realization other than engineering?

OPTIONS

But perhaps it is possible to consider the question in a nonrhetorical manner. After all, is it not reasonable to ask, Why be an engineer?

Yet it is important in the asking of such a question—which is not an engineering question—to appreciate the achievements of the tradition of engineering philosophy of technology as *Bildung*. At the same time, it is equally crucial to recognize in this engineering tradition a fundamental break with the tradition of philosophical reflection that links back to Socrates. It is equally possible to argue for the utility of philosophy to engineering, without thinking philosophy as abandoning its heritage within engineering. The argument would be that in some deep and not to be glossed over sense, engineering philosophy of technology is not fully philosophical. Rather than serving as a fundamental questioning of engineering, the philosophical engineering that emerges within engineering itself functions more as an apology, an attempt to extend the engineering approach to encompass not just the making and using of artifacts but all of human experience. As such it constitutes an ideology if not a totalizing vision of human experience. It is *philosophy for engineering* rather than *philosophy of engineering*.

By contrast, for Socrates the life of philosophy is a repeated effort to recover the naiveté of inquiry that recognizes limitations to all human knowledge and action. It is to become once again a child, but a mature child, living with wonder at all that is. In service to the god who spoke through the oracle at Delphi, proclaiming that none was wiser than Socrates, Socrates proceeded to interrogate politicians, poets, and artisans about their presumed wisdom.

His conclusion was that only he is wise "who, like Socrates, knows that his wisdom is limited."

In the archetype of Socrates, philosophy constitutes, as it were, a kind of learned ignorance—a learned ignorance that serves to infect any claim to knowledge with a qualification and caution regarding its expansion. For Socrates this qualification of separation, distance, or detachment is introduced by means of a philosophical questioning that aims to free one from being captured by confidence so that confidence is replaced by modesty and an appreciative assessment of the virtues of nonaction over action—even in the presence of a manifold of demands for action. From such a philosophical perspective, are there any options for engineering to become philosophical from without?

A response to this question may begin with four observations. First, the modern attitude that takes world transformation as a fundamental human obligation is manifest, as has been indicated, in the physical, biological, and the social sciences as well as in engineering. But in the sciences—physical, biological, and social—there has nevertheless emerged a degree of self-reflection and criticism that is, if not wholly lacking, at least much more attenuated in engineering. Social scientists especially have, since the mid-twentieth century, become increasingly prone to ask questions of social phenomena—but also to reflexively question themselves about why they are asking such questions. Reflexivity can also be found to some degree among scientists who seek to admit and on occasion question the paradigms within which they work. But in engineering the status of reflexivity remains underdeveloped.

Second, although the engineering community has admirably developed ethical guidelines for professional practice—more so, to some extent, than the sciences—these ethical codes are largely restricted to issues of process rather than product. It is much easier to identify how to do things right than what are the right things to do.

Third, one may also observe that in the humanities as well there has been a progressive retreat from substantive measures of the good in favor of process or proceduralism. Perhaps the greatest criticism of post–World War II liberalism in both Europe and North America—a criticism that has come from the Marxist left and the religious right—is that in its commitment to process, liberalism is unable to advance a substantive vision of the good. In the words of one recent critic (Leon Kass), liberal proceduralism tends to focus on the fair distribution of things at the expense of inquiring whether the things being distributed are themselves worth distributing. Even when the food may be contaminated, we focus on worries about "why my portions are so small." If the humanities and philosophy are to be of benefit to engineering, they may also have work to do on themselves. There may be weaknesses in philosophy as the queen of the humanities that will make it difficult to introduce any real distancing into engineering.

Finally, fourth, for engineering to become philosophical from without it will be necessary for scholars and practitioners in the humanities, for those who would take up this task, to develop a genuine appreciation of engineering. Philosophy is often weak not only in itself but also in its appreciation of what is really taking place in engineering. Too often philosophers approach engineering with a caricature of the profession and its practices that would not be recognized by engineers themselves. Is it possible to become genuinely knowledgeable in engineering without being sucked into it?

There is a paradox here. The foundation of Socratic philosophy as distinct from ideology is distance or detachment. If engineering were philosophical, it would need to be pursued with some degree of detachment, one that allows the engineer to rise above the details of technical proficiency about doing things right, in order to consider deeper issues related to doing the right things. With Socrates such distancing comes through questioning, a questioning that bears in the primary instance on public affairs and the nature of the good. Can this kind of questioning be redirected toward engineering itself in ways that engineers can practice it and still function as engineers? Alternatively, does the practice of Socratic detachment in the engineered world simply leave one at the mercy of engineering?

Prescinding from such questions, consider two examples of contemporary efforts at the reconfiguration of engineering education so as to introduce engineering students to possibilities of detachment—right in the midst of their dedication to technical things. The first instance constitutes what may be termed a strong without, the second a without that nevertheless emerges from within. That is, in these two instances of an attempt to develop philosophical engineering from without, the without at issue can be implanted from without or it can (as it were) be a without that emerges from within. Both examples come from my home state of Colorado and are associated with two of its leading establishments of engineering higher education.

The first example is the Herbst Program of Humanities for Engineers in the College of Engineering and Applied Science at the University of Colorado. In parlance typical of the United States, the Herbst Program is termed a "great books program." It was founded in 1989 as the result of a related great books experience by a Colorado engineering alumnus, Clarence Herbst. In the early 1980s, Herbst became involved in the great books program that takes place during the summer at the Aspen Institute. Founded in 1950, the Aspen Institute is a not-for-profit organization based in Washington, DC, with campuses in Aspen, Colorado, and on the Wye River in Maryland. It is dedicated, according to its website (aspeninstitute.org), "to fostering enlightened leadership and open-minded dialogue [focused on] nonpartisan inquiry and an appreciation of timeless values." One primary means to this end is its Socrates Society Seminars. Socrates Seminars convene small groups of persons who aspire to advanced leadership training for intensive periods of reading and discussing great books of the past with leading scholars of the present. For Herbst himself, the Socrates Society Seminars experience was so rewarding that he offered to fund a related great books experiment in his former engineering college.

As it has developed since, a select group of students, in small 12-person seminars, take courses centered around a close reading of books by such figures as Plato, Aristotle, Marcus Aurelius, Vitruvius, Augustine, Dante, Machiavelli, Shakespeare, Galileo, Bacon, Milton, Descartes, Hobbes, Pascal, Newton, Swift, Goethe, John Stuart Mill, Darwin, and Einstein. Enrollment is selective, with the number of students involved in any one semester numbering under 200 (out of an undergraduate student body of approximately 2,500). Nevertheless, with leading faculty drawn from St. John's College (in Maryland and Santa Fe), the premier great books liberal arts institution of higher education in the United States, the program attracts dedicated students and has a leavening influence in an otherwise extremely technical curriculum.

As Athanasios Moulakis, the founding director of the Herbst Program, has written, "Engineering attracts bright people and equips them to do important things, but it does little to help them understand the human condition or, indeed, the fullness of their own humanity" (1994, p. 1). To complement engineering knowledge about how to do things right, it is useful to stimulate engineers to reflect on what things it is right to do. The hypothesis of the Herbst Program is that one of the best ways "to do this is through patient and dedicated teaching that sets out with the modest goal of helping students learn to read important texts that deal with aesthetic and ethical issues and to talk about their reactions among themselves" (p. 6). The effect is to introduce into engineering a modest philosophical distance, among engineers who thereby come to recognize the limitations of their technical knowledge and skills. The goal is Socratic engineers who are no longer so sure about what they engineer.

The second example is a program in humanitarian engineering with which I have been involved at the Colorado School of Mines. This is a more recent experiment, having been initiated in 2003 with a grant first from the William and Flora Hewlett Foundation, and then extended in 2006 with a grant from the U.S. National Science Foundation. One way to describe this experiment is that it asks students to take seriously the classic modern definition of engineering as the systematic skill "of directing the great sources of power in nature for the use and convenience of [humans]." How can this be the case when, in so many instances, there are such class differences in the distribution of uses and benefits? Who is really benefiting from the use and convenience that is being mass-engineered today?

Many faculty and students are surely attracted, no doubt some more than others, to what they think of as the humanitarian value of engineering. When the CSM program was established, some faculty objected that the existence of a program called "humanitarian engineering" implied that chemical or petroleum or mining engineering were anti-humanitarian. Yet as engineer-philosopher Byron Newberry has pointed out, if all engineering is humanitarian this would lead one to "expect engineers to devote considerable time to, and to be quite vocal on, the subject of what does or does not constitute a benefit to humanity, particularly with respect to technology [—something which] is generally not the case" (Newberry, 2007, p. 109). There would seem to be a disconnect in the engineering ethos—a disconnect that deserves critical examination. In Newberry's words, again, engineers are

> people who, on the one hand, are not necessarily uncaring about the effects of their work on the larger society, and who generally are interested in doing—and believe they are doing—the right thing. On the other hand, they perhaps exhibit some ill-considered complacency, and perhaps do not have sufficient depth and breadth of perspective for situating their work within the larger whole, or critically evaluating it from non-engineering perspectives.
>
> (p. 112)

In Newberry's deft turn of phrase, the engineering ethos is one of "proximate instrumentalism." Engineers typically exhibit "an instrumental attitude about technologies they are close to [—an instrumentalism derived] not from any philosophical ideology, but [as] a byproduct of familiarity and expertise." Such, it may be suggested, is a deeper description of engineering ethos than the commitment to heuristics offered by Koen. It is in an effort to stimulate reflection on such easy and unquestioned commitments to both humanitarianism and instrumentalism that it may be possible to cultivate within engineering itself something approaching a philosophical perspective.

Take simply the issue of humanitarianism. Humanitarianism is a much more complex historical phenomenon and ideology than is generally appreciated. As it arose in the nineteenth century, humanitarianism is characterized by efforts to treat human beings equitably without regard to class, nationality, or other group identifications, especially with regard to issues of health and material welfare. It is also important to distinguish humanitarianism from some closely related modern terms such as humanism, the humanities, and human rights—and to consider possible relations to premodern ideals such as cosmopolitanism. Additionally, humanitarian action overlaps differentially with work for social justice, development, peace building, and democratic governance. In each case, analysis easily reveals conceptual and practical issues that are not easily engineered away. Can such questioning help to awake in engineers a kind of philosophical sensitivity that would otherwise be found wanting? It is by means of such questioning of its internal assumptions that the without may emerge from within.

CONCLUSION

In conclusion, then, I would argue that the dream in the head of humanists of a *Bildung* in engineering may be explored in multiple ways. Indeed, because of the tensions between alternative senses of both *Bildung* and philosophy, and the dangers to which history and experience have exposed us, such exploration may itself be a form of *Bildung* and a philosophy peculiarly appropriate to engineering.

Moreover, there are many paths beyond the alternative approaches that have been only too briefly indicated here. To complement the great books of the European tradition there are, for instance, the great books of Asian traditions. Surely the sutras of Buddhism and the texts of Kongzi, Laozi, and their followers, have something to teach with regard to wisdom in the present. To supplement humanitarianism one might venture reference to the human potential movement. In addition, to include in engineering curricula intensive examinations of such issues as those associated with the tension between science and religion might have a salutary, distancing influence. It is one achievement of the project that we are here to celebrate—the project of discovering and introducing *Bildung* and philosophy into engineering—to have stimulated the consideration of such multiple explorations.

The cultured engineer project is a significant effort not just at enhancing engineering education but more generally at enhancing our collective life in a world of advancing and intensifying artifice. If we are to learn to practice the examined life—a life that Socrates describes as the only one that is truly worth living—it will require thinking in a context different than that of the Greek *polis*. Today, in a world of human-induced global climate change, genetically modified foods, and nanotechnological design, we are all engineers or the beneficiaries of engineering. May we also all aspire to become not just engineers or the consumers of engineered products—but thinking, philosophical engineers or engineering fellow travelers. Surely this is the ultimate implication of that dream in the head of humanities scholars concerning a *Bildung* in engineering.

PHILOSOPHICAL ENGINEERING: FIVE THESES

Finally, as a contribution to transforming this dream into a waking reality, I offer for consideration the following manifesto:

1. If engineering education is to be more than training in high-level trade school technical skills, then it must include philosophy and the humanities. For more than 2,000 years in the European tradition, philosophy, the liberal arts, and humanities have been the basis for learned professions such as medicine and law. Engineers would be selling themselves short not to aspire to a similar breadth and depth in their education.
2. Philosophical reflection in such areas as ethics is crucial to support engineers in their professional practice. Engineers are regularly challenged by such problems as conflict of interest, confidentiality, and honesty. They are often asked to make decisions about the design and construction of products, processes, and systems that implicate public safety, health, and welfare. All such situations call for technical decision-making to be complemented with ethical judgment.
3. Advances in engineering often provide significant ethical and philosophical challenges to consumers, users, and citizens. Dramatic examples include nuclear power and other forms of energy generation, transportation, and communication systems that may disrupt established ways of living, products and processes that induce environmental transformations, and technological devices that alter patterns of perception and human physiologies. For engineers to properly exercise their responsibilities as professionals to assist employers and the public in making informed choices with regard to such issues requires a measure of philosophical knowledge and understanding.
4. In a globalizing world, engineers are increasingly ambassadors between cultures. To be able to bring their technical knowledge and skill to bear in the most effective and culturally sensitive ways will often require an ability to appreciate the complexities of intercultural interactions. A knowledge and philosophical appreciation of one's own history and culture is the best foundation for appreciating that of others and successfully negotiating such complexities.
5. Engineers are properly leaders in the design and creation of the world of artifice that supports human life and mediates among humans and between humans and the rest of nature. This mediating artifice is not a neutral means but constitutes multiple decisions and choices about goodness, beauty, and justice. More than physicians who must help their patients understand the meaning of health or lawyers who must counsel

their clients about the demands of justice, engineers are called on to help educate their employers and the public about the implications and consequences of design alternatives. Only a substantial grounding in philosophy and the humanities can prepare engineers to realize the highest dimensions of such professional responsibilities.

CHAT WITH EPICTETUS

– Hello. This is your online chat bot. How may I help you?
– Hello. This is Epictetus, from Rome.
– Epictetus, what a surprise! What can I do for you? Are you coming to America for a lecture?
– No, nothing like that. I don't travel much anymore. But I'm concerned about the practice of acceptance in your modern world and about the need maybe to update or revise my philosophy.
– You mean stoicism. Important topic. Amazingly, there have been efforts to revive it here. Modern life, especially modern engineering and technology, are very well accepted. It's only some fringe philosophers who don't quite go along. Even if no one really anticipates events, everyone still just accepts what happens. I'd say in some sense we're all pretty good stoics in this regard.
– Well, that's what seems to me the problem, what leads me to think I may need to rethink some doctrines. Stoicism now seems to lead to acceptance of the non-natural, whereas for me it was acceptance of nature. Also, I used to argue that the pursuit of self-interest, as human beings were in fact not self-sufficient, required commitments to the common good. But now I understand that's been questioned by the so-called economists among you.
– That's right. Especially by free market and libertarian economists. They have lots of statistics in their favor.
– Did you say "Stoics in their favor"?
– No. *Statistics* in their favor. Deals with data analysis, in this case of human behavior. Data has become the new big thing, even called "big data."
– Well, I'm not sure about all that and what could be so big about just data. But one thing. . . . Interrupt: Your VPN connection has dropped.

Essay 7

Convivial Software

An End-User Perspective on Free and Open Source Software

The ideals of free and open source software arose within the technical community, although under significant nontechnical influences (Turner, 2006). While there have been debates among those who would distinguish the free (sometimes *libre*) and the open source movements, since the late 1990s the two regularly fly under the unified acronyms FOSS (free and open source software) or FLOSS (free *libre* open source software), thus reducing the importance of such differences with regard to the present argument. Beyond internalist debates, considerable interest has also developed concerning implications for the nontechnical or end-user world. Lawrence Lessig (1999) and Steven Weber (2004), for instance, have analyzed the open source movement from the perspectives of law and political economy, respectively; Henry Chesbrough (2006) has argued the importance for new models of business. Computer scientists, software engineers, and associated hackers have also regularly promoted nontechnical benefits across a spectrum of societal dimensions. The free software community has seen the bazaar of open software as productive of both technical and end-user goods in ways that proprietary market software production is not (see, e.g., the collections by DiBona et al., 1999; Feller et al., 2005; DiBona et al., 2006). In all such cases, the general approach has been to provide a narrative description of some aspect of the free and open source phenomenon, exploring how what may have initially appeared as nonstandard behaviors led to the social construction of new norms with implications for legal, political, or economic orders in ways that are potentially beneficial to society.

Indeed, on the basis of such narratives, openness has migrated from software into a plethora of related terms such as "open access" publication, "open content" development (as in *Wikipedia*), "open innovation," "open courseware," and even "open source culture." Remarkably, however, there has been little if any linkage between open source ideals and Karl Popper's political philosophy of the open society (Popper, 1945), and only quite limited discussion of relations to Robert Merton's classic sociology of how science functions when guided by an appropriate set of scientific norms (Merton, 1942), now denominated as "free" or "open science" (see Kelty, in Feller et al., 2005; Kelty, 2008). Relations between open source methods and the ontology of open systems (see, for example, Bertalanffy, 1968) is another connection deserving exploration.

The present argument will nevertheless take a different tack. The goal here is to begin with the end user and from this perspective to reflect on the free and open source software movement, not so much as it were from the inside out as from the outside in. To do so will involve placing the movement in larger historico-philosophical perspective. Then, going beyond the technical details, the aim will be to enhance and support the free and open source movement, by showing how it is heir to a tradition of professional-ethical idealism and potentially related to important issues in the history of science, technology, and society relations. A normative argument will also be advanced for assessing free and open source software in something more than its own conceptions of freedom and openness but with an appeal to norms that remain inherently complementary.

The approach here has obvious affinities with end-user and user-centered software design analyses (see, e.g., Norman and Draper, 1986; Adler and Winograd, 1992; Kaasgaard, 2000), although it has been argued (see, e.g., Lieberman et al., 2006; Seffah et al., 2005) that such design methods sometimes treat end users as no more than another factor in an ongoing technological construction—that is, as customers or consumers—rather than as citizens or true ends for software functionality. Instead, my argument has more in common with the user-centered perspective of Robert Johnson (1998) or Peter-Paul Verbeek (2005) in their analyses of technological artifacts in general; both Johnson and Verbeek, however, adopt rhetorical and post-phenomenological stances, respectively, which are kept at a distance. Perhaps the closest affinities are with value-sensitive design discourse (see, e.g., Friedman et al., 2006).

The effort to focus on software from an end user's perspective will also lead to the coining of a new term. From an end user's perspective what is important is not so much the direct availability of source code and the creative freedom to manipulate it as what may be called *program* or *technical conviviality*, phrases deeper and richer in meaning than "user friendly." Related terms that might be proposed include "software without frontiers" or even "humanitarian software"—indicating a desire to break the boundaries of software sovereignty or technicalist nationalism. There is more with regard to openness than program transparency and rights to play in the fields of the code.

From a nontechnical perspective, the invention and development of software is no more than another instance of engineering and technology broadly construed. Software is simply a new example of technology, and the effort to assure that technology is developed and utilized in a socially responsible manner has a significant history. The argument will thus begin with observations about the history of technology. This will lead to critical reflections on the development of professional engineering ethics and to a discussion of the alternative technology movement. Finally, it will conclude by indicating some policies or criteria to consider when imagining the design of convivial software.

TECHNOLOGICAL INVENTION IN A SOCIAL CONTEXT

The creation of software is a kind of invention. But it is invention with specific problematic features. Writing and adapting source code is an activity more than once removed from end-user utilization and one that requires high levels of abstract, analytical thinking. It is useful from a historical perspective to consider such features.

Human beings cannot live without technics. By necessity they fabricate clothes and shelter, and they must secure and prepare food. Thus it is reasonable to redescribe *homo sapiens* as *homo faber*. But just as the thinking of *homo sapiens* is not limited to one kind of rationality—since thinking includes the creative use of language in the arts and the humanities as well as in the analytic and investigative skills of mathematics and science—so too the making of *homo faber* takes many forms. Although necessity may be the mother of invention, she is mother to a large family with many children.

According to philosophers Alfred North Whitehead and José Ortega y Gasset writing independently of each other in the early decades of the twentieth century, the history of technology went through a critical watershed in the late nineteenth century, when the process of inventing became rationalized. What Whitehead called "the invention of the method of invention"—that is, the invention not just of another technical artifact but of techniques to promote invention itself—was "the greatest invention of the nineteenth century," an event that fundamentally transformed technics and the technics-society relationship (Whitehead, 1925, p. 141). It replaced craft technics with systematically pursued technology through engineering.

Thomas Edison, who has become a kind of inventor archetype—having been awarded a record 1,093 patents—played a key role in this process by inventing what he called an "invention factory" (see Israel, 1998). With regard to the motivations driving his invention factory, or research and development (R&D) work, there exists a revealing story. Edison's first invention was an electric vote recorder for a legislative assembly, developed when he was only 21 and working in telegraph communications—that is, the computer industry of his day. In place of the roll-call or written vote systems, which were slow and time-consuming, all members of the legislature would be supplied with electric switches on their desks. This would allow votes to be counted and the results determined quickly.

But this invention was a flop. It did not sell. The Massachusetts legislature, to which Edison offered his electric vote counting device, was not interested. In fact, it was positively opposed. Politicians often wanted to learn how colleagues voted before casting their own votes, or wanted time to reconsider as written votes were manually tallied. In a roll-call vote, members were free to "pass" until others had voted, in order to learn how a vote was going before finalizing their own decisions. Bargaining was possible even in the midst of voting. This possibility would have been curtailed by electric voting. In response to this failure to sell his new electric vote tallying machine, Edison determined never again to invent something unless he knew that the people for whom it was intended really wanted it. From now on, he promised himself, he would "devote his inventive faculties only to things for which there was a real, genuine demand" (Dyer and Martin, 1929, vol. 1, p. 103). Edison was not enamored of technology practiced solely for what Samuel Florman (1976), in an apologia for creative engineering, would subsequently call the "existential pleasures" of technological experience.

What this story reveals is not just an aspect of Edison's business plan but something about the new systematic invention process itself. The process makes possible what might be called alienated inventing. Karl Marx had previously identified alienation as a key feature of the industrial mass production process. In the large-scale factory structured around extensive division of labor, workers were separated from the full experience of making and from the products of their piecemeal fabrication. Now, in the research and development factory, alienation had moved from technological production into the production of technology. It would give rise to what has become known as the problematics of technology transfer, especially transfer from laboratory to market.

It is difficult to imagine traditional artisans inventing unwanted or truly superfluous products. Traditionally, artisans are so culturally and socially embedded that their inventive skills are just naturally applied to culturally and socially meaningful goods. Their contributions to what might be mistaken for anticipations of a consumer economy were ornamentation and aesthetic decorations that enhance cultural integration by uniting technology, religion, politics, and art. Newness came into existence not suddenly through systematic or conscious pursuit but slowly by give-and-take accretions over extensive periods of time.

In research and development laboratories, however, the great process of separating or disaggregating the elements of culture that is a defining characteristic of modernity (Polanyi, 1944) took hold of the invention process and dis-embedded it from traditionally integrated sociocultural orders. As one critical social theorist has put it, prior to the modern period,

> Technology was associated with a way of life, with specific forms of personal development [and] virtues. . . . Only the success of capitalist de-skilling finally reduced these human dimensions of technique to marginal phenomena.
> (Feenberg, 1995b, p. 18)

The Renaissance, Reformation, and Enlightenment introduced distances among, for example, art, religion, politics, and economics—all of which became semiautonomous social activities, whose integration henceforth had to be constructed, not assumed. So too the invention process was now uprooted from all implicit connections with a sociocultural nexus and on a macro level decontextualized, which is not to deny that at individual or micro levels technical activities remain enmeshed with and influenced by individual motives and interests. Because Edison had no tacit or intuitive knowledge of political life, he invented something that did not fit in with that life—and vowed in the future to do what came to be called market research prior to any inventing.

For a time, the new invention of invention method could rely on the residual memory of contextualized human need. Edison's major inventions did not require any systematic market research to be successful. Electric lights, voice recording, and motion pictures all responded to almost mythical human desires—although many of these primordial desires and possible device-meeting responses had also been mythologically criticized. His great techno-economic innovation of the electric power system was a direct response to popular enthusiasm for electric lights. But henceforth, especially in particulars, cultural and social contexts of use would increasingly have to be investigated in advance by means of empirical research—or created by means of advertising. In the process such research and advertising had the effect of forming a new socioeconomic order in which inventing could take on a kind of self-sustaining character.

Shortly after Whitehead, Ortega (1939) rightly noted, without reference to Edison, that once there is in place a dis-embedded method of invention, although inventors may initially tend to become alienated from the traditional lifeworld they subsequently begin to transform human affairs through their infusions of inventiveness. Inventors do have a tendency to form a culture of interest in invention for its own sake, to think of themselves as nothing more than inventors, sucked into the vortex of the existential pleasures of invention—at once depriving themselves of the common pleasures of social existence. They turn "art for art's sake" into "technology for technology's sake," at the same time inventing a new macro sociocultural order. Divisions between expert and citizen, producers and consumers—dimensions deeper than a mere electronic digital divide in product accessibility—open up at the feet of technoscientific specialists and behind the backs of democratic end users. Such divides can only be bridged or reunited through the systematic development of ways to relate the different worlds.

Comparisons with software development are no doubt obvious. The hacker culture has a tendency to become turned in on itself. Hackers are defined by their technical skills and their delight in problem solving, as well as by their commitments to share solved problems. Reinventing the wheel is not only a waste of time and energy, it is not fun. What is fun is inventing something new and then sharing the excitement. Software engineers write code that is technically sweet and pleasurable.

Having posted Linux kernel version 0.01 without fanfare a month earlier, here are the words of Linus Torvalds, another 21-year-old inventor, when he announced version 0.02 (October 5, 1991):

> Do you pine for the nice days of Minix-1.1, when men were men and wrote their own device drivers? Are you without a nice project and just dying to cut your teeth on an [operating system] you can try to modify . . . ? No more all nighters to get a nifty program working? Then this post might just be for you.

In the midst of his technical delight there is surely a measure of separation from end-user interests. In an earlier e-mail (August 25, 1991) Torvalds wrote, "Any suggestions are welcome, but I won't promise I'll implement them." (For these e-mails and more, see Torvalds, 1992.) In a later book, he elaborated on "the meaning of life" by distinguishing three basic human motivations: survival, social relations, and entertainment. "Taking survival for granted," Torvalds argued that Linux "brought people both the entertainment of an intellectual challenge and the social motivations associated with being part of creating it all" (Torvalds, 2001, p. 249)—the existential pleasures of creating software "just for fun."

For Torvalds, of course, the implicit invitation is also for other technically sophisticated users to take pleasure in doing some implementation themselves. But this is a technical energy and excitement that commercial powers readily seek to harness and direct. As even the greedy Microsoft has on occasion reminded, the proprietary commercialization of software did not originate solely out of greed. Yet commercialization has its own downside.

The market must itself be created and designed to benefit the common good because it is in constant danger of being captured by private or special interests. The free and open source movement is one attempt to shift the balance of power in the market economy, but it is a shift that may call for supplemental adjustments among those who participate in this type of engineering creativity and its entrepreneurial follow-ons.

THE ENGINEERING IDEAL

During the same period that Edison was inventing the method of invention, engineers were inventing their profession—an engineering ethos. Edison was not himself a professionally trained engineer; like many hackers today, he was an autodidact. So were many engineers of the time. And just as Edison created a method of invention, so other engineers were in the process of creating a method for producing engineers, one that would systematize the process of technical self-education.

The invention of professional engineering and its correlate, a professional engineering ethos, later articulated in part as engineering ethics, exhibits its own complex history. Here it is sufficient to note how the invention of engineering education was a response to the mass need for engineers brought about by mass production and mass invention. As part of his creation of the invention factory, Edison himself began to employ increasing numbers of engineers and to support engineering education. Moreover, the emergence of professional engineering ethics may be read as a parallel response to Edison's technology transfer problem, the difficulty of coordinating technological invention and design with end-user interests. Instead of the commercial responses of market research and advertising—or Edison's own personal commitment—the engineering professional began to construct institutional commitments to address such dangers.

The emergence of engineering ethics may be conveniently sketched in terms of a three-phase argument. The first phase occurred during the 1700s and the 1800s as civil engineering emerged from the military. As engineers emerged out from under the shadows of the military to become civil engineers, they established professional engineering societies. In England, as an outgrowth of informal dining clubs, there emerged the Society of Civil Engineers in 1771, then the Institution of Civil Engineers in 1818, which was granted a royal charter in 1828. The Institution of Mechanical Engineers followed two decades later.

When granted its Royal Charter the Institution of Civil Engineers asked one of its members, Thomas Tredgold, to formulate a definition of engineering. The result is the now classic description of engineering as "the art of directing the great sources of power in nature for the use and convenience of [humans]." But the truth is that engineers do not work for human beings in general; most work for private corporations, and thus easily become captives of the organizations that pay their salaries. This is different from such professionals as physicians and lawyers, who mostly run their own companies—or often work directly for the state. The ethical results are obvious: Engineers lack the level of professional autonomy enjoyed by physicians and lawyers. Indeed, having arisen out of the military, there is a tendency for engineers to adopt some form of military ethics. The ethics of soldiers is obedience to authority. The implicit ethics for civil engineers tended toward company loyalty. Engineers quietly assumed, despite the ultimate ideal of serving humanity, that humanity was best served by obedience to their employers.

A second phase occurred in the early 1900s. Professional engineering societies drafted the first explicit professional engineering ethics codes. These codes simply made explicit what was already implicit: the importance of company and professional loyalty. But the inadequacy of this position was immediately felt. Physicians had a moral obligation to promote health. Lawyers at some fundamental level were guided by the moral ideal of justice. Neither physicians nor lawyers saw themselves as able to be told what to do by patients or clients; instead, they sought to mediate to patients or clients a substantial ideal that would otherwise be denied them and to which the professionals saw themselves in ultimate service. Was there no substantive ideal to which engineers, as well, were in service?

During the early 1900s in the United States proposals were debated concerning the possibility that for engineering the substantive ideal was efficiency. The problem with this proposal was twofold. First, it failed to acknowledge the extent to which efficiency is context-dependent, that is, varies with what outputs and inputs are to be considered in calculating any output-input ratio. Second, it tended to justify some form of technocratic governance, thus undermining democracy.

Finally, in a third phase the inadequate ideals of loyalty and efficiency were replaced by a new formulation of engineering responsibility. Immediately following World War II, professional engineering ethics codes began to argue for "public safety, health, and welfare" as the primary professional engineering ideal. Over the course of five decades, from the late 1940s to the late 1990s, this argument became a consensus view. A philosophical restatement of the trajectory and its projection is that engineers have a duty *plus respicere*, to take more than engineering into account (Mitcham, 1994, p. 164).

This consensus is reflected in the "Software Engineering Code of Ethics and Professional Practice" (1999), which declares that software engineers shall adhere to eight principles "in accordance with [a general] commitment to the health, safety and welfare of the public." (Software engineers just make a minor shift by exchanging the order of safety and health.) Indeed, the first of these principles is that "software engineers shall act consistently with the public interest."

The free and open source software movement constitutes a major opportunity for the exercise of this engineering ideal of responsibility. Additionally, since one of the more common ways to teach engineering ethics is through case studies, the free and open source software movement can be a good illustration of efforts by engineers to take ethics into account. It can easily be argued that open source is a necessary means for protecting public safety, health, and welfare in general, and for practicing at least two of the eight principle specifications. Principle three of the ACM "Software Engineering Code of Ethics" (1999), for example, states that "software engineers shall ensure that their products and related modifications meet the highest professional standards possible." As Eric Raymond has argued in *The Cathedral and the Bazaar* (1999), in many instances this requires open source code availability. Principle five further declares that "software engineering managers and leaders shall subscribe to and promote an ethical approach to the management of software development and maintenance." It is difficult to see how this is possible without open source code and the freedom to alter and adapt such code. Thus a strong argument can be made that the "Software Engineering Code of Ethics" entails support for the free and open source software movement. Indeed, it could be argued that the even more general "Association for Computing Machinery (ACM) Code of Ethics and Professional Conduct" (1992) points in the same direction.

But to make public safety, health, and welfare a key value also suggests the importance of the public—and raises again the question of the role of the end user. In the various fields of applied ethics—of which computer and software ethics are only two instances—there have been subtle if measurable movements toward involving the public in technical decision-making. (See, for example, discussions of collaborative design, as examined in Scrivener et al., 2000.) There are, of course, strong resistances to this idea from many quarters. After all, the argument goes, how can nonexperts tell experts what to do? Will public interference not undermine professional technical autonomy? Is there not a danger of politicizing science and engineering?

Despite the dangers, there are nevertheless three overlapping arguments for promoting the proper involvement of the public in technical decision-making (Mitcham, 1999). A first is that public involvement promotes democracy and public intelligence. Insofar as citizens let experts decide for them, they cease to function as active members of the body politic; but insofar as they work with experts to help make decisions about the proper goals for scientific research and technological development, they learn about science and engineering, their powers and their limitations, and enhance their lives as democratic citizens.

A second argument is that public participation may also promote better technical decision-making. All knowledge is not technical knowledge. Local knowledge, nonexpert knowledge, what is often called indigenous knowledge, when appropriately utilized, has the power to enhance even the technical aspects of technical decisions. One good example comes from HIV/AIDS research (see Feenberg, 1995a, chapter 5). It was HIV/AIDS activists who helped increase funds for such research and redirected the research away from basic work on immunology and toward more immediately practical therapies for those afflicted with the disease. Without the involvement of nontechnical HIV/AIDS activists, not as many lives would have been saved by the related scientific research.

Finally, a third argument is simply that people have a prima facie right to influence those decisions that affect them. Although there are clearly instances in which this right may be trumped by other considerations (see Nozick, 1974, p. 268ff.), the public implementation of technological design decisions is arguably not among them. Many scientific and engineering decisions have major impacts on our ways of life. Democratic revolutionaries once legitimately proclaimed, "No taxation without representation." Today, as Langdon Winner (1991) and Steven Goldman (1992) have argued, they may just as legitimately argue, "No invention without representation." This extends the principle of free and informed consent as generally accepted in medicine to engineering (see Martin and Schinzinger, 1983).

One of the clearest instances of a shift from "technical professional knows best" to "technical professional consults best" has occurred in the medical profession. Traditional medical ethics emphasized the physician as decision maker for the patient. Today this model is in transition to one in which the physician helps educate patients so that they are able to make free and informed decisions about the course of treatments they may or may not undergo. Physicians may be experts but are not by that reason alone the final authority. The practice of medical expertise requires collaboration with end-user patients who themselves determine how that expertise is finally to affect their lives.

THE CONVIVIAL TECHNOLOGY IDEAL

The Edison vote recording machine anecdote illustrates the problem of technology transfer from laboratory to market. In this context, the separation between invention and end-user need is intensified and rationalized by notions of technical autonomy and the existential pleasures of cutting-edge invention, with the responses of marketing and advertising not so much revealing the problem as obscuring it. The development of personal and professional engineering ethics, insofar as professional ethics calls for commitment to a public good, is another way to address this obscured issue.

But the problem of alienation from end-user concerns is refocused by another type of technology transfer, that from one country or society to another. Since the middle of the twentieth century this second type of technology transfer has played an increasingly important role in world history.

On January 20, 1949, in his inaugural address, Harry S. Truman, the first post–World War II president of the United States, declared that having used science and technology to win the greatest military conflict in history, the United States now had a new responsibility: "We must embark," Truman declared, "on a bold new program for making the benefits of our scientific advances and industrial progress available for the improvement and growth of underdeveloped areas." With these words and subsequent actions, Truman not only inaugurated a presidential term but also the "age of development" (Sachs, 1992) Ever since, the United States and its allies have systematically attempted to transfer technology from developed to underdeveloped countries, in order to bring all societies into a common framework. Twenty-first-century discussions of economic globalization are but the most recent permutations of the development ideology.

Yet by the mid-1970s it had become obvious that the initial development strategies for transferring advanced technologies from one country to another often failed for reasons similar to those experienced by Edison with his vote tallying device. What donor countries wanted to give, recipient countries often resisted taking or tried ineffectively to take. Water and electrical power systems broke down; new roads and machines were not maintained. Gaps between the rich and the poor were not reduced but widened.

Four responses emerged: One response was to promote social science research into the problem in order to enhance the development bureaucracy. The resulting products often functioned as international marketing studies.

A second response was to promote education as a required adjunct to international technology transfer. The education in question, however, often functioned more like advertising than anything else. In many cases it simply facilitated immigration of a newly trained technical elite from poor to rich countries, a kind of social capital exploitation in parallel to natural resource exploitation.

A third response was simply to cut back on development aid, in the belief that recipient countries had to do it for themselves. As one book proclaimed, "Underdevelopment Is a State of Mind" (Harrison, 1985)—referring, of course, to the state of mind of the poor, not the rich.

The fourth response was proposals for what were variously denominated as intermediate, appropriate, or alternative technologies. Intermediate technologies were designed to function in between manual tools and complex high-tech machines. One example: replace animal transportation not with the internal combustion engine but the bicycle. Such intermediate technologies were also described as more appropriate to the cultural and social contexts because they introduced smaller, less destabilizing technical changes, and their functioning was easier to understand. The workings of a bicycle are much more transparent than those of an internal combustion engine.

In a reflection back into the high-tech world itself, especially after the energy crises of the 1970s and in the face of mounting issues of environmental pollution, such technologies were seen as appropriate not just to underdeveloped or developing countries but also to highly developed countries. Now they were called alternative technology. The soft energy paths of renewable energies—wind and sun instead of coal, oil, and nuclear power—were argued to be better for both the poor and the rich, and a new basis for creating common frameworks of development (Lovins, 1977).

The most influential statement of the intermediate and appropriate technology movement was E. F. Schumacher's *Small Is Beautiful* (1973). Although Schumacher's book is the most widely known, another provides a deeper analysis: Ivan Illich's *Tools for Conviviality* (1973). For a short period of time the term "convivial tools" or "convivial technology" supplemented those of appropriate, intermediate, and alternative technology. Although the term has been sidelined by history, it is revived here to suggest again broader ways of thinking about the free and open source software movement.

In *Tools for Conviviality* Illich opened by noting the following about many technologies: they begin as means to some specific end but often wind up subverting that end through what he calls "counterproductivity." Illich's case study example was medicine. What started out as a means to health has become the cause of a new kind of illness: iatrogenic or physician-caused illness and disease. According to Illich, a first watershed in the history of scientific medicine occurred when it came to pass that physicians actually did something beneficial for their patients more than 50 percent of the time. This watershed occurred during the early decades of the twentieth century. But a second watershed occurred when it came to pass that more than 50 percent of the time physicians treated patients for illnesses or diseases to which medical practice itself contributed. Such illnesses and diseases range from the results of overt malpractice to the negative side effects of drugs, infections caused by treatments, diseases caught in hospitals, diseases that did not even exist until the evolution of bacteria in response to antibiotics, and more. According to Illich, this second watershed had been reached or was on the verge of being reached at the time his book was written.

For Illich, the proper response to the counterproductivity of the second watershed is moderation: not more medicine, but less. Convivial technology is properly limited technology, technology that does not so overwhelm with its presence that people think they cannot live without it. This is an argument that has been reiterated in the course of more than one critical examination of the contemporary healthcare system (see, for instance, Callahan, 1998).

As a response to the challenge of transferring technology from one country to another, then, Illich's argument called for the exercise of critical distance. Perhaps not all technology should be transferred. Prior to attempting a transfer, rather than doing a marketing study of what people wanted, it would be better to do an assessment of the technology itself. Not all technologies are inherently beneficial. Many technologies bring with them political implications if not unintended negative consequences that deserve careful consideration (see Winner, 1980). What is more important than technology itself is to create a modestly ironic adaptation from Illich, its conviviality quotient. To what extent does a technology make possible or perfect human living (*vivere*) with (*con*) its artifactual presence?

The appropriation and adaptation of the term "conviviality" in this context might be questioned as failing to appreciate fully a Kantian-like personalism at the core of Illich's normative stance. At one point, for instance, Illich writes that he means "conviviality" to indicate "autonomous and creative intercourse among persons, and the intercourse of persons with their environment; and this in contrast with the conditioned response of persons to the demands made upon them by others, and by a man-made environment." He considers "conviviality to be the individual freedom realized in personal interdependence and, as such, an intrinsic ethical value." Further, "in any society, as conviviality is reduced below a certain level, no amount of industrial productivity can effectively satisfy the needs it creates" (Illich, 1973, p. 11). One way of interpreting this claim is to see Illich as affirming individual autonomy of the will over and against any heteronomy, especially in the use of tools or technologies. Human beings are diminished in individual human flourishing insofar as their use of technologies is determined by engineer designers, capitalist producers, advertising marketers, or any other external managers. Like Michel de Certeau (1980) and Feenberg, Illich argues on the basis of and celebrating creative consumer transformation of technologies.

Although there is clearly some appeal to individual autonomy in Illich's normative argument, it is a mistake to emphasize this at the expense of its communitarian element. According to Illich's anthropology, humans are human insofar as they live with others. In the quoted passage, for instance, Illich references autonomous creativity in relation to "intercourse among persons, and the intercourse of persons with their environment." In his first justification of the term, Illich writes that the convivial society is one "in which modern technologies serve politically interrelated individuals rather than managers" and that "convivial" is meant as "a technical term [designating] responsibly limited tools" (Illich, 1973, p. xxiv). The emphasis is on "*interrelated* individuals," not simply individuals. In the only other text with an extended discussion of the term, Illich likewise emphasizes, along with autonomy, that "convivial tools" as those "which facilitate the individual's enjoyment of use-values—without or with only minimal supervision by policemen, physicians, or inspectors" (Illich, 1978, p. 42). But again, Illich's anthropology is one that understands individuals as living with others, convivially, among friends. His position is less sympathetic to libertarian individualist rights and more to groups rights as argued, for example, by Michael Ignatieff in *The Rights Revolution* (2007). A convivial assessment of technologies could thus be postulated as one that considers critically, to venture a provocative analogy, to what extent a tool-human grouping might tend to be broken apart and users separated from promised or projected use values simply as a result of artifactual design.

As one example of such an assessment, consider how the shift from tools to machines to computerized devices tends inherently to diminish human-artifact integration. Tools require human energy input and guidance. A hammer is held in the hand, has energy imparted to it from the muscles of the arm, and is guided by the informed eye, which in turn re-forms the motion of the hand. The hammer functions as a part of the body is embodied.

A machine, by contrast, receives energy and movement from nonhuman sources, although the human remains as an immediate guiding operator. The automobile is powered by an internal combustion engine but requires a human driver. The form of conviviality or "living with" manifested in the car is quite different than the experience presented by the hammer. The car begins to take on a kind of independence or semiautonomy—if the driver unhands the steering wheel, the automobile can continue along the road on its own. Shortly, we can expect to have autonomous automobiles. The car is less an extension of the human body and more an energized artifact standing over against the body if not a capsule that temporarily encloses and transports the body.

In computerized machines this trajectory toward living more *away from* than *with* reaches a new level. Computerized devices such as automated teller machines are powered not by human energy but by electricity and guided not directly by human beings but by stored programs. Human guidance exists behind the scene, of course, but so deeply hidden or removed that the artifact appears to take on an independent life. To live with automatic teller machines becomes less and less a direct bodily engagement, as with tools, and more and more a living in the midst of, as with plants and animals. Technical conviviality exhibits distinctive experiential structures.

A further investigation of this issue of technical conviviality can be found in the work of Albert Borgmann, another philosopher deeply concerned with the interactions of technology and culture. According to Borgmann (1984), the separation between inventor and end user is reified in the material artifact, in the form of a separation between the machinery of a device and the commodity it delivers. The commodification of culture is not just the creation of an economy of quick-drop shopping but one in which the goods shopped are commodities whose operations are fundamentally opaque.

Without the details, one can understand how a mechanical, analog clock works. Without the details, one has hardly any clue about what is going on in an electronic, digital watch. Both deliver a commodity we can call well-tempered time. But in the former, the mechanism of delivery is imaginable even when not fully understood. In the latter, the mechanism disappears behind a veil of technical sophistication. The digital divide becomes an ontological feature of high-tech artifacts.

Something similar happens in the shift from typewriter to word processing computer. Even if one cannot completely master the mechanics of a mechanical typewriter, its structure and how it works—the transfer of motion from keyboard to precast typewriter key—is readily comprehensible. In living with the typewriter one's intelligence extends itself with ease into the world of artifice. One does not feel dumb or stupid, or come up against an opaque construction that belittles one's common cognitive abilities. Instead, looking at and using the typewriter gives one a certain satisfaction. One quickly develops a sense of competence and confidence in its presence.

In the presence of word processing computers, by contrast, almost the opposite seems to be a norm: one feels stupid, even angry, unable to understand what is really going on, and often unable to cope. Would it be possible for the free and open software movement to address such a problem? Could the free and open source movement design a word processing program that, like the mechanical typewriter, promoted rather than inhibited technical conviviality? And would such software not be inherently more available for transfer from one country or context to another, in the way that bicycle technology is more transferable than automotive engineering?

CONCLUSION AND IMPLICATIONS

The history of technology and Edison's invention of the method of invention have been used to point up a problem in technology transfer from laboratory to market. The problem arises from a temptation—a temptation from which the software community is not immune—to become an elite removed from the nontechnical, end-user life that it is meant to serve.

The argument then reviewed the development of engineering ethics as a partial response to the problem. The contention was that the professional-ethical commitment to public safety, health, and welfare supports the free and open source software movement.

Finally, reflection turned to the challenge of development and the issue of technology transfer from rich nations to poor, in order to highlight ways in which the free and open source movement might be thought of as another form of the intermediate, appropriate, and alternative technology movements.

By way of conclusion, it is appropriate to offer a brief statement of possible policy implications. As noted at the outset, from the end-user perspective what is important is not so much the direct availability of the source code as program or technical conviviality, phrases richer in significance than "user friendly." Such is the background for proposing the concept of convivial software. From the perspective of end users, there is more with regard to free expression or openness than technical source code transparency.

To provide concrete illustration of this *more*, perhaps it can be suggested that convivial software be judged in terms of three criteria:

1. **Stability**: The need for continuous upgrades detracts from technical conviviality. It is like having friends who insist on continuously reinventing themselves. It is difficult to get comfortable with ever-new friends and software. A missing good in software is the stability of the old—forcing one to repeatedly upgrade. As a personal example, since I first started using Word Perfect in the mid-1980s I have been forced into the upgrade and relearning cycle at least five times—on average, once every three or four years. Compare this with my typewriter. I bought a typewriter when I was in high school and used it continuously, with only minor repairs and cleaning, for 30 years. The shift from typewriter to word processor was a positive improvement in writing tools. No upgrades in software have ever come close and, in fact, many have been negative.
2. **Transparency**: One cannot readily figure out how a word processor works. This is true at multiple levels. At one level, opacity occurs in the toolbar icons, which the context-sensitive help notes seldom resolve, and in the customizing of default configurations. At another, it concerns what happens behind the scenes, as it were, with formatting. Word Perfect at least allows users to reveal formatting codes, and then edit them. Microsoft Word takes this ability away from its users and is thus much more mysterious and difficult to adjust. At still a third level of transparency, users might like to understand what happens when a word processing program locks up, requiring the computer to be turned off and rebooted. The error messages are completely opaque. (Indeed, it is difficult to image that the error messages mean very much to the technical expert without some reference

materials at hand.) Would it be possible to design a word processing software program that could be understood without earning a degree in computer science and engineering?
3. **Simplifiability**: Most word processing programs include many more features than any one user actually needs. The toolbar is cluttered. I never use "Quick Format" or "Picture Draw" capabilities and have met few people other than technical users who do. When users need such features, they could install specific programs. I would also willingly dispense with "Zoom" and "Change View." But it is not possible to get rid of these features or even to hide them as a way to simplify the screen.

In response to the obvious criticism that one does not have to use all the features built into a word processing program, yes, this is true. But they clutter up not just the screen but one's technical life with artifacts. Some of us would like to live with a computer screen that has the aesthetic feel of Shaker or Scandinavian furniture. Why is this not a digital option? Is it not one that the free and open source software movement for the first time might make possible for all?

In summary, then, the argument here is that the free and open source software movement might well adopt and promote an ideal of convivial software that is stable, transparent, and simplifiable. To restate in other words, consider the ideal freedoms zero through three as described by the Free Software Foundation in 1986:

Freedom 0: The freedom to run the program for any purpose.
Freedom 1: The freedom to study how the program works and change it to make it do what you wish: algorithmic transparency.
Freedom 2: The freedom to redistribute copies so that you can help your neighbor.
Freedom 3: The freedom to improve the program and release your improvements (and modified versions in general) to the public so that the whole community benefits.

Might it not be possible to complement such freedoms with three others, two negative and one positive?

Freedom 4: Freedom from upgrades
Freedom 5: Freedom from technical study as a prerequisite to basic understanding
Freedom 6: Freedom to simplify (not just improve)

Upgrade free, transparent, and simplified software would be able to help mediate the technology transfer divides between laboratory and market and between developed and developing countries—and maybe even transform our own development. Free and open source software has the prospect of being able to contribute not just to technical power but also to the good life.

Finally, the pursuit of the concept of conviviality in open source software might well have implications for other forms of open source culture—from open access to open science. That is, rather than focusing on the internal structure of knowledge production and then working out to the social implications, it might be reasonable on occasion to reverse the analysis, to work from what would be most useful to humans back toward embedding end-user concerns in the knowledge production processes themselves. The normative concept of conviviality, bringing a notion of how best to live to bear on the means of living, may have benefits that go beyond computers and information technology.

ADDENDUM: THE SPEED TRAP

The argument for not updating software has so many problems it is difficult to know how to proceed. On top of which is our facticity of being born into a well-entrenched and apparently rational cultural commitment to speed.

From a modest effort to engage the issue (Mitcham, 2018), one of the most widely shared values in our world is an entrenched commitment to speeding things up. From the earliest years of the Industrial Revolution, efforts to accelerate the production of goods and services have been seen as multidimensional benefits. Increasing speed in productivity increases worker pay—provided workers fight for it!—and the availability of goods. From the nineteenth through the twentieth centuries, science and engineering have sponsored accelerations in transport and communications as public goods. In the twenty-first century, increasing the speed of knowledge production and technological innovation are largely unquestioned ideals that are further asserted to be the most effective means for addressing social problems such as poverty and disease as well as environmental challenges such as pollution, biodiversity loss, and climate change. Who wants to go back to the dark ages of economic and political stagnations?

Translated to software: Who wants to go back to MS-DOS or WordStar? Or to the Internet before the Web? And given the proliferation of malware and hacking made possible by the ease and speed of digital communication systems, rapid updating appears to be a categorical imperative. What other option is there?

Yet recognition of how deeply we are stuck in the speed trap should occasion philosophical reflection. Here and there are small efforts, although they are not being advanced with any great rapidity. One of the most neglected issues for any philosophy of engineering and technology is how to regulate and moderate the speed of innovation. There is something deeply destabilizing about the current (and increasing) pace of life.

As a naïve counter to those who want to run ever more rapidly into the future: Who wants to be going so fast that it becomes impossible to live in the present? Why can't updating and innovating be practiced with dedication to stability instead of creative destruction?

Essay 8

Comparing Approaches to Philosophy of Engineering

Including the Linguistic Philosophical Approach

(with Robert Mackey)

What follows is philosophically incomplete. One feature of this incompleteness is that we do not fully agree about what should be included—or even what is included. This essay is thus a provisional presentation of what has emerged from an extended discussion that often has the character of a disagreement. In addition, the paper is methodologically naive, accepting as given certain meta-philosophical distinctions that are in fact controversial—and failing to include all that might be considered. In the current context of the relative underdevelopment of the philosophy of engineering, we nevertheless hope that our effort may encourage others to consider a greater range of possibilities than might have been imagined. To anticipate our conclusion, we want to argue for the pursuit of pluralism in the philosophy of engineering—as a way to transform engineering itself. The goal is not just to understand engineering but also to change it. But in order to advance such a thesis it will be necessary to indicate more about the kind of pluralism we have in mind.

1. INTRODUCTION

Philosophy is composed of different branches and practiced in different schools or traditions. The branches include at least epistemology, metaphysics, and ethics. What might be called a conventional philosophical pluralism would thus simply emphasize the importance of each of these branches, along with others such as political philosophy and aesthetics. Philosophy in the comprehensive pluralism of all its branches, however, also comes in distinct historical schools or traditions.

Historically, for instance, it is common to distinguish between Platonism and Aristotelianism, Augustinianism and Thomism, rationalism and empiricism, and more. In the contemporary world the two most widely recognized distinctions are between the phenomenological school of continental Europe and the analytic school of Anglo-American provenance—although the geographies are less significant than their differing methodologies. In considering possibilities for pluralism in the philosophy of engineering it is thus also useful to reflect on how this new regionalization of philosophy might take shape within different philosophical traditions. A pluralism that includes not just the different branches of philosophy but is open as well to contributions from different schools of philosophy is a richer and more robust pluralism.

With slightly more specificity than the phenomenological/analytic differentiation, it is possible to identify six currents in contemporary philosophy: (1) phenomenological philosophy, (2) postmodernist philosophy, (3) analytic philosophy, (4) linguistic philosophy, (5) pragmatist philosophy, and (6) Thomist philosophy. These six currents or schools are not mutually exclusive and often overlap. To this particular set one can easily add a number of others such as Marxist philosophy and critical social theory, feminist philosophy, or environmental philosophy. Yet for purposes of initial inventory and orientation in relation to possibilities for a philosophy of engineering, our particular set of contestable metaphilosophical distinctions should be sufficient. Beyond the review of opportunities in the six schools considered here, however, one approach will be outlined in slightly more detail. This detail will focus on what the philosophy of engineering might look like from the perspective of linguistic philosophy.

To these introductory qualifiers, let us add two more. First, by beginning with philosophy rather than engineering, we acknowledge a bias in favor of philosophy. Our approach might well be complemented by considering the philosophy of engineering from the perspectives of a widely diverse number of types of engineering: civil, mechanical, electrical, chemical, electronic, nuclear, social, computer, biological, and more—perhaps also in relation to their various historical or cultural forms. But our bias is one in favor of a philosophy of engineering for philosophers rather for engineers, although we hope that some engineers might find it useful. Second, engineering is seldom fully and carefully distinguished from technology. Indeed, we believe that the engineering-technology distinction is itself a philosophical problem—one that will be dealt with in different but complementary ways in various philosophical traditions. With all these qualifiers in mind, let us now provide summary presentations of six contemporary currents or approaches to philosophy and how they might engage with engineering.

2. SIX BASIC TYPES

Let us reiterate that the distinctions among these six basic types of contemporary philosophy can be debated. In an increasingly globalized and interconnected world—with fusions in music, cuisine, and literature—we could expect crossovers to occur in philosophy as well. Nevertheless, differences remain. Acknowledging the porosity of boundaries, the argument for a pluralism that accommodates different schools of philosophy may still sketch how each might approach the phenomenon of engineering.

1. A phenomenological approach to the philosophy of engineering would take engineering and/or some of its diverse manifestations as phenomena calling for more careful description and attentive, critical reflection than they may have previously been accorded. Human practices tend to become sedimented with received interpretations, and the phenomenological effort is constituted by a return "back to the things themselves" (Edmund Husserl), to inspect with a stance of fresh interest what has become routine or occluded by worn assumptions and clichés. This effort to disclose with freshness the engineering experience is well illustrated in the work of Don Ihde. Although Ihde seldom refers directly to "engineering"—preferring instead the terms "technics" and "technology"—what he says about technology can in many instances apply to engineering.

Consider, for instance, Ihde's (1990) distinction of embodiment and hermeneutic relations between humans and technology. Embodiment relations are exhibited by such simple tools as the blind person's cane or the dentist's probe. As Ihde says, for the sightless person the world begins at the end of the cane; the cane is experienced as incorporated into the sensing and perceiving body. Hermeneutic relations are, by contrast, exhibited by such simple instruments as the thermometer. The thermometer, as an instrument "out there" apart from the body, has to be read or interpreted. Such phenomenological distinctions of embodiment and hermeneutic relations are manifest in engineering practice. Many of the once-ubiquitous drafting instruments, such as pencils and triangles, functioned as if they were extensions of the hand. Today, however, sensors and computational devices that function only insofar as they are read or interpreted have become increasingly prominent elements in engineering practice.

One historical associate of phenomenological philosophy, existentialism, deserves special notice. As is well known, Samuel Florman's *The Existential Pleasures of Engineering* (1976) is an effort to adapt this school to a description of the engineering experience. That Florman did not find dread or "fear and trembling" to be distinctive features of the engineering experience places him in good company with Ihde—as both exhibit a confident optimism that it is important for any philosophy of engineering to acknowledge although not necessarily endorse.

2. Postmodernism most commonly if ambiguously refers to the work of such figures as Michel Foucault, Jean-François Lyotard, or Jacques Derrida. In each case philosophical thinking typically undertakes to historicize or destabilize its subject matter, often by playful or ironic means. A postmodernist philosophy of engineering might, having noted some of the variegations of its alleged history and different contextualizations, question whether there even really is such a thing as engineering.

One wry distinction might be drawn between postmodern and postmodernism. Postmodern thought seeks to demythologize or demystify a belief or practice, to help us see—almost after the manner of descriptive phenomenology—how things really are. What is disclosed in the demystifying process is something one did not previously suspect. Foucault's disclosure of how everything from prisons to knowledge production can serve to maintain and promote power relations is such a demystification, one that finds power everywhere, in places most people have not previously located it. In a structurally analogous manner, Billy Vaugh Koen (2003) has attempted to demystify engineering as heuristics and then argued for the presence of heuristics in all aspects of human behavior.

Postmodernism carries the postmodern approach one step further. Whereas postmodern thought might presume that disclosure can at some point come to a conclusion or be finalized, postmodernism is more skeptical and undertakes an ironic and repetitive pursuit of one demystification after another. Certainly the ironic and playful style typical of postmodernism is also present in Koen's ironic and playful prose.

3. An analytic philosophy of engineering is easily if inadequately illustrated by the work of Mario Bunge. (Bunge calls his philosophy scientific rather than analytic, but for us this is a distinction without a difference.) From the 1960s to the present, and with repeated verve, Bunge has advanced analytic distinctions between science and engineering by arguing for recognition of special forms of logic and knowledge in engineering, even when he wants to describe engineering as a kind of applied science, and by proposing an engineering transformation of ethics. Without necessarily accepting all of Bunge's positions, one could nevertheless argue that a proper philosophical analysis of engineering would attempt to rethink engineering in terms broader than those of engineering itself while considering how engineering methods might be used to rethink many other aspects of human experience.

In brief, an analytic philosophy of engineering would seek to resolve engineering into its constitutive elements, which it would attempt to clarify conceptually. In addition it would criticize and attack misconceptions and false beliefs such as those associated with such general theories as those of technological determinism or the technological fix—often by arguing their incoherence or dependency on category mistakes. We take it that Joe Pitt (2000 and 2011) and his effort to practice Wilfrid Seller's "finding one's way around" in the world of engineering and technology provides complements to Bunge's exemplification; the Dutch school of analytic empiricism that began with technology but in ways that emphasized the centrality of engineering is a further development (see Kroes and Meijers 2001). As the philosopher Deborah Johnson (2001) has described her own efforts with regard to computer science and engineering, the analytic philosopher seeks to reject fuzzy conceptualization and hype—whether positive or negative in character—surrounding and associated with engineering and technology.

4. Linguistic philosophy takes the analysis of language as its starting point and, as in the later Ludwig Wittgenstein, suggests that clarification of language use can enable us to see through certain conundrums that have accumulated in both popular and professional philosophical thought. What more precisely this might mean with regard to engineering remains to be explored. This approach to the philosophy of engineering will subsequently be considered in more detail.

5. Pragmatism in the last quarter of the twentieth century became divided into epistemological and social philosophical schools. The epistemological school, as represented by such figures as Willard Van Orman Quine and Donald Davidson, has paid little if any attention to engineering, perhaps because engineering does not engender epistemological or logical puzzles the way science does. Instead, the possibilities for a pragmatist philosophy of engineering can be most clearly discerned in the social pragmatism of such figures as John Dewey, Paul Durbin, and Larry Hickman. Durbin, in fact, edited a book titled *Critical Perspectives on Nonacademic Science and Engineering* (1991) that brought together the works of people commonly associated with the philosophy of technology to address questions concerning the relation between engineering and social reform. In the efforts of Dewey to understand science itself as instrumental, and to use especially the social sciences to advance what is often called piecemeal social engineering, we see another possibility for a pragmatic, social philosophy of engineering. Hickman's version of the pragmatist approach is to call for a philosophy of technological culture—but which might with equal legitimacy be termed a philosophy of engineering strongly interacting with culture.

At the foundation of the social reformist school of pragmatism is a view of society as an evolving, humanly constructed order. In this sense a pragmatist philosophy of engineering could make common cause with or be implicitly allied with work in the social constructivist studies of science and technology. Bruno Latour (1988), Wiebe Bijker (1995), and Harry Collins and Trevor Pinch (2002)—in quite different approaches to social constructivism—have all mentioned engineering, at least in passing. Additionally, engineer and philosopher Louis Bucciarelli's *Designing Engineers* (1994) and *Engineering Philosophy* (2003) may be read as exemplifying social constructivist interpretations of engineering practice and pragmatist efforts to use philosophy in engineering, respectively.

6. Thomist philosophy, probably the least recognized of the traditions reviewed here, understands itself as a continuation of Aristotelian philosophy, which it nevertheless encloses in an ontology of supernatural creation and a cosmological vision of Platonic inspiration. In the first half of the twentieth century, Etienne Gilson and Jacques Maritain cut Thomism loose from many scholastic encrustations, especially in metaphysics and epistemology; in the last half of the century Alasdair MacIntyre did much the same in ethics. Thomism typically defends a realist approach to science while, as in the work of Bernard Lonergan, occasionally adapting and incorporating elements from both Kantian and phenomenological traditions.

Although it is primarily a treatise in realist epistemology, Lonergan's study of *Insight: A Study of Human Understanding* (1957) adumbrates in the margins what might be described as a realist, Aristotelian-based philosophical interpretation of engineering. For Lonergan, insight is a heuristic movement from intuition to explicit cognition and operates in both science (knowing) and technology (making), the latter of which becomes engineering through the process of conscious elaboration. Thus it is that Thomism has adopted an epistemological approach representative of modern philosophical tendencies. More generally, however, a full-bodied Aristotelian-Thomistic philosophy of engineering should attempt to understand engineering in relation to all the fundamental branches of philosophy as Aristotle himself began to articulate them: epistemology, metaphysics, and ethics.

3. TOWARD A LINGUISTIC PHILOSOPHY OF ENGINEERING

Having itemized six possible types of or approaches to the philosophy of engineering and offered selective references for each, it is now appropriate to take up the challenge suggested by our comment regarding the underdevelopment of a linguistic philosophy of engineering. Let us attempt to map out some potentials for this neglected approach. In the process it may be possible to indicate more fully some necessary aspects of any philosophy of engineering.

To talk about something called "linguistic philosophy" may appear to reference an anachronism. Twenty-five years after his effort to define "the linguistic turn," Richard Rorty (1992) declared linguistic philosophy to be indistinct from analytic philosophy—if not dead. Yet what continues to be called an analytic tradition emphasizes a need to reflect critically on the ways of talking about topics of philosophical interest (see, for example, Losonsky, 2006). Even if the slogan that "the problems of philosophy are problems of language" no longer has the naive appeal it once did, it is useful to consider in what ways a reflection on language within a regionalized field of human practice might contribute to advancing a philosophy of that practice.

Our task here has, however, become more problematic than was initially imagined. As with many projects in philosophy, what began with plausible (it seemed to us) intuition, turned progressively difficult—if not confusing. Yet even though we are reduced to a somewhat episodic set of comments, and disagreements, with further disagreements among ourselves regarding their perspicacity, we continue to believe in the viability of the effort. Indeed, even the difficulties we experienced have provided us insight into issues in the philosophy of engineering for philosophers, if not for engineers. Consider, then, three comments on possibilities for a linguistic approach to the philosophy of engineering.

As a first comment, it can be noted that regional or specialized philosophies such as the philosophy of science, the philosophy of art, or the philosophy of religion typically include analyses of their subjects from the perspectives of the main branches of philosophy—that is, epistemology, metaphysics, and ethics. To appreciate what a linguistic philosophy of engineering could look like, one might therefore consider how epistemology, metaphysics, and ethics have been manifested in linguistic philosophy and then see if the methods of linguistically oriented epistemology, metaphysics, and ethics can be projected into engineering. This may be a somewhat pedestrian, but not for that reason inappropriate, approach to a linguistic philosophy of engineering.

A second comment is that any examination of what linguistic philosophy may have achieved in epistemology, metaphysics, and ethics, will depend on some understanding of how linguistic philosophy—sometimes called ordinary language philosophy—arose as a movement in England following World War II. It originated in opposition to such other philosophical schools as Hegelianism and logical positivism and was at one point closely associated with the thought of the later Wittgenstein, especially his posthumously published *Philosophical Investigations* (1953). Other major practitioners have included Gilbert Ryle, J. L. Austin, and P. F. Strawson. One practice common to these philosophers is an attempt to pay careful attention to linguistic usage on the grounds that "the meaning of a word is its use in the language" (*Philosophical Investigations* I, §43). Language is understood as including a number of "language games" so that learning how to correctly use words in any of those games is to learn their meaning. Philosophical problems about the meaning of knowledge (epistemology), reality (metaphysics), or goodness (ethics) are argued to arise when one incorrectly plays the corresponding language game. Philosophy functions like an umpire in the games of language, able to resolve if not solve disputes among players.

Thus one linguistic philosophical approach to the philosophy of engineering would be to think of engineering as a particular language game. More specifically, one may note that philosophy of engineering—like virtually all other specializations of philosophy—might well begin with a definition of engineering. But what engineering is might be better determined by how the word "engineering" and its cognates and associated terms (such as invention, innovation, design, technology, science, etc.) are used, especially in relation to each other. From a linguistic philosophical perspective, it would be appropriate to begin not so much with our experiences of engineering but with the words we use to talk about such experiences.

This view has been interestingly advocated by Andrew Light and David Roberts (2000). In their interpretation of Wittgenstein's project, what is central is a "descriptivist thesis": the method of describing or giving an account not so much of things themselves as of our diverse descriptions of them. They suggest that one way to avoid the too quick fall into normativity that they argue has vitiated so much philosophy of technology is to call up with greater self-consciousness the manifold "forms-of-life and descriptive grammars bound up with various families of technologies" (p. 139). What Light and Roberts would practice with regard to technology might be as beneficially prosecuted with regard to engineering. Adopting their tactic of introducing suggestive substitutions into a Wittgenstein text would yield, for instance, the following linguistic effort to appreciate yet philosophize engineering complexity:

> And for instance the kinds [of engineering] form a family in the same way. Why do we call something ["engineering"]? Well, perhaps because it has a—distinct—relationship with several things that have hitherto been called [engineering]: and this can be said to give it an indirect relationship to other things we call by the same name. And we extend our concept of [engineering] as in spinning a thread we twist fibre on fibre. And the strength of the thread does not reside in the fact that some one fibre runs through its whole length, but in the overlapping of many fibres.
> (see *Philosophical Investigations* I, §67)

From this perspective, engineering is not some one language game—the game of efficiency, as Lyotard (1984) has termed it—but many.

At this point, however, the descriptivist project presents a certain conundrum. For Wittgenstein, seeking linguistic clarity aims to rid philosophy of its confusions.

> For the clarity we are seeking is indeed complete clarity. But this simply means that the philosophical problems should completely disappear.
> (*Philosophical Investigations* I, §133)

Yet for philosophical problems to disappear they must first have appeared. In other regionalizations of philosophy problems have appeared within nonphilosophical practices, and it is their appearance that gives rise to regionalized philosophy. In religion, for instance, the belief in a nonempirical God gives rise to such questions as "How can I know that God exists?" or "How can I know God?" The philosophy of religion seeks to address such questions in ways different from those in religion itself. In art, the question of the nature of art itself is posed by such artistic schools as surrealism and dada—again feeding into that regionalization of philosophy known as the philosophy of art. In science, relativity theory and quantum physics present both epistemological and ontological paradoxes that engender questions for the philosophy of science. But what are the questions in engineering that could give rise to a philosophy of engineering? It sometimes seems that philosophers have to go hunting for such problems and are alone in talking about them.

Third comment: Ian Hacking observed in *Why Does Language Matter to Philosophy?* (1975) that there are two common positive responses to his question that have linguistic philosophy operating across all philosophical schools both synchronically and chronologically. One response is that any philosophy worthy of the name must involve the pursuit of linguistic clarity through a refinement of common speech; thus one finds efforts to clarify linguistic usage by the construction of definitions that avoid ambiguity, equivocation, contradiction, and paradox and thereby to create a technical language for philosophy, in all philosophies from Plato and Aristotle to the present.

Another, contrasting response criticizes technical languages and appeals to common speech, maintaining that the path to clarity is one of sensitive and subtle reflective appreciation of the inherent richness of common speech. As Hacking summarizes,

> There are two well-known minor ways in which language has mattered to philosophy. On the one hand there is a belief that if only we produce good definitions, often marking out different senses of words that are confused in common speech, we will avoid the conceptual traps that ensnared [others]. On the other hand is the belief that if only we attend sufficiently closely to our mother tongue and make explicit the distinctions there implicit, we shall avoid the conceptual traps. One or the other of these curiously contrary beliefs may . . . be most often thought of as an answer to the question Why does language matter to philosophy?
>
> (p. 7)

Hacking himself actually thinks these are minor if not trivial answers to the question, and in an analytic review of the development of that special form of philosophy known as the philosophy of language—from what he calls "the heyday of ideas" (in the seventeenth century) through "the heyday of meanings" (twentieth century) to "the heyday of sentences" (a parallel development in the twentieth century)—proposes the emergence of what he terms a "lingualism" in which language itself has become the primary phenomenon if not reality. On this basis it might be possible to examine engineering language—that is, the technical language of engineering—as its own special phenomenon.

Such a lingualist approach to the philosophy of engineering would consider the kinds of words used and the ways they are used within engineering. At the same time, any attempt to reflect critically on linguistic usage within engineering needs to consider what such reflection could add over and above the technical use itself. This last is in fact a question that has been raised about linguistic philosophy as a whole.

4. CONCLUSION

A linguistic philosophy of engineering is only one possible approach to the philosophical engagement with engineering, but it is an underdeveloped approach with serious promise. At the same time, it is not an approach that we would recommend to the exclusion of others, including approaches not mentioned here. Instead, as indicated at the outset, we wish to argue for pluralism in the philosophy of engineering—a pluralism that could even help transform engineering.

Before offering arguments for pluralism in the philosophy of engineering it may be appropriate to indicate briefly why we think engineering deserves to be transformed. There are at least two reasons. First, engineers often complain that they do not receive proper respect and/or that engineering does not attract sufficient interest among students, especially women and minorities. Second, nonengineers have criticized engineering as insufficiently well developed as a profession—as lacking the professional maturity of, say, medicine or law. A philosophy of engineering may offer a path to the transformation of engineering that would constitute a response to such complaints and criticisms. But this would only be likely, we suggest, if the philosophy of engineering eschewed the ostensible narrowness of engineering itself by the practice of pluralism.

Let us conclude, then, with three basic arguments for pluralism in the philosophy of engineering—a pluralism that may at the same time contribute to a measured transformation in engineering itself. The three arguments are from philosophy, from interdisciplinarity, and from responsibility.

First, as we have argued, the linguistic approach is not the only possible approach, and there are undoubtedly strengths and weaknesses in any approach. The most robust engagement of philosophy with engineering will, it is thus reasonable to propose, entail engagement from more than one philosophical perspective. This is likely for a couple of reasons. Since philosophy itself is not some one thing, philosophy *tout court* can engage engineering only through the more specific engagements of its distinctive traditions. Because philosophers themselves represent different schools as much as they represent philosophy, the rich possibilities for a philosophy of engineering will only be realized through different philosophies of engineering; the more philosophies, the better.

Second, traditions or schools of philosophy are somewhat like disciplines. As is often argued today, problems that need to be addressed through the use of disciplinary knowledge do not themselves come with disciplinary boundaries. More often than not problems can be successfully addressed only by marshaling knowledge from more than one discipline, that is, inter- or multidisciplinarily. In like manner, the problem of creating a philosophy of engineering would surely benefit from drawing on different disciplinary-like approaches to philosophy.

Third, one of the major challenges facing engineers and nonengineers alike concerns trying to figure out what constitutes the ethical responsibility of those who practice and those who benefit from engineering. But ethics itself involves critical reflection from more than one perspective. So even and especially in this branch of philosophy—which must surely be part of the philosophy of engineering in a conventionally pluralistic sense—there is reason to argue for pluralism in a more robust sense. This more-robust pluralism in the philosophy of engineering is one that would be open to contributions from different schools of philosophy.

What might be the result of such a pluralistic philosophy of engineering, in which linguistic philosophy of engineering would play a significant role? How might this lead not just to a philosophical understanding of engineering but to changing it? Perhaps it would be more accurate to say that we aim to change engineering by attempting to understand it philosophically, that is, by introducing engineering into philosophy and therefore, however indirectly, by insinuating a measure of philosophy into engineering. At the same time, what is most remarkable with regard to relationships between engineering and philosophy is the absence of engineering in philosophy. In no major introduction to the philosophy of science and in only the most limited number of studies in the philosophy of technology does engineering play any significant role. To include engineering in philosophy may thus have some implications for philosophy as well.

But what, it remains reasonable to ask, might be the kind of changes that philosophy could introduce into engineering? Consider the analytic and protolinguistic philosopher Bertrand Russell's concluding defense of philosophy in *The Problems of Philosophy* (1912). According to Russell, philosophy raises more problems than it solves.

> As soon as we begin to philosophize ... we find ... that even the most everyday things lead to problems to which only very incomplete answers can be given. Philosophy, though unable to tell us with certainty what is the true answer to the doubts which it raises, is able to suggest many possibilities which enlarge our thoughts and free them from the tyranny of custom. Thus, while diminishing our feeling of certainty as to what things are, it greatly increases our knowledge as to what they may be; it removes the somewhat arrogant dogmatism of those who have never traveled into the region of liberating doubt, and it keeps alive our sense of wonder by showing familiar things in an unfamiliar aspect.
>
> (p. 157)

The philosophy of engineering, insofar as it is able to highlight problems within the customary ways of thinking about engineering—both within engineering and without—may introduce into engineering a kind of liberating skepticism and wonder regarding engineering that could be especially beneficial to a world such as ours, which is increasingly dependent on and identified with engineering. This in turn might even make engineering more attractive to some who have shied away from undertaking it while mollifying to some degree others who have criticized it.

Part Two

Contingency (vs Necessity)

(after Lydia Davis)

We might avoid catastrophe.
But we will not avoid catastrophe.
So we will look the other way.

Essay 9

A Spectrum of Ideals in Engineering Ethics, Simplified

Engineering is a profession of relatively recent origin. By contrast, the classic secular professions of medicine and law have premodern roots. If one dates the birth of modern natural science as a social institution from the founding of the Royal Society in 1660, engineering as a civilian profession may be historically anchored in John Smeaton's informal convening of the Society of Civil Engineers a hundred years later in 1771. Smeaton initiated a trajectory of institutionalization that carried forward to the royal chartering of the Institution of Civil Engineers (ICE) in 1828. Additionally, just as one distinguishing feature of modern natural science (in its specialized knowledge production process) is the regular publication of research, which began in a systematic way with the *Philosophical Transactions* of the Royal Society in 1665, a key aspect of engineering (in its distinctive artifact making process) is the patenting of inventions, which began to be formally protected by national statute in the 1790s (1790 in the United States, 1791 in France).

Publication of words purporting to describe reality and the patenting of physical inventions aiming to remake reality: The practices of science and engineering have evolved, in historian Edwin Layton's (1971) well-turned phrase, as "mirror-image twins" of uniquely modern forms of thinking and making, evidencing as well historically unique interacting ideals of broad ethical and political significance. Both science and engineering arose at particular times and places and from there have spread out across the world. Their ideals in turn are pursued with means that acquire their salience from an ability to contribute to the ideals themselves; means are informed by results.

In the context of the Royal Society commitment to knowledge production, there early on emerged an informal constellation of behavioral norms associated with the new epistemic methodology. The fruitful practice of its methods of empirical experimentation and mathematical theorization depended on and socialized scientific community members in such behavioral ideals as honesty in the reporting of results and aspiring to ignore the social status of producers in assessing them. Science tended to create a transnational, cosmopolitan community open to members from all countries who were willing to adopt its epistemic methods and interpersonal ethos. In ideal form it constitutes what H. G. Wells (1928) termed an "open conspiracy."

Insofar as the end is knowledge production, honesty in reporting results is going to be more salient than the Aristotelian virtue of physical courage. The behavioral virtues of doing science will be a selective subset of the general human virtues, operationalized for a specific context. Nodal points for an incomplete history of this ethics of means in science would include at least the following:

- the emergence of peer review at the *Philosophical Transactions* under editor Henry Oldenburg;
- the calling out of hoaxes and the deformations of "trimming, cooking and forging" by Thomas Babbage (1830);
- analysis of social norms in the scientific community by Albert Bayet (1907) and Robert Merton (1938 and 1942);
- idealistic defense of science as a model for political order by Michael Polanyi (1962); and
- public exposures of fraud and misconduct in science that have led since the 1980s to formulation of the concepts of responsible conduct of research (RCR) and responsible research and innovation (RRI).

Merton identified what he called the "ethos of science" with four key ideal social practices: communalism (C) in the sharing of results, universalism (U) or rejection of any argument from appeal to the social status of the producer, disinterested (D) commitment to the success of the scientific enterprise as a whole rather than limited personal gain, and organized skepticism (OS) as willingness to entertain systematic questioning. Although scientists often fail to enact these CUDOS ideals (Greek *kudos*, praise or renown) they still regularly criticize themselves insofar as they recognize shortcomings—and in the contemporary world have sometimes worried about institutional pressures that weaken or threaten them.

One simple reading of the historical development of the ethos of science is as the conceptual modulation of the end (the production of reliable or certified knowledge) in fluctuating, iterative adoption of a set of behavioral norms as means for enhancing community practice; individual elements differentially rise to prominence in response to the pressures of different scientific tasks, historical contexts, and social situations. Hence the strong debates that have readily centered on Merton's original claims, both within and without the scientific community, at different times and places. Note, too, that discourse about this ethos will on some occasions appropriately take place not only within the scientific community but also in dialogue with the larger society in which science necessarily inheres.

1. HISTORICAL DIALECTICS OF ETHICS AND ENGINEERING

The historical development of an ethos of engineering may be described in similar terms, mirror imagining the case of science. One difference is that the aim of engineering is less well defined than that of science. Although there are serious debates about the epistemology of science, these have been subject to extensive philosophical examination and there is a relatively well-agreed-upon concept—at least in soft or folk philosophical terms—describing its distinctive features. We all know, even if there seem to be exceptions to any proposed final definition, that scientific knowledge differs from other types of knowledge. Science combines empirical evidence with mathematical conceptualization to produce reliable or certified cognition.[1] Within this framework, the elements of the ethos of science can reasonably be assessed in terms of their ability to contribute to reliability in the knowledge produced.

With regard to engineering, however, there is less consensus regarding what an engineered artifact is and how it might be same or different from all other types of physical objects. Do engineered artifacts exhibit some kind of greater reliability than nonengineered artifacts? What is distinctive about engineering methods? The issue of what makes an engineered entity engineered has until quite recently (since the early 2000s) been subject to little extended philosophical inquiry. One consequence is that discussions of ethics in engineering often attempt at least two related but different tasks: to identify the form of making that distinguishes engineering and thus its end and to describe (and justify in terms of end realization) appropriate behavioral norms for engineering practice. Again, as in science, the particular virtue set that is appropriate for emphasis or codification in professional engineering will be a selective subset of the virtues, now oriented toward or operationalized for another special end and context. In engineering, for instance, the Aristotelian virtue of wittiness (*eutrapelos*) contributes little to effective engineering practice.

Historically, the end goal of engineering was first given a name in the ICE charter as artifacts of human "use and convenience." (See the Appendix, "On Engineering Use and Convenience," for an extended examination of the 1828 charter and this concept.) One reading of the history of ideals in engineering ethics is as a dialectic of assumptions or interpretations regarding this end and the social as well as technical means to its realization. Any complete unpacking of such complex relationships is beyond the scope of the present essay. But as a modest effort to think more deeply about ethics in engineering, what follows is a simplified narrative of the dialectic emergence of three key ideals and some associated principles, together with brief comments on their strengths and weaknesses. The ideals can be lumped under three headings: (1) loyal obedience, (2) efficiency, and (3) public safety, health, and welfare. The aim of this somewhat less-than-nuanced account is not so much a detailed or fully satisfactory history as a framework that may be useful for furthering philosophical dialogue and reflection.

FIRST THESIS: OBEDIENCE TO AUTHORITY AND COMPANY LOYALTY

Engineering as a civilian profession in the West arose out of the military. The fourth-century Roman soldier-historian Ammianus Marcellinus in his *Rerum gestarum* (XXIII, 4) describes military *ingenia* such as the ballista and battering ram. A thousand years later, the soldier-constructor of such devices had come to be called an *ingeniator*. The original "engineer" was thus someone who designed military fortifications and/or operated "engines of war." William Shakespeare refers to Achilles as "a rare engineer" (*Troilus and Cressida*, act 2, scene 3, line 8) and writes of a soldier "engineer/Hoist with his own petard" (*Hamlet*, act 3, scene 4, line 206).

The initial institutions of engineering education were also in bed with the military. Some examples are as follows:

- In Russia: Czar Peter the Great's Academy of Military Engineering (Moscow, 1698)
- In Austria-Hungary: Emperor Joseph I's Estates School of Engineering (Prague, 1707)
- In France: the *Bureaux des Dessinateurs du Roi* (1744), which became the *École des Ponts et Chaussées* (Paris, 1747); the *École des Mines* (1783); the *École Polytechnique* created by the National Convention of the French Revolution (1794)
- In the United States: the Military Academy at West Point founded by President Thomas Jefferson (1802)

The *École des Ponts et Chaussées* and *École des Mines* were established to educate members of a regimented national corps of road builders and mining specialists, respectively; the *École Polytechnique* was placed at the service of Napoleon Bonaparte shortly after he became brigadier general of the Revolutionary Army of the Republic. The Military Academy at West Point was the first school in the United States to offer engineering degrees; General Sylvanus Thayer, creator of the West Point curriculum, subsequently endowed the Thayer School of Engineering at Dartmouth College. Within such contexts the general duty of engineers, as with all soldiers, was oriented toward defending the nation-state in a hierarchical institution with a disciplinary ethos of loyalty and obedience.

In the military, loyalty to fellow soldiers and obedience to command authority dominates behavior norms. Soldiers who do not bond with the corps are unlikely to survive on the battlefield. Insubordination of any form (from failing to make reveille to desertion) is subject to censure, punishment, and condemnation—in the extreme, even execution. Soldiers are required to obey orders even when they might dislike or disagree with them—even when they might judge the orders to be unrealistic, irrational, perhaps even immoral. *Raison d'état* trumps ethics not only at the level of international affairs. Such is a defining feature of the role of members of the military as a social institution, with few exceptions across cultures or throughout history. Only in the twentieth century in the West have efforts been made to qualify this obligation with the caveat that soldiers are not required to obey unlawful (or patently immoral) orders. The exception, however, is easier to state than to operationalize. In any immediate instance, disobedience is likely to be punished; only after the fact might disobedience be exonerated or (rarely) praised.

During the same period as the founding of the first professional engineering schools, a few builders of "public works" began to distinguish themselves from their military forebears with the term "civil engineer"—a designation that continues in some languages to denote all nonmilitary engineering. The creation of this civilian counterpart to military engineering initially gave little cause to alter the basic ethos. Civil engineering was simply peacetime military engineering for "use and convenience" and engineers remained duty-bound to obey their employers, whether a civilian branch of government or a private corporation. Such an obligation was further rationalized and enforced by an industrial capitalism in which corporate owners exercised state-enforced legal authority over employees, whom they were free to fire at will for poor performance of any kind, especially disobedience. Obedience (sometimes complemented with the honorific term "loyalty") was a pervasive if not explicitly stated role morality for engineers as employees; this was the case even when technical expertise provided them a modicum of independence, especially when directing major projects of public infrastructure or private industrial construction. Even consulting engineers assumed subservience to larger purposes such as national competition. Much less than scientists were engineers inclined toward cosmopolitanism; engineers were married to the state.

Along with the emergence of civilian engineering, the late eighteenth and early nineteenth centuries witnessed the growth of public associations of professionals (such as physicians, lawyers, and engineers) in what Alexis de Tocqueville described as social (rather than economic) organizations intermediate between family and state—and as the foundational institutions of civil society.[2] During the same period, technical knowledge became increasingly rationalized in various semiautonomous disciplines, reflecting what sociologists identify as a key feature in the logic of modernity—structural differentiation.[3] Intermediate associations of engineers thus formed to reflect disciplinary differentiations of civil, mechanical, electrical, chemical, and other branches of engineering.

Historically, ethics too became an issue for differentiation. The American Medical Association at its founding in 1847 drafted a "Code of Medical Ethics," making a link to the premodern "Oath of Hippocrates" (fifth century BCE) and framing obligations to patients and the public good in terms of relieving suffering and securing health. The American Bar Association, three decades after its 1878 founding, in 1908 adopted a "Canons of Ethics" that framed obligations to clients and society as promoting justice. During the opening decades of the twentieth century, professional engineering societies likewise began to formulate ethics codes that, naturally enough, tended to make explicit what had previously been implicit: loyal obedience. Although all three regionalized ethics codes highlighted obedience, the medical and legal professions made themselves subject to individual patients and clients in the furtherance of general ideals (health and justice). By contrast, engineering codes placed engineers under the control of corporate employers without any explicit reference to a technical ideal. Indeed, it is somewhat surprising that no reference was made to the ideal of "use and convenience" included in the original ICE definition.

Primary examples are the codes of ethics of the American Institute of Electrical Engineers (AIEE, later to become the Institute of Electrical and Electronic Engineers or IEEE) adopted in 1912 and of the American Society of Civil Engineers (ASCE) and the American Society of Mechanical Engineers (ASME), both of which were adopted in 1914. Each of these three codes was less than a page in length and stressed that "the engineer should consider the protection of a client's or employer's interests his first professional obligation" (to quote the AIEE) or required the engineer to act simply "as a faithful agent or trustee" (ASCE language).

With regard to the ASCE, Sarah Pfatteicher has insightfully uncovered conflicting influences active in the emergence of its code and in the internal discussions that extended from the 1870s to the early 1900s. As she put it,

> The first code of ethics adopted by the ASCE was intended to describe, rather than guide, the behavior of ASCE members. . . . Early codes of ethics were

intended to document and publicize existing standards of behavior (largely for the benefit of potential employers), not to establish ideals toward which ASCE members might strive.

(Pfatteicher, 2003, p. 21)

This descriptive code admonished members to be true to existing practice and "to be loyal to their clients, their fellow engineers, and their profession" (p. 29). Paradoxically, although one goal of this early code construction was to enhance public recognition and a degree of autonomy, because of the prominence given to business interests and company loyalty, the practical effect was to undermine independence. In other words, professional engineering—insofar as it articulated loyalty as a primary value—tended to promote a kind of self-imposed tutelage to capitalist corporate employers.

One criticism of this historical narrative deserves acknowledgment. Michael Davis (2002) challenges the idea that engineers initially took loyal obedience as their primary obligation. Such a view, he argues, ignores historical context and the role of interpretation required by any law or code of conduct. Davis's argument makes important points, but historical context also supports a loyalty narrative. Repeatedly in various early-twentieth-century engineering society proceedings, there is an emphasis on some form of loyalty as primary. Indicative of the intense interest in the issue during that time was the publication of a rather remarkable 15-page annotated bibliography on engineering ethics by the Carnegie Library of Pittsburgh, with the following introduction:

> Numerous requests have come to the Technology Department for material on Engineering Ethics. A brief list on the subject, prepared several years ago, has been supplemented from time to time in response to more recent requests, but though the list as now published represents a rather systematic search, it does not include all that is available in this department. The list is confined pretty closely to ethics for engineers, but it includes a few articles on the ethics of other professions.

(McClelland, 1917, p. 3)

Among the numerous references in this bibliography is one to a proposal leading up to adoption of the AIEE code, in which it was clearly stated that "the electrical engineer should consider the protection of his client's interests as his first obligation" (Steinmetz et al., 1908, p. 1422); and in a discussion preparatory to the ASME code, it was proposed that "the engineer should consider the protection of a client's or employer's interests his first obligation, and he should avoid every act contrary to this duty" (Baker et al., 1913, p. 29).

Another publication five years after the bibliography is a symposium of seven papers dedicated to "The Ethical Codes of the Engineers," in a theme issue on "The Ethics of the Professions and of Business" of the *Annals of the American Academy of Political and Social Science* (Cooke, 1922; Rice, 1922; Newell, 1922; Hering, 1922; Clausen, 1922; Wilgus, 1922; and Christie, 1922), further highlights the issue with numerous comments on the early codes and tensions between engineering and business. An appendix to the symposium includes what were apparently the five existing codes at the time.

That loyalty was considered a special problem to be confronted in engineering ethics education (and in political and social science) into the 1980s is further confirmation of the important role it has played (see Baron, 1984). Davis is clearly correct, however, for reasons he somewhat overlooks. Engineers often believed such loyalty was in the public interest because of their strong beliefs that capitalist corporations were themselves public benefactors.

THE PRINCIPLE OF LOYAL OBEDIENCE

There is undoubtedly value in the related principles of loyalty and obedience. Loyalty is a widely recognized virtue under many circumstances. Without loyalty it is hard to imagine any social relationship, whether personal friendship or larger group solidarities, being sustainable. Some serious degree of obedience is also fundamental to the reasonable functioning of social groups, perhaps even more so in organizations that construct and operate large technical projects. The safe operation of a nuclear power plant depends on workers at all levels of an organizational hierarchy following a well-defined set of rules and procedures. The same goes for any number of other engineering projects and activities, from aircraft design and operation to the electric power grid and communication networks. There is thus prima facie reason to endorse the idea that engineering ethics should include a significant component of loyalty and obedience.

The problem is that loyalty and obedience can be captured and misused by corrupt authorities for illegitimate ends. Any obediential ethics runs the risk of leaving its adherents subject to manipulation by unjust forces. Only when and to the extent that the authority to be obeyed is good, is obediential ethics fully justified. In the Abrahamic religions, where obediential ethics plays a prominent role, it is based on belief in the absolute goodness of the authority to be obeyed. One obeys God because God is wholly good. In any other instance obedience requires qualification. Even in the military it is now common to say that one is obligated to carry out only legitimate or just orders. Physicians and lawyers, too, are obligated to respond positively to the wishes of their patients and clients, only to the extent that patients and clients desire health and justice. An initial attempt to meet the weakness of an unqualified loyal obedience—and to articulate a regulative ideal for engineering comparable to those of health in medicine and justice in law—can be found in the technocracy movement.

SECOND THESIS: TECHNOCRATIC EFFICIENCY

An emergent ideal related to professional ethics but highlighting something other than obedience and company loyalty was an ideology of leadership in technological progress through pursuit of the ideal of engineered efficiency. Historian Edwin Layton finds this ideology for engineering, which exercised a strong influence on engineering self-understandings, manifested as early as 1895 in an ASCE presidential address by George S. Morison, one of America's premier bridge-builders. Making implicit reference the ICE definition of engineering, Morison envisioned the engineer in pseudo-religious terms as the primary agent of technological change and human progress:

> We are the priests of material development, of the work which enables other men to enjoy the fruits of the great sources of power in Nature, and of the power of mind over matter. We are the priests of the new epoch, without superstitions.
>
> (Morison, 1895, p. 483)

There is, as well, an unconscious echo here of the *Nouveau Christianisme* of Henri de Saint-Simon (1760–1825).

Two decades later, reaffirming this theme, a Stanford University civil engineer and leading ASCE member published a paper on "The Philosophy of Engineering" proposing that "the new engineer" has a "higher duty [than simple engineering] of reconciling business and the people, of being a peacemaker between those who produce our economic goods and those who consume them at what seems like too high a cost" (Parsons, 1914, pp. 39–40). A discussant following went further, proposing that "engineers in general [should become] the dominant agency for the general uplift of humanity" (p. 53).

During the first third of the twentieth century in the United States this ideal of engineering promoted efficiency and uplift contributed to the Progressive Era "efficiency movement" and eventually spawned a positive social vision of technocracy in which engineers would be given political and economic power for public benefit. Because of his criticism of the "withdrawal of efficiency" by captains of industry (when they reduced production in order to prop up prices) and praise of "efficiency engineers" and "scientific-management experts" (as the creators of modern technology and thus its proper managers), political economist Thorstein Veblen is commonly associated with this vision. The title of his *The Engineers and the Price System* (1921) succinctly referenced what he saw an inherent tension between engineers, who aimed for efficiency in both industrial systems and consumer products, and captains of industry, who pursued monetary profit. Engineers as "the indispensable General Staff" of industry should take charge. But while Veblen promoted the potential ideal of engineering governance, he was ironically skeptical about its realization, given a lack of class consciousness among engineers and their general psychology of subservience.

Some engineers who inspired him, especially from the ASME, nevertheless appeared as possible members of a prospective revolutionary wedge. Mechanical engineer Morris L. Cooke, as a public works manager in Philadelphia from 1911 to 1915, was radicalized by the resistance of public utility owners to increasing efficiency in electric generation and distribution, insofar as they might suffer some reduction in profitability. For Cooke, engineered efficiency was more useful and beneficial to society than deference to commercial interests.

Mechanical engineer Frederick W. Taylor's *The Principles of Scientific Management* (1911) attacked the problem of efficiency from a different perspective. Whereas Veblen worried about inefficiencies introduced from the top by corporate managers, Taylor stressed the problem of inefficiencies from the bottom caused by deficiencies in worker ambition, skill, or discipline. Corporate leaders may have resisted Veblen's engineer-experts, but they welcomed Taylor's: the right technocracy could be good for business.

Among studies of the ideal of technocratic efficiency that highlight the connections with engineering are William Aiken (1977), which focuses on the American context, and Richard Olson (2016), which emphasizes an international synergy between scientism and what might better be called "engineering-ism." In a broad historico-philosophical study, Jennifer Alexander's *The Mantra of Efficiency* (2008) calls attention to how the pursuit of efficiency is simultaneously benefit and harm: increasing efficiency of action requires increasing regulation of actors. Additionally there is the paradox of efficiency as simultaneously performance within strict limits and performance that exceeds limits:

> On the one hand, it was an accounting principle, comparing what went into a machine or process with what came out and measuring how much of the original input had been conserved. On the other hand, efficiency was creative and dynamic, yielding growth through careful management and bringing as its reward not merely conservation but increase.
>
> (Alexander, 2008, p. 170)

THE PRINCIPLE OF EFFICIENCY

Efficiency never functioned as a principle in engineering ethics in the same manner as loyal obedience (or any other of the subsidiary cannons). It still played a key role in debates about the engineering ethos and can be interpreted as representing an effort to find a conception of the good internal to engineering analogous to that of health in medicine.

Indeed, there are reasonable arguments in defense of the ethical value of efficiency and the politics of technocratic governance. Certainly the subordination of production to short-term money making without concern for the good of the products being designed and manufactured is not desirable in the long run, and inefficiency or waste can be described as types of imperfection if not evil. Moreover, in a highly complex technoscientific world it is often difficult for politicians or average consumer-citizens to know what is in their own best technical interests. And whether efficiency can be adequately promoted either by the consumer pull of imperfect markets or by a push from technical professionals under the thumb of corporate interests remains seriously questionable. Still, it is hard to imagine complex public utilities like water systems or air transport systems managed by anyone other than engineers.

Yet to govern in pursuit of technical efficiency remains problematic on at least three counts. First, efficiency as an ideal only functions within well-defined boundary conditions that are often surreptitiously set by non-engineers and may exclude important factors. Limitations in perspective can occur even when engineers set their own boundary conditions for conducting input-output analyses, given a training and mind-set that emphasizes problem quantification. Second, the pursuit of technical perfection in the form of maximum output for a given input is not always the best use of societal resources—as when, for example, cars are increased in efficiency at the expense of designing and constructing public transport systems. Third, technocratic decision-making by engineers is obviously in tension with democratic decision-making. In response to such objections there developed a third distinct idea of engineering ethics, that of the public good understood in terms of public safety, health, and welfare.

THIRD THESIS: PUBLIC SAFETY, HEALTH, AND WELFARE

The World War II mobilization of science and engineering for national purpose in the United States and the North American postwar recovery contributed to a temporary suspension of tensions between technical and economic ends, efficiency and profit, that had been highlighted by the technocracy movement. But the anti-nuclear weapons movement of the 1950s and 1960s, in conjunction with the consumer and environmental movements of the 1960s and 1970s, brought tensions again to the fore and provoked some engineers to challenge national and corporate or business direction. In conjunction with a renewed concern for democratic values—especially as a result of the civil rights movement—this led to a new interpretation of the engineering use and convenience ideal.

Matthew Wisnioski's *Engineers for Change: Competing Visions of Technology in 1960s America* (2012) is a narrative analysis of this era extending and complementing Layton's (1971) for an earlier period of ferment. As Wisnioski's summarizes the issue of professional engineering ethics in this historical context (quoting from an engineer's letter to the editor in a 1965 issue of *Mechanical Engineering*), "The key task facing engineers was rather to identify responsibilities 'corresponding to the clear obligations for doctors to save life, and for lawyers to defend the accused'" (Wisnioski, 2012, p. 72). Existing ethics codes were outmoded insofar as they treated corporations as individual clients and placed "devotion to client above the public or individual conscience" (p. 73). More specifically, the recodification of engineering ethics emphasizing public responsibility can be documented by developments in three commonly acronym-denominated professional organizations: ECPD-ABET-AAES, NSPE, and IEEE.

In the ECPD-ABET-AAES

In 1947 the Engineers' Council for Professional Development (ECPD)—founded in 1932 as an organization of organizations (not of individuals), and charged in part to develop an ethics code acceptable to its constituent societies—adopted an ethics code that made it a leading duty for engineers "to interest [themselves] in public welfare" and to "have due regard for the safety of life and health of the public." Revised in 1963, 1974, and 1977, this code eventually formulated the first of seven "fundamental canons" as follows: "Engineers shall hold paramount the safety, health and welfare of the public in the performance of their professional duties."

In 1980 the educational supervising activity of the ECPD was restructured into the Accreditation Board for Engineering and Technology (now simply called ABET) to certify engineering degree programs. ABET assumed the final ECPD revision of its code, along with an extended "Suggested Guidelines for Use with the Fundamental Canons of Ethics." In this form the ABET code influenced engineering education, insofar as ABET slowly began to stress the importance of professional ethics in university engineering curricula.

The restructuring of ECPD educational activities into ABET took place parallel with the restructuring of ECPD interdisciplinary professional development activities into a new American Association of Engineering Societies (AAES). One of the perennial problems of professional engineering development in the United States has been fragmentation in the professional engineering community, a dispersal of social capital that dilutes public influence. Unlike ABET, however, the new AAES did not assume the ECPD code but in 1984 officially adopted its own "Model Guide for Professional Conduct," which sought to provide a unifying framework for all existing disciplinary codes. This AAES guide, in revisions, likewise progressively stressed the importance of safety, health, and public welfare.

In the NSPE

A further illustration of the post–World War II emergence of the importance of social responsibility in engineering ethics was a code developed by the National Society of Professional Engineers (NSPE). Like the ECPD, one of the original objectives of the transdisciplinary NSPE (founded 1934) was "the establishment and maintenance of high ethical standards and practices." Unlike the ECPD, which was an organization of organizations, the NSPE is an NGO of something like 50,000 individual members, all of whom have gone through a special vetting process to become Professional Engineers (PEs). According to its mission statement, the NSPE "promotes the ethical and competent practice of engineering, advocates licensure, and enhances the image and well-being of its members."

Although an ethics code was proposed as early as 1935, none was formally adopted until 1946, when the NSPE endorsed the new ECPD code even before the ECPD formally did so. With the 1963 revision of the ECPD code, however, the NSPE moved to create its own code. The evolution of this distinctly NSPE code led by 1981 to adoption of a short list of "Fundamental Canons," the first of which is to "Hold paramount the safety, health and welfare of the public."

In the IEEE

Still a third example of the rise in social responsibility characteristic of U.S. engineering ethics codes in the second half of the twentieth century can be found in the Institute of Electrical and Electronics Engineers (IEEE)—which emerged in 1963 from the unification of the AIEE and the Institute of Radio Engineers (IRE, founded 1912) and by the early 2000s had become the largest professional engineering NGO in the world, with more than 300,000 members worldwide.

In the early 1970s, the IEEE undertook to write a new code of ethics. In the preamble to its initial code of 1974, the IEEE declared,

> Engineers affect the quality of life for all people in our complex technological society. In the pursuit of their profession, therefore, it is vital that engineers conduct their work in an ethical manner so that they merit the confidence of colleagues, employers, clients and the public.

The fourth article of the code itself specified that IEEE members have a responsibility to "protect the safety, health and welfare of the public" and even to "speak out against abuses in those areas affecting the public interest."

In 1990, following significant debate in the late 1980s about the way properly to amend it, the code was simplified and public responsibility was elevated to the first of ten principles. IEEE members committed themselves "to accept responsibility in making engineering decisions consistent with the safety, health and welfare of the public, and to disclose promptly factors that might endanger the public or the environment."[4]

PUBLIC SAFETY, HEALTH, AND WELFARE AS PARAMOUNT

This third distinct engineering ethics ideal meets many of the objections that can be raised against the first two and has been widely adopted by the professional engineering community both in the United States and elsewhere. It clearly seems like a more-adequate interpretation of the original end of engineering as use and convenience. It also allows retention of desirable elements from prior theories. For instance, loyal obedience remains but within a larger, more encompassing framework. Now the primary loyalty is not to some individual or corporation but to the public. Leadership in technical development likewise remains but is explicitly subordinated to the common good interpreted as safety, health, and welfare.

Yet at least two new questions arise with this principle: First, how is an engineer supposed to know what is best for the public, whether with regard to safety, health, or welfare? Each of the four key terms in this principle is somewhat problematic, and their conjunction makes things more so. The situation is not unlike the situation that arises with the utilitarian ethical principle of maximizing the greatest amount of good for the greatest number. Which is it: holding paramount the greatest amount of public safety . . . or public health . . . or public welfare? Is this perhaps a rank ordering of goods? If not, how does one aggregate such different goods, which often conflict with each other? And which of many possible publics (employers, workers, investors, and local inhabitants) are to be considered? As John Dewey (1927) argues, publics are fluid and do not exist independently of the problems that constitute them.

Second, does the concept of public safety, health, and welfare in its vagueness not easily devolve into a moral cliché or demand expansion in multiple dimensions? Public safety, health, and welfare is a little like patriotism or responsibility. It can mean so many different things to so many different people that it winds up either functioning like a panacea that papers over differences or as a topic that stimulates disagreements. It also readily seems to invite inflationary inclusions of all sorts of other public goods. After all, what about competing public goods such as privacy, security, stability, development, justice, peace, and more? Could one not argue that one or more of these is implicated in, say, public welfare (understood as more than economic wealth)?[5]

ENVIRONMENTALISM AND SUSTAINABILITY

Going beyond a narrow (directly human centered) conception of the principle of public safety, health, and welfare, the 1980s witnessed efforts to broaden its interpretation to include some kind of protection of the natural environment. This expansion was influenced by the emergence of environmental ethics as a regionalized ethics parallel to engineering ethics.[6] As American engineer Aarne Vesilind and New Zealand philosopher Alastair Gunn have observed, however, the integration of environmental ethics into engineering ethics was complex and contentious, driven as much by external political forces as by internal professional concerns (Vesilind and Gunn, 1998, chapter 3).[7]

In 1990 the IEEE "Code of Ethics" was apparently the first professional society to adopt an explicit reference to the environment, but in a weak form by presenting the obligation "to accept responsibility in making engineering decisions consistent with the safety, health and welfare of the public" as including a responsibility to disclose dangers to "the public or the environment." Then in 1997 the ASCE endorsed the political emergence of sustainability as a conceptual restatement of environmentalism. Its first fundamental canon became, "Engineers shall hold paramount the safety, health and welfare of the public and shall strive to comply with the principles of sustainable development" (ASCE, 2007), although it was not until 2009 that ASCE formally defined sustainable development as "the process of applying natural, human, and economic resources to enhance the safety, welfare, and quality of life for all of society while maintaining the availability of the remaining natural resources." Following the ASCE lead, a National Academy of Engineering report on the "engineer of 2020" likewise subsumed environmentalism under a sustainability rubric (NAE, 2004).[8]

A PARTICIPATION PRINCIPLE

Concern for the public good, including the environment and sustainability, does not necessarily involve any citizen participation in decision-making. A technocracy/democracy problem remains. The engineer committed to public safety, health, welfare, and sustainable development may make decisions about technical issues in an authoritarian manner at odds with democratic procedural ideals, based on a strictly technical analysis and evaluation of risks associated with some product or process. Vesilind and Gunn even go so far as to argue, not unlike Morison, that as "engineers are often the only people with the knowledge of potential environmental harm and the professional authority to command attention [that] this gives them *more* environmental responsibility" than others and that "engineers are in a unique position to make a difference in the care and nurturing of our planet" (Vesilind and Gunn, 1998, pp. 128 and 135).

Yet recognition that technology often brings with it not only benefits but also costs and risks argues for granting all those affected some input into technical decisions. On the basis of a proposal to understand engineering as social experimentation, the philosopher-engineer team of Mike Martin and Roland Schinzinger (1983, chapter 3) argued for adaptation of the biomedical ethics model of human subjects research that requires free and informed consent of all participants.[9] Deploying a political analogy, political theorist Langdon Winner (1991) and philosopher Steven Goldman (1992) have defended a principle of "no innovation without representation." Participation has become a widely adopted ideal but one incorporated into engineering ethics only to a limited extent.[10]

One important qualifier is that participation need not imply veto power but only intelligent and relevant involvement in decision-making. Equity (a stake in the game) is not the same as equality (an equal stake). In accordance with a participation principle, the role of the engineer as technical specialist becomes not so much that of independent adviser or decision maker as one participant in an educational dialogue and contributor to various regulatory processes within appropriate democratic structures and guidelines (see Jasanoff, 1990, especially chapter 10). The ideal of responsibility for the public good, especially as extended with the participation principle, has nevertheless been circumscribed by skeptical debate concerning theoretical justifications and practical implications.

2. ELABORATING: SELECTIVE NORTH AMERICAN CASES AND ISSUES

The fundamental professional ethical problem for the engineering community has been how to develop an autonomy that would enable engineers to practice a professional commitment to the primacy of public safety, health, and welfare without undermining appropriate professional or company loyalties or subverting democracy. The problem is that engineers are often placed in positions that call for the public criticism of other engineers or are granted the power to make technical decisions for the public that can easily lead to the promotion of nonpublic interests. The principal-agent problem and special (personal) interests are always with us. One dramatic illustration of the way that nonpublic interests can distort public work occurred in Los Angeles in the 1930s. Another influential "conflict of interest" case occurred at the Hydrolevel Corporation in the 1960s. With regard to technical power, however, in comparison with scientists, engineers are more "on tap than on top."[11] This fact has been driven home by a number of case studies of disasters associated with design flaws resulting disasters such those involving Goodrich A-7D airbrakes,[12] cargo bay doors on the DC-10 (Fielder and Birsch, 1992), gas tanks on the Ford Pinto (Birsch and Fielder, 1994), and others. Two of the most widely influential such disaster cases concern the Bay Area Rapid Transit (BART) system from the 1970s and the space shuttle *Challenger* explosion in the 1980s.

THE JAKOBSEN, PAYNE, AND ASCE CASE (1930s)

In southern California in January 1932 two engineers, Bernhard F. Jakobsen and James H. Payne, were notified that they had been expelled from the American Society of Civil Engineers for failing to abide by the organization's professional ethics code. The particular principles that had been breached were apparently canons two and/or five which declared,

It shall be considered unprofessional and inconsistent with honorable and dignified bearing for any member of the American Society of Civil Engineers:
. . .
2. To attempt to injure falsely or maliciously, directly or indirectly, the professional reputation, prospects, or business, of another Engineer.
. . .
5. To review the work of another Engineer for the same client, except with the knowledge or consent of such Engineer, or unless the connection of such Engineer with the work has been terminated.

The context is complex, and part of the story inspired elements in the 1974 neo-noir film *Chinatown*. But according to Jakobsen the crucial issue was that he and Payne had assisted the *Los Angeles Daily News* and the County of Los Angeles in exposing fraudulent work related to a proposed Los Angeles County Flood Control District dam on the San Gabriel River. The charge was that Payne and Jakobsen were guilty of "bringing on a brother engineer injury to his reputation" (Jakobsen, 1955, p. 3)—despite the fact that some of those so exposed were eventually charged with and convicted of criminal behavior and even served jail time.

Jakobsen and Payne appealed, but their appeal was denied. A subsequent editorial in the *Los Angeles Times* (February 12, 1937) concluded,

> The decision of the governing board of the American Society of Civil Engineers not to reinstate as members B. F. Jakobsen and J. M. Payne, who performed a distinguished public service by calling attention to the San Gabriel Dam scandal, will be greatly regretted here by those who know the inward facts of that situation. . . . Jakobsen and Payne were substantially correct, and . . . for a merely technical infringement—in good cause—of the society's code of ethics, they ought not to be punished.

Despite numerous petitions from various local governmental bodies—along with more than 20 years of demands from Jakobsen himself—the ASCE failed to recognize the injustice of his expulsion and properly reinstate him. As Jakobsen concluded in own self-published memoir of the affair,

> Professional ethics in the field of engineering that precludes exposure of graft on public work is of importance to both engineers and citizens, because it poses the question: Does the engineer's duty to protect his fellow engineer transcend his [duty] as a citizen?
>
> (Jakobsen, 1955, p. 7)

Yet remarkably, indicating something of a schizophrenic character of the engineering profession, Jakobsen was able to maintain his registration as a civil and structural engineer in the State of California and continued to run a consulting engineering firm.

Without making any reference to the Jakobsen and Payne case, in its own 2007 history of ASCE ethics code development, the ASCE website acknowledges,

> The canons of [the 1917] code deal only with an engineer's interactions with other engineers and with clients and make no mention of an engineer's duty to the public. It would not be until 1976 that the board adopted an express statement of the engineer's fundamental ethical duty to "hold paramount the safety, health, and welfare of the public.
>
> (https://www.asce.org/question-of-ethics-articles/dec-2007)

However, in a separate, subsequent comment on the Jakobsen–Payne case, the ASCE admitted to retaining the essence of canon two in its current revised code, acknowledged that "this canon likely served as the basis for the . . . expulsion" but maintained that "it is questionable whether the actions in this case would meet today's threshold for violating this canon" (asce.org/question-of-ethics-articles/dec-2017).

THE HYDROLEVEL CASE (1960s)

In the early 1800s, in response to an increasing number of public fatalities resulting from steamboat boiler explosions, the U.S. government awarded the first research contract to the Franklin Institute, an educational and research organization founded in 1824 "for the promotion of the mechanic arts." This research disproved a number of then-current theories (e.g., that water at high temperature in a boiler can decompose into hydrogen and oxygen and then explode) and pointed up the need for better materials standards in boiler construction, regular maintenance, and more adequate safety devices (such as high-pressure release valves).

In response, by the mid-1800s Congress enacted regulatory legislation.[13] As they evolved, the regulatory agencies that were established became the enforcers of technical standards that by the early 1900s had become the responsibility of the ASME—in a typically American outsourcing of a regulatory activity. In the process research engineers, in a benevolent technocratic manner, had clearly taken on a responsibility for helping to protect public safety and welfare.

By the mid-1900s, however, this responsibility and relationship had become rather ingrown. In the late 1960s research at a small engineering firm, Hydrolevel Corporation, developed a new type of low-water fuel cutoff device for steam boilers that threatened the business of McDonnell and Miller, Inc. (M&M), the primary supplier of such devices. When an appeal was made to the ASME for an interpretation of section HG-605a (a 43-word paragraph) in its 18,000 page *Boiler and Pressure Vessel Code* that would certify the new Hydrolevel design, ASME members who were also involved with M&M acted to secure a negative response. The result was a lawsuit that went all the way to the Supreme Court, which in 1982 ruled that ASME had violated the Sherman Anti-trust Act. (Hydrolevel eventually went out of business because it could not market its new product.)[14] A commitment to protect public safety and welfare through technical ideals and engineering autonomy had been used to protect the welfare of a private engineering-based firm.

Essay 9

THE BAY AREA RAPID TRANSIT (BART) CASE (1970s)

In the late 1950s metropolitan San Francisco decided to create the Bay Area Rapid Transit (BART) network, designed to be the most advanced metro in the world—one that would eliminate both operators and conductors in favor of an automatic train control (ATC) system. The proposal involved an early effort to design a real-world unmanned transport vehicle. Construction began in 1964 and at the end of 1971, almost three years behind schedule and considerably over budget, BART was finally nearing a first stage of completion.

During this time, however, Holger Hjortsvang, an engineer working on the ATC, became seriously concerned about its design and testing—especially the attempt to deal with problems in a complex project on a piecemeal rather than a systemic basis. Beginning in 1969 he expressed these concerns to management, arguing for more extensive, collaborative testing. By late 1971 he found himself supported by two newly hired engineers: Max Blankenzee, a senior programmer analyst, and Robert Bruder, an electrical-electronics construction engineer, both of whom were also involved with the ATC. For months the three engineers expressed their concerns to management both orally and in writing, only to have them consistently ignored.

Finally, seeking to have their concerns addressed, in early 1972 the engineers contacted a member of the BART District Board of Directors and provided papers documenting their case. Very shortly there were unexpected newspaper stories on the problems (a board member, despite promising the engineers confidentiality, had leaked to the media), followed by a February meeting of the board that yielded a split vote of confidence in BART management. Management then undertook to identify the sources for certain critical documents provided to the board and at the beginning of March fired Hjortsvang, Blankenzee, and Bruder.

The engineers appealed for support to the California Society of Professional Engineers (CSPE), arguing that they were only attempting to live up to an ethics code obligation to hold "the public welfare paramount" and to "notify the proper authorities of any observed conditions which endanger public safety and health." In June the CSPE submitted a report to the California State Senate largely supporting the engineers. Then in October, in dramatic confirmation of the engineers' concerns, an ATC failure caused a BART train to overrun a station, injuring four passengers and an attendant.

Prior to the BART case, the primary way in which professional engineering societies had acted to enforce ethics codes had been to discipline engineers for disloyalty to an employer—often by expelling them from a professional society. The BART case was the first in which a professional engineering society publicly backed its members and criticized a specific firm. A further supportive report by the IEEE, the largest engineering society in the world, led to the creation of an Award for Outstanding Service in the Public Interest—which was first given in 1978 to the three BART "whistle-blowers."[15]

THE *CHALLENGER* DISASTER (1980s)

In 1986 the need for engineering independence was again brought to professional and public attention by the space shuttle *Challenger* disaster. Once more a major high-tech, government-funded project was years behind schedule, considerably over budget, and thus subject to strong management pressures to meet new and unrealistic deadlines. As came to light afterward, Roger Boisjoly, a mechanical engineer at Morton-Thiokol in charge of design for the field joints on the solid rocket booster, had been questioning the safety of O-ring seals for almost a year. The night prior to the January launch, Boisjoly and other engineers had explicitly opposed continuing the countdown, only to have their decision overridden by senior management.

As a result of their testimony before a presidential commission during its post-disaster investigation, these engineers came under severe pressure from Morton-Thiokol to defend the company. Instead, Boisjoly became an outspoken advocate for both greater autonomy in the engineering profession and the inclusion of engineering ethics in engineering curricula, thus helping to promote development of engineering ethics courses in engineering colleges throughout the United States.[16]

WHISTLE-BLOWING AS A DUTY TO PUBLIC DISCLOSURE

Uniting these three cases is not only the practical problem of developing or promoting the right kind of engineering autonomy but also what might be called a principle of public disclosure—one clearly allied with that of public participation. Supporting such a principle is the argument that public good is served by a duty to disclose both the full process of technical decision-making and any shortcomings in relationship to safety, health, welfare, and the environment—especially to those who might be affected. Transparency is an important ethical value. How much transparency can be debated?

With the Hydrolevel case, for instance, a more forthright disclosure of personal interests might well have led to a different outcome early in the technical standards definition process. With both the BART and *Challenger* cases, public disclosure was a principle underlying the actions of the engineers involved—although with Hjortsvang, Blankenzee, and Bruder disclosure took place before the accident, whereas with Boisjoly it took place after the fact. In all three instances, it is reasonable to argue that the engineering design work deserved disclosure because in the Hydrolevel case it supported safety with enhanced efficiency of operation and in the BART and *Challenger* cases it was research by especially Hjortsvang on the ATC and by Boisjoly on the O-ring seals, which founded their beliefs that safety was being compromised.

Compare, too, the situation of physicians and lawyers with engineers, in regard to respective responsibilities to those for whom they provide professional services. One aspect of such responsibilities for physicians and lawyers is a responsibility to protect patient and client confidentiality. The physician is obligated not to reveal the state of health of a patient to others without patient consent; an attorney is not permitted to share knowledge about the guilt of a client without client consent. Exceptions occur, however, when patient illness endangers public health and when clients reveal plans to commit a crime. In such cases, where the professional has knowledge about what might happen in the future, as opposed to what has happened in the past, and to others, the obligation to uphold confidentiality weakens if not vanishes. As engineers often know not just the past but the future, a prima facie duty to public disclosure is likely to be present.

But the hypothesis of a duty, under certain conditions, to disclose engineering knowledge must not be too quickly affirmed without recognizing a possible tension between duties to truth or knowledge and duties to welfare or the good. Earlier it was remarked that the duties to knowledge are characteristic of science, whereas duties to good use are more prominent in engineering. In the free market economic system, corporations are also generally granted legal protection for trade secrets—that is, granted the right not to disclose knowledge or information of a certain type—in the belief that this will promote invention and therefore societal welfare. But knowledge that is publicly disclosed can no longer be protected under trade secrecy laws, so that corporations have a reasonable interest in requiring that engineers not reveal knowledge except when necessary to secure a patent or pursue other business purposes. Engineers, unlike scientists, typically do not publish their work. The resultant tradition of engineering secrecy is thus in tension with any duty to disclose knowledge related to issues of safety, health, and welfare.

Finally, there is a problem that knowledge can confuse or be misinterpreted. More information is not always an unqualified good (see Mitcham and Frodeman, 2002, pp. 88–89). In a world of the attenuating of traditional knowledge vetting processes and social media appropriations and disseminations of personal interest or manipulative gain, duties to disclose become ever more complex.

CONCLUDING NON-DIALECTIC POSTSCRIPT

There is no simple conclusion to this narrative of ideas in engineering ethics. The question of professional engineering ethics is one of auto-conceptualization and self-knowledge in an ongoing dialogue with institutional self and circumstances. Like all ethical discourse, it is always only beginning, completed, and affected by history.

It may nevertheless be appropriate to recall and modestly extend earlier observations about the dialectics of ethics and engineering. Ethics is a reflective effort to identify the spectrum of goods immanent to life as a whole and to apportion them properly across different realms of experience. There are more goods in the world than can be realized at any one time and place by any one person or community. At the same time, communities—including professional communities—are defined and constituted by the goods around which they unite. The spectrum of ideals in engineering ethics reveals successive efforts to identify the good or goods around which engineering as a profession is best united.

In the first instance this good is, of course, technical: the designing and making with a historically unique method or technique (which has, elsewhere, for want of a better term, been described as involving rationally anticipatory miniature construction) material goods of a morphologically distinctive type. In a second instance, however, efforts are made within the technical community to identify other goods, sometimes called virtues, which may support or otherwise enhance the technical good. The particular virtue set emphasized or codified in professional engineering will be a selective subset of human virtues, now oriented toward or brought to bear for a special end and context.

This selection will be strongly influenced by the conception of the spectrum of obligations that the engineering community might attribute to itself and assessments of their relative importance. The constellation of relationships for any profession conceptually stretches across (a) the profession itself, (b) clients or employers, and (c) the larger society. The idea of loyalty tends to give some priority to (a); of obedience to (b); of efficiency to a less-social aspect of (a); and of safety, health, and welfare to (c). All codes, implicitly if not explicitly, make some reference to all three relationships and thus invoke one or more of their associated virtues but not necessarily in the same rank order. The early codes typically placed (b) and/or (a) above (c)—although, of course, because, if pushed, the engineers involved would have argued that this is the best way to realize (c). Later codes have ranked (c) as primary, in an implicit criticism or relativizing of previous rankings. A model code proposed by the engineer ethicist Stephen Unger, for instance, ranked (c) first and even parsed it out into two components: "responsibility to society" and working "in a responsible manner . . . only with honorable enterprises" (Unger, 1982, pp. 163–64; 1994, pp. 281–82). Engineers are obligated to use only good means to relate only to the good elements of society.

Even though there is unlikely to ever be any hard, permanently satisfying dialectical synthesis, this interaction might still be amenable to and benefit from a Hegelian analysis[17]. Certainly the influence of context on this conceptual reordering within engineering ethics deserves to be analyzed with much more nuance than so far exists. Especially is this true as engineers and nonengineers alike are faced with globalization in its many forms, almost all directly or indirectly related to engineering.

3. TOWARD A SOFT DIALECTICAL SYNTHESIS

Despite the unlikelihood of any hard synthesis, a soft one may still be worth formulating. The absence of hard power in ethics does not justify a complete dismissal of soft power. The Pope may not have any military divisions, but he still has influence.

Return to the BART case: Hjortsvang argued for a more complete testing of the ATC system. He worried that piecemeal testing could fail to reveal the full scope of problems. Fixes for isolated malfunctions can interact to create new system malfunctions. The fact that functional failure might then also harm users is, in some sense, secondary. Even without potential user harms, piecemeal testing simply can never fully assure full system functionality. Piecemeal testing is thus bad engineering. Hjortsvang wanted to create an interdisciplinary team to oversee ATC design and development in a more holistic manner.

Another and deeper version of the need for a part-to-whole testing has been called "the paradox of information technology." Beyond a certain point human beings "will never be able to model (and thereby check)" in all relevant ways (particularly speed and complexity) data processing operations. Indeed, "the possibility of controlling information processing systems diminishes in proportion to the introduction of modeling or checking instances" as these actually further complicate a program.[18]

A general statement of the modeling paradox goes like this: The utilization of a simplifying model in a technical design process also complicates the process, as it introduces a new factor to be considered, namely the relation between the model and the phenomenon modeled. To attempt to model this new relationship will only further complicate things. More simply put, there is no way to test a model by modeling it. One can only attend to the reality modeled (real-world phenomena), attempt as carefully as possible to take into account all relevant factors, and then pay attention to see if things work the way the model has indicated things should.

In the early 1990s the National Academy of Engineering held a symposium on "Engineering as a Social Enterprise" arguing the need for a sociotechnical systems interpretation of the profession. Symposium organizer, aeronautical engineer Walter Vincenti pointed out how engineers regularly have to deal with technical systems and are thus familiar with how such systems must be subdivided for analysis. "In the sociotechnical model, the entire society is visualized," according to Vincenti, "as a vast integrated system, with varied social and technical areas of human activity as major interacting subsystems." And although

> this subdivision is made so that each subsystem can be analyzed in quasi isolation . . . analysis must be carried out . . . —and this is the crucial point—with attention at all times to the interactions between and constraints on the subsystems and to the eventual need to reassemble the system.[19]

Only such a broad attempt to take everything into account can address problems raised by unexpected weaknesses in engineering simplification.

A DUTY *PLUS RESPICERE* TO TAKE MORE INTO ACCOUNT

While implicit in interdisciplinary engineering practice, environmental engineering, systems engineering, and other approaches that see engineering as part of larger sociotechnical systems,[20] the principle involved here is seldom formulated as such. Its justification can be extended from the technical to the ethical. From the perspective of engineering practice, the fundamental imperative in the face of failure amid complexity can be phrased as "Take more factors into account." Without an attempt to follow this imperative, failures will indeed become "normal accidents."[21]

The obligation to take more into account, as a general counterbalance to model simplification, has a moral dimension, not just because it can on occasion avoid some specific harm. As in the BART case, more complex testing will also cause some harm (e.g., greater financial costs). The moral dimension of taking more into account is realized when it links engineering practice into general considerations of and reflections on the good. In this sense the duty to take more into account may be formulated as a duty *plus respicere* (from the Latin *plus*, more and *respicere*, to be concerned about).[22] This might even be interpreted as an engineering version of the motto "Think globally, act locally."[23]

The exact character of ethics, like that of design, has been widely debated in ways unnecessary to rehearse here. Suffice it to suggest that there is a strong argument to the effect that truly ethical behavior is based on as wide a reflective base as possible. One reason why altruism is superior to selfishness is that it involves a broader perspective and takes more into account, of others as well as oneself. In the historico-philosophical development of ethics codes in professional engineering, one defense of the moral superiority of social responsibility is that it entails more generous or inclusive reflection, even including company loyalty and technical efficiency, which are then placed within a more expansive framework.

Stating the issue as pointedly as possible, to take more into account in engineering will include taking ethics into account. The problem of the remoteness and subtlety of ethical factors in engineering is obvious. This itself might even be said to constitute a distinct ethical problem. An imperative to reduce such remoteness becomes part of a duty *plus respicere*.

To repeat: Civil engineer George Bugliarello has argued that the social responsibility of the engineer should include upholding human dignity, avoiding dangerous or uncontrolled side effects, making provisions for possible technological failures, avoiding the reinforcing of outworn social systems, and participating in discussions about the "why" of various technologies. For Bugliarello, "engineering can best carry out its social purpose when it is involved in the formulation of the response to a social need, rather than just being called to provide a quick technological fix."[24] Such is but another version of the imperative *plus respicere*.

PRACTICAL GUIDELINES FOR EXERCISING A DUTY *PLUS RESPICERE*

The proposed duty *plus respicere* is admittedly a loose and somewhat shapeless deontology. It might even be described as a "soft" ethics, in analogy with soft or alternative technology—perhaps as an example of the poetry that must complement mathematics[25] or of what has been called "weak thinking."[26] But to push this weak synthesis a bit further, perhaps it is legitimate to suggest that the following questions could serve as useful guidelines for self-interrogation by design engineers. Applying the basic argument concerning the danger of idealization, a research or design engineer should ask the following questions:

- Does the idealization or modeling utilized in this particular design process per chance ignore some factors that, while irrelevant to the boundary conditions of the technical problem, are important to wider concerns?
- Are the models being utilized sufficiently complex to include a diversity of nonstandard technical factors?

- Would it be possible to take other factors into account? What might their implications be?
- Does reflective analysis include explicit consideration of ethical issues?

Beyond such general questions, engineers might also ask themselves:

- Have efforts been made to consider the broad social context of the engineering practice, project, and end-use implications, including impact on the environment?
- Have likely end-user assumptions been critically examined?
- Has the engineering project been undertaken in dialogue with personal moral principles and with the larger nontechnical community?
- Are there peripheral implications of the engineering project that deserve to be given more direct consideration?

In summary, the scientist should ask: Is this knowledge significant? The engineer should inquire: Is this project worthwhile—and have all relevant factors been taken into account?

TRUMAN AND HEGEL

Most people don't realize that President Harry S. Truman was a great admirer of Georg Wilhelm Friedrich Hegel, the greatest of German philosophers, one who had *aufgehoben*-ed (as Truman liked to put it) all that preceded him.

Because of Hitler and World War II, Truman had to keep his admiration under his hat, but a few people knew. In fact, Truman once secretly snuck out of the White House to attend a Hegel seminar at the Pentagon and read a brief paper on the logic of the Encyclopedia of the Philosophical Sciences disguised as a question.

It was a big hit—one repeated shortly with the atomic bomb, a truly Hegelian device.

But what if the bomb were approached not as a thesis or antithesis but as a Heideggerian-like revelation? What might it reveal or disclose? What if biodiversity loss and climate change were not just problems?

NOTES

1. Latour (1979 and elsewhere) and others add instrumented actor networks as crucial.
2. See Alexis de Tocqueville, *Democracy in America*, vol. I (1835), chapter 12.

3. For sociologists "structural differentiation" describes how modern social orders are characterized by separations among, for example, religious, political, and economic institutions in ways not typical of premodern societies. Among those who have developed the concept of structural differentiation most fully are Talcott Parsons and his student Neil Smelser; see, for example, Smelser, *Social Change in the Industrial Revolution: An Application of Theory to the British Cotton Industry, 1770–1840* (Chicago, IL: University of Chicago Press, 1959).

4. For a much more detailed and nuanced analysis of the IEEE 1974 code, see Tang and Nieusma (2017).

5. Gunn and Vesilind (2003) attempt in the form of an eccentric fiction with commentary to explore some of these questions. With regard to the issue of peace, see Vesilind (2005).

6. For a useful overview of the historical emergence of environmental ethics, see Nash (1989). A complementary study is Worster (1994).

7. An earlier version of this book was published as Gunn and Vesilind (1986). There are some small confusions with their account. According to Vesilind and Gunn, in 1975 the ASCE code simply adopted verbatim the Fundamental Principles from the ABET model code. But (a) according to ASCE itself, the new code was adopted in 1977; (b) although the ASCE code states, like Vesilind and Gunn, that it adopts the principles from ABET, ECPD did not become ABET until 1980; and (c) the first principle is not in fact the same as the ECPD principle. The ECPD first principle states that engineers will use "their knowledge and skill for the enhancement of human welfare." The ASCE first principle states that they will use "their knowledge and skill for the enhancement of human welfare and the environment." (Another effort to bring environmental ethical issues into engineering ethics is Gorman et al., 2000.)

8. For an argument complementary to Vesilind and Gunn on the emergence of sustainability as a theme in engineering, see Beder (1998). For analysis of the contested character of the notion of sustainability, see Mitcham (1997).

9. The argument is further developed in subsequent editions of their influential textbook (1989, 1996, and 2005). For another articulation of the analogy, see Shrader-Frechette (1991), pp. 72–74, 86–87, and 206–14.

10. For brief review of this discussion, see Mitcham (1999).

11. Winston Churchill is often quoted as describing the proper role of scientists (by which he really means engineers) as being properly "on tap, but not on top." See Randolph S. Churchill, *Twenty-One Years* (London: Weidenfeld, 1964), p. 127.

12. "Airforce A-7D Brake Problem," U.S. Congress, Hearings, Subcommittee on Economy in Government of the Joint Economic Committee, Senator William J. Proxmire, Chair, August 13, 1969; Vandivier (1972); and Vandivier (1975). Selections from the first item and a complete version of the third are reprinted in Baum (1980), pp. 136–38 and 139–54, respectively.

13. For details on this history, see Burke (1966). Ward (1989) provides further information on the role played by the insurance companies. Unfortunately, Cross (1990) is neither well written nor especially informative. For more general analysis of American efforts in the deployment of technical knowledge for the democratic regulation technology, see Jasanoff (1990), where "science" is used, as is often the case throughout the literature, to cover engineering and technology.

14. For a more robust analysis of the Hydrolevel case, with references to the relevant legal documents and historical studies, see Wells et al. (1986).

15. For documentation, see Anderson, et al. (1980) and "The BART Case" in Unger (1994), pp. 20–27. Tang and Nieusma (2017) provide further analysis.

16. For documentation, see Boisjoly (1991). For interpretations different than Boisjoly's, see Vaughan (1996); Pinkus, et al. (1997); and McDonald (2009).

17. Feenberg (2017) is one contribution. A revisiting of Hegel that could provide further inspiration for such an analysis is Robert B. Pippin, *Hegel's Practical Philosophy: Rational Agency as Ethical Life* (Cambridge: Cambridge University Press, 2008).

18. Zimmerli (1986), p. 296.

19. Vincenti (1991), p. 2.

20. See, for example, Bijker et al., eds. (1987).

21. See Perrow (1984).

22. Note, also, that *respicere* is composed of *re-*, intensifying prefix + *specere*, to look at or behold, the latter of which is related to the Greek σκέπτω (from whence comes the English "skepticism").

23. This slogan, which has become a popular motto of the environmental movement, was apparently coined independently about the same time by two French social critics. See René Dubos and Jean-Paul Escande, *Quest: Reflections on Medicine, Science, and Humanity*, trans. Patricia Ranum (New York: Harcourt Brace Jovanovich, 1980 [French original 1979, from interviews recorded in September 1978]), p. 105; and Jacques Ellul, *Perspectives on Our Age*, ed. Willem H. Vanderburg, trans. Joachim Neugroschel (New York: Seabury, 1981 [from interviews recorded early in 1979]), p. 27.

24. Bugliarello (1991), pp. 77 and 81.

25. See Rosen (1988), especially the title essay, pp. 1–26.

26. See Vattimo (1988).

Essay 10

The Concept of Sustainability
Origins and Ambivalences

According to Xenophon's account, Socrates brought philosophy down from the heavens to ask questions such as "What is piety?" "What is impiety?" "What is good?" "What is shameful?" and "What is justice and injustice?"[1] In this questioning, philosophy at once ruptured a mythopoetic connection between heavens and the earth while supporting and criticizing ideals that provide guidelines for the conduct of human affairs. In our time this double-sided inquiry that promotes and questions social behavior can appropriately be brought to bear on the notion of sustainability in its most this-worldly form, sustainable development, an ideal that now infuses much contemporary discourse about and public policy with regard to science, engineering, and technology—and has even been introduced into professional engineering codes of ethics as an extension of the principle of safety, health, and welfare.

Sustainability is an ideal that, like love or patriotism, points toward something necessary and even noble but can also become an obfuscating cliché and misused by ideologues and dreamers. It is worth noticing that the word "sustainability" as an abstract noun did not appear until the 1970s.[2] As an abstract concept, sustainability is especially subject to ambiguities and disagreements about its precise implications—ambiguities spread across economic, medical, political, ecological, and related interpretations.[3] Faced with such a situation, philosophy can attempt to promote recognition and understanding of its truths through a paradoxical process of question enhancing reflection on four basic themes: historical and philosophical background, immediate origins, sustainable development and its near neighbors, and criticisms.

1. A HISTORICAL AND PHILOSOPHICAL BACKGROUND FOR SUSTAINABILITY

The historical and philosophical background against which the ideal of sustainability has emerged can be indicated by contrasting it with two basic theories about the character of historical change. These two views see history as fundamentally either cyclical or progressive. But the idea of progress in turn can be either toward a perfect future or away from an imperfect past.

The primordial experience of time and history is as cyclical. The sun rises and sets, then rises again. The moon returns every 28 days. The seasons come and go and come again. Anything that at first looks like it is not part of some cycle tends to be interpreted as forming part of a larger, perhaps unseen cycle: millennial cycles, cycles of great-many-thousand-year *yugas*, maybe even just one great big cycle from creation and fall to redemption as new creation.[4] The inherently conservative uses of such a view of history are recent claims by critics of carbon emissions regulation that climate change is just another large-scale and natural-cyclical phenomenon.

The idea of one great cycle is perhaps the decisive step toward a theory of progress. When the whole cycle of history becomes so large that one cannot see the beginning or the end—so immense that one can only remember the beginning and hope in the end—what becomes more important than any cyclical return to a previous state is the more immediate change taking place on one limited arc within the cosmic circle. As the past is left ever further and further behind and the future becomes more and more an imaginative projection, attention becomes increasingly focused on the present and its requirements. Within the restricted frame of reference thus defined, historical movement toward a possible future can now be thought of as progress. The idea of progress originates in and presupposes a restriction of perspective or horizon.

The initial or early modern theories of progress were of this type: some kind of singular great movement toward a future conceived as recovery of a remote past or perhaps potential perfection, through our efforts in the present. In Christian theology, for instance, an apocalyptic second coming is conceived as recovering the state of perfection indicated by the Garden of Eden narrative. But as the second coming of Christ was indefinitely postponed, late classical and medieval Christians began more and more to focus on the practice of the spiritual life in the present—a phenomenon that finds expression, for instance, in the rise of Christian monasticism. Then as an element in the rise of modernity the Christian mythology of spiritual progress toward transcendent salvation by means of faith was criticized and replaced by an ideology of progress toward this-worldly happiness and satisfaction by means of power and control over nature through science, engineering, and technology.

In the political theology of science, the future is imagined as realizing that state of perfection implied by natural human potentials for health, well-being, and autonomous freedom. Thus was the idea of progress toward some final state that renews a beginning state superseded by one of progress toward a future achievement unique to itself. But insofar as this finality or end becomes questionable or difficult to specify—not wholly unlike the second coming—then the idea of any termination or final state evaporates and progress becomes indefinite, more away from an imperfect past than toward a perfect future.

This is what has happened, for instance, with the ideology of modern scientific knowledge—not to mention that of engineering prowess. Originally science was presented by its modern founders as able to provide final and definitive knowledge about the world. In Alexander Pope's heroic couplet, "Nature, and Nature's laws lay hid in night. / God said, *Let Newton be!* and all was light." But with recognition of the inability of any scientific theory to withstand the onslaughts of further investigation or the introduction of further refinements—that is, with recognition that no scientific theory is ever final—the idea of indefinite (if not infinite) improvement was substituted for the realization of an epistemological perfection. Or, as one satirist wrote "In continuation of Pope on Newton": "It did not last: the Devil howling 'Ho!/ Let Einstein be!' restored the status quo."[5]

The idea is that science and engineered technology get "better and better" in comparison with the past but not necessarily closer and closer to some determinate ideal. Scientific theories may be more comprehensive today than yesterday, but it remains difficult to imagine what a completely comprehensive theory would be like—and whether it is even possible to strive for such a theory. The ideal of a comprehensive theory or discipline is replaced by one of complementary theories and multidisciplinarity.

The theory of evolution in biology further exemplifies this idea of indefinite progress away from a past that fails to point toward any definite future. Some animals can smell; others can see. Some can smell and see. What the perfect sensorium would be like—who knows? But having five senses is nevertheless better or more inclusive than having only one. A multiplicity of senses, indeed an increasing number of senses, denotes progress.

The perfect human being, the perfect society as well, are difficult to specify—but surely we can recognize that this one or that one is better than some other one. This is the implication of Winston Churchill's well-known observation that democracy is definitely not the perfect form of government. In fact, "democracy is the worst form of government except all those other forms that have been tried from time to time."[6] With the fall of Soviet communism, the same might be said for capitalist economic and technological development, especially in light of the widespread scandals associated with the post-socialist, high-capitalist decades. Capitalist development of science and engineering is not perfect and indeed has got to be the worst economic system there is—except for all the others that have been tried from time to time.[7]

It should be noted, however, that this backward-looking or escapist theory of progress entails a stigmatization of the past. The motor of progress so conceived is captured in the popular question, "Do you want to go back to a time when?"—with little appreciation of what that "time when" might have really been like. With regard to the sustainability question, it is possible that at some deep level there is simply the fear of an allegedly unsustainable and inhuman past of scarcity that must at all costs be avoided. Sustainability may not be so much a positive ideal as an attempt to escape the unsustainable. The ghost of Thomas Hobbes hovers in the shadows.

2. IMMEDIATE ORIGINS OF THE CONCEPT OF SUSTAINABLE DEVELOPMENT

Implicit in the modern idea of progress as an indefinite and continuous superseding of the past, rather than as an approach toward some well-defined future, is the notion that it is without limits. The early modern theory of progress, such as can be found in the Marquis de Condorcet and Immanuel Kant, actually had its own kind of limits: some kind of perfect state. But the theory of indefinite progress lacks any such projected termination. Perhaps, for example, one could keep adding on senses forever: we certainly create instruments that register more and more aspects of reality through X-ray machines, radio receivers, ultrasound and magnetic resonance imaging devices, and more. Perhaps such extensions of perception could just go on indefinitely, and even be incorporated into our physiological sensorium, after the manner of the science fiction dreams of some transhumanists and artificial intelligence enthusiasts. Androids and cyborgs become our siblings if not ourselves.

By contrast, it is the discovery and articulating of limits to progress that is a distinctive character of what has been called the postmodern historical consciousness. Or perhaps one should say, more cautiously, that one theme in the conflict between modern and postmodern interpretations of history is precisely this question of the limits to progress. For postmodern theorists, this discovery and its articulation has taken multiple forms but early on was often associated with research sponsored by a group called the Club of Rome and first published in a book titled, appropriately enough, *The Limits to Growth* (1972).[8]

The Club of Rome was an informal, international association of scientists, business executives, scholars, and public officials that coalesced in the late 1960s and early 1970s around the charismatic Aurelio Peccei and his concerns about the global development system.⁹ Its intentionally vague character as a kind of transnational, intellectual think tank contributed to its ability to attract a good deal of media attention. The fundamental argument of the Club of Rome and that "report to" it called *The Limits to Growth* is that technological development and population increase simply cannot continue as they have for the last 200 or 300 years.

The theory of progress that compares the present with the past, and does not consider the future except as an open-ended possibility for further growth and improvement, is a source of increasingly intractable problems. Comparing the present only with the past and noticing how much better things are now than they were then, fails to consider that in the near future things may well get worse—worse even than they were in the past or worse in dramatically different ways. Moreover, the catastrophe will occur precisely as a result of the action of the very forces that make the present appear better than the past—and which continue to be supported because they do so.

To put the point in graphic form (see Figure 10.1): the simple and obvious point of *The Limits to Growth* is that exponential growth cannot continue indefinitely. Exponential increase of one kind or another has been characteristic of modernity since the seventeenth century. Exponential growth— whether in population, food, industrial production, energy consumption, CO_2 emissions, and so on (dark solid line in graph)—will either terminate in a catastrophe (thin solid line) or level out into a logistic curve (broken line). Some wits have even described the exponential portion of this graph as the "club" of Rome, as that with which Peccei and colleagues were wont to hit others over the head.

The argument of *The Limits to Growth* was that we are going to go over the cusp of catastrophe if we do not take conscious action to create a curve of logistic accommodation to resources. Even in the follow-up book, *Mankind at the Turning Point* (1974)—which tried to put forward a positive vision of the future as one of possible "organic" rather than "undifferentiated" growth—the message continued to be phrased in largely negative terms and tone.[10] Human beings have to stop what they are presently doing. The implicit message is that we should replace growth with a no-growth or steady-state economy.

The shift from emphasizing what should not be done to stressing what should and can be done is constituted by the shift from a discussion of "limits to growth" to "sustainable development." This shift in the framework of the discussion was initiated by two other reports:

- the *World Conservation Strategy* (1980) of the International Union for Conservation of Nature and Natural Resources; and
- *Our Common Future* (1987) of the World Commission on Environment and Development.

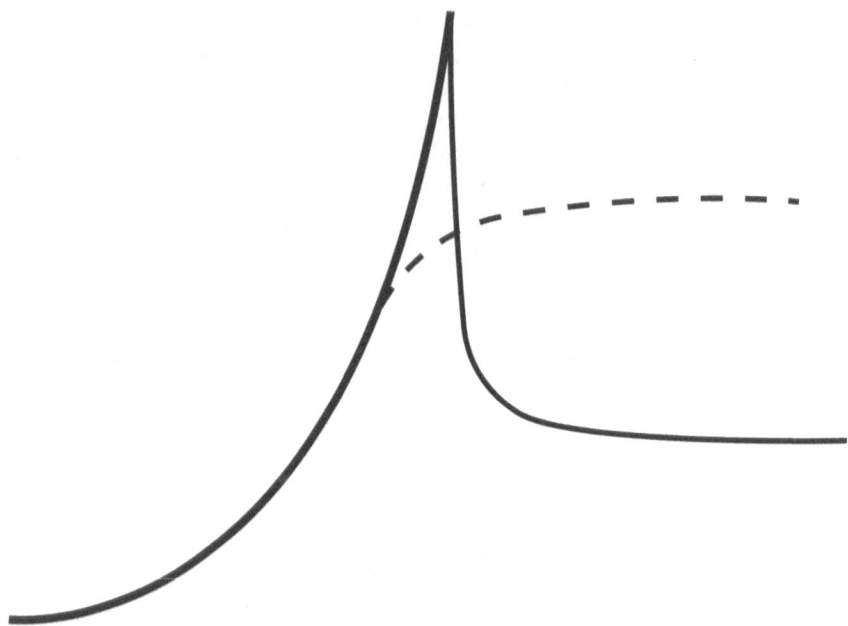

Figure 10.1. From exponential growth to collapse or steady state

Because of their importance for all subsequent discussion and their influence on such international programmatic documents as *Agenda 21*, the product of the 1992 Earth Summit in Rio de Janeiro, it is worth considering both documents in some detail.

The first of these, the *World Conservation Strategy*, is exactly what its name implies, a strategy document aimed at government policy makers, conservationists, and development practitioners.[11] It consists of a spare 50 pages of text divided into an introduction followed by three major sections on the general objectives of conservation, a set of priorities for national action, and a set of priorities for international action. Each of the three sections is further broken down into six or seven two-page subsections with between eight and fifteen numbered paragraphs. This is clearly a policy and action manual organized around thought bites if not sound bites.

Development is defined (section 1, paragraph 3) as "the modification of the biosphere and the application of human, financial, living and non-living resources to satisfy human needs and improve the quality of human life." Moreover, "for development to be sustainable it must take account of social and ecological factors, as well as economic ones; of the living and non-living resource base; and of the long term as well as the short term advantages and disadvantages of alternative actions."

At the same time, conservation is defined (section 1, paragraph 4) as "the management of human use of the biosphere so that it may yield the greatest sustainable benefit to present generations while maintaining its potential to meet the needs and aspirations of future generations." "Conservation, like development, is for people" (section 1, paragraph 5) and not for the things conserved.

Such definitions lead to the conclusion that "development and conservation operate in the same global context, and the underlying problems that must be overcome if either is to be successful are identical" (section 20, paragraph 1). For the *World Conservation Strategy* recognition of the "limits to growth" is, as it were, the precondition to further and continuing growth. The limits and conservation must be incorporated into further growth or development strategies.[12]

Our Common Future (1987) is more discursive, with an influence that is difficult to overestimate.[13] As the first chapter of *Agenda 21* acknowledges,

> In 1987, the U.N. World Commission on Environment and Development linked the issue of environmental protection to the seemingly unrelated topic of global economic growth and development. Headed admirably by Norwegian Prime Minister Gro Harlem Brundtland, this commission produced a stunning report entitled *Our Common Future*, which carefully documented the status and future of the global economic and ecological situation. Perhaps the most lasting accomplishment of the Brundtland Commission, however, was to thrust the concept of "sustainable development" into the mainstream of world debate. Although this concept had been the focus of discussion in the world scientific community for some time, its introduction into international dialogue elicited an almost instant response.[14]

Our Common Future constituted the third in a series of reports from so-called "independent commissions" headed by prominent European social democratic politicians. The first was *Common Crisis* (1980), from the Independent Commission on International Development Issues chaired by Willy Brandt, former prime minister of West Germany. The second was *Common Security* (1982), from the Independent Commission on Disarmament and Security Issues, chaired by Olof Palme, former prime minister of Sweden. All reports were addressed not only to governmental and developmental specialists but also to the general public.

Our Common Future, building on the approach of the *World Conservation Strategy*, reemphasized the shift from limits to sustainability. Gro Brundtland, prime minister of Norway and chair of the World Commission on Environment and Development, was faced with competing interests. On the one side were environmentalists, who argued the limits to growth or no-more-growth position to meet the threat of pollution, protect natural resources, and respect the rights of future generations. On the other were economists, especially of the Third World, who argued the need for development, more growth, to alleviate poverty in the present and to make it possible for these nations to play their proper roles in international affairs. The Brundtland report bridging of these conflicting interests proposed neither simply limits nor simply development but *sustainable development.*

To quote the widely influential definition of the Brundtland report, sustainable development "meets the needs of the present without compromising the ability of future generations to meet their own needs" (p. 8). As *Our Common Future* immediately explains,

> The concept of sustainable development does imply limits—not absolute limits but limitations imposed by the present state of technology and social organization on environmental resources and by the ability of the biosphere to absorb the effects of human activities. But technology and social organization can be both managed and improved to make way for a new era of economic growth.
>
> (p. 8)

The change that takes place after unlimited growth is thus defined not as a situation of no-growth or no-more-growth but as sustainable growth. The Brundtland report is proposing a possible third option in the transformation of the exponential curve (see Figure 10.2): not catastrophe (thin solid line) or leveling off (broken line) but continued more moderate growth (dotted line)—sustainable growth, that is, growth that does not eventuate in catastrophe.

It is something like this concept of sustainability that appears to animate a host of post-Earth Summit books on sustainable development.[15]

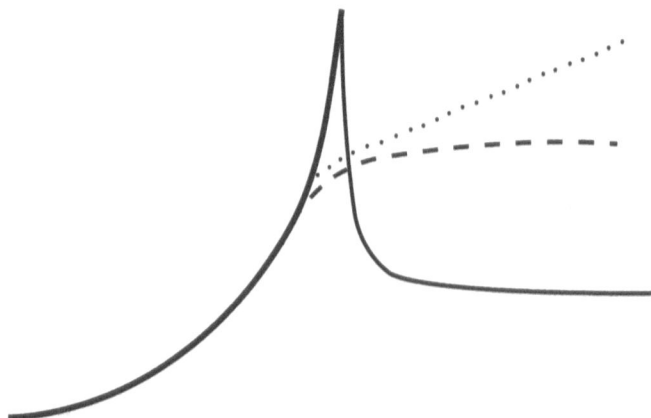

Figure 10.2. Collapse, steady state, or sustainable growth

The implications for engineering deserve to be highlighted, especially for the mutation of civil engineering that has become environmental engineering.

3. SUSTAINABLE DEVELOPMENT AND SOME NEAR NEIGHBORS

It is not clear, however, to what extent sustainability has to mean continued growth. The concept entails some studied or creative ambiguity, which is precisely what makes it useful for bridging the gap between no-growth environmentalists and progrowth developmentalists. In order to explore this ambiguity, it is useful to consider some of the near neighbors of the concept of sustainability and its permutations.

The concept of sustainability in relation to economic or technological growth and development was strongly influenced by post–World War II theories of economic development. Here two classic references are President Harry Truman's inaugural address on January 20, 1949, and economist (and later political adviser) Walt W. Rostow's influential 1956 article on "The Take-off into Self-sustained Growth."[16] In his address, Truman attacked the "false philosophy of communism" and committed the United States to a fourfold opposition: support of the United Nations, programs for "world economic recovery," military defense of "freedom-loving nations," and "a bold new program for making the benefits of our scientific advances and industrial progress available for the improvement and growth of underdeveloped areas."[17] Point four in effect enrolled engineers as leading contributors to an anticommunist American foreign policy agenda.

Rostow elaborated his argument in a subsequent book on *The Stages of Economic Growth: A Non-Communist Manifesto* (1960), distinguishing five successive stages:

1. Traditional society,
2. Preconditions for take-off,
3. Take-off,
4. Drive to maturity, and
5. High mass consumption.

In traditional society there is no dynamism of techno-economic growth, which instead occurs only more or less by chance or accidentally. In traditional and pre-take-off societies, growth is often slow, the result of external factors and influxes of capital or technologies and development interests from outside the society. When these external inputs falter, so does growth.

According to Rostow, for growth to take off and become what he terms "self-sustaining,"

> The forces making for economic progress, which yielded limited bursts and enclaves of modern activity, expand and come to dominate the society. Growth becomes its normal condition. . . . New techniques spread in agriculture as well as industry. . . . [T]he basic structure of the economy and the social and political structure of the society are transformed in such a way that a steady rate of growth can be, thereafter, regularly sustained.[18]

As Hannah Arendt noted at this same time, however, such "unhindered development"—as exemplified by the "economic miracle" of postwar Germany, in which "the expropriation of people, the destruction of objects, and the devastation of cities [turned] out to be a radical stimulate"—undermines the uniquely human world of cultural and political orders in which humans have traditionally found themselves at home. In the new order of insistent production and consumption, "not destruction but conservation spells ruin, because the very durability of conserved objects is the greatest impediment to the turnover process, whose constant gain in speed is the only constancy left."[19] Two decades later, in an effort to respond to the problematics of what economist Joseph Schumpeter termed "creative destruction"[20] attempts to spell out alternative approaches emerged in two areas: one dealing with energy, the other with agriculture.

In the energy crisis of the mid-1970s, the argument for the replacement of dependency on nonrenewable resources (coal, oil, and gas) with utilization of renewable resources (water, wind, solar) and the adoption of what Amory Lovins termed a "soft energy path" pointed toward a kind of sustainability. As Lovins and colleagues argued, energy production and use cannot continue indefinitely at the present rate of increase. Increased pollution and the eventual depletion of resources will not allow it. The self-sustained growth in energy production is not indefinitely sustainable. Rostow's last stage of high mass consumption is a prelude to disaster.

Two noteworthy features of Lovins's argument are that he does not see energy use as growing at an exponential rate and that he postulates not just a leveling off but an actual decline in energy production and use. These ideas, condensed for display in two graphs, were first presented in *Foreign Affairs* (October 1976) article, which became the central chapter of *Soft Energy Paths* (1977).[21] One controversial implication of Lovins's argument concerned the extent to which his soft energy paths and energy conservation proposals call for more general alterations in ways of life in advanced industrial societies. For Lovins himself, basic changes need not occur; more efficient energy use is presented as actually increasing the sustainability of increasing material standards of living.

It may also be noted that Lovins and the alternative or sustainable energy coalition had an influence on U.S. governmental policy. Extending the 1974 Solar Energy Research, Development, and Demonstration Act, in 1977 President Carter founded the Solar Energy Research Institute, which evolved into the National Renewable Energy Laboratory in Golden, Colorado (Lovins's home state). In relation to energy, "renewable" became another term closely linked to "sustainable."

In agriculture, the idea of "sustainability" (if not the exact term) has deeper roots. The notion that farmers should not just use the land, wear it out, and move on, but should instead conserve and sustain it, became a major theme in U.S. governmental policy and agricultural education with establishment of the Department of Agriculture in 1862 and the associated founding of the major land grant universities.

The first theory about how to achieve this conservation was simply through "scientific farming" as promoted by agricultural extension agents attached to the new agricultural universities. Building on the work of British and German chemists such as Sir Humphrey Davy (1778–1829) and Justus von Liebig (1803–1873), agricultural science proposed that the regular utilization of inorganic fertilizers, especially nitrogen, would indefinitely increase and prolong land productivity. Agricultural science evolved in the early 1900s into (and would from the beginning have been more accurately termed) "agricultural engineering" (first degree program at Iowa State University in 1907). But nitrogen utilization establishes a farming dependence on the industrial production of chemical fertilizers. Agricultural engineering tends to replace farming with agribusiness—and to serve as a gateway to genetic engineering and biotechnology.

A counter-vision of conservation or sustainability developed in conjunction with a critique of industrialized agriculture as tending to promote unnecessary dependence on industrial development. This critical movement originated with Sir Albert Howard, a British agricultural advisor to India in the 1920s, who argued that Indian farmers could maintain their independence and increase yields by means of decentralized fertilizer production through organic composting.[22] But more immediately influential in North America was the writer and back-to-the-land pioneer J. I. Rodale (1898–1971), who in 1942 coined the term "organic farming" when he founded what would become the journal *Organic Gardening and Farming*. With arguments that foreshadowed some Club of Rome ideas, Rodale's proposals for organic farming included criticisms of monocropping or undifferentiated agriculture. Indeed, this critical or alternative agriculture movement became widely identified with the work of J. I. Rodale, his son Robert Rodale (1930–1990), the Rodale Institute, and associated publications such as *Prevention* (1950–present) and *New Farm* (1979–present).

For Robert Rodale, especially, the idea of organic farming needs to be expanded beyond the realm of agriculture:

> In recent years, events have forced the concept underlying organic research to expand.... There is an energy crisis, and the nonorganic way of growing food uses too much energy. Our environment has become contaminated by agricultural chemicals, many of which pollute the normal sources of food. The ozone layer of the upper atmosphere ... is threatened with destruction by the overuse of chemical nitrogen fertilizers.... Because of these and similar conditions, there is now a need for a concept of organic research that will support an effort going far beyond the basic goals of natural soil improvement and improved food quality.[23]

As a result, during the latter half of the 1980s Rodale's preferred term became "regeneration," and he worked to define methods of farming that not only sustained the land but actually regenerated it. Similar attitudes and methods can, he argued, be carried over into society.[24]

In the 1970s again E. F. Schumacher's *Small Is Beautiful* (1973) and the "intermediate technology" movement gave added scope and life to organic farming arguments.[25] The intermediate technology movement itself devolved into the appropriate and then the alternative technology movements, which in turn spawned the term "alternative agriculture." For example, during the 1990s in the pages of the academic, peer-reviewed journal *American Journal of Alternative Agriculture* (founded 1986, renamed 2003 as *Renewable Agriculture and Food Systems*), alternative agriculture often used "sustainable agriculture" as a functional synonym.

For over a century the primary strategy of North American agriculture had been to increase yields and therefore profits per person-hour of labor by the increased utilization of machines, chemicals, and hybrid plants and animals. Yet this had the effect of making farmers increasingly dependent on external, high-technology and high-cost inputs as well as the vagaries of a globalizing economic system. Coupled with growing public concern about environmental degradation in the 1980s, there emerged arguments for alternative, low-technology inputs. Although a reduction of high-tech methods might reduce yields, net farm income could in some cases be maintained or even raised through corresponding reductions in high-cost of inputs as well as the development of specialized markets.

The influence of this contrast between high-tech, high-cost inputs and low-tech, low-cost inputs was reflected in a 1988 appropriations bill for the Agricultural Productivity Research component (Subtitle C) of the 1985 Food Security Act that explicitly specified "low-input" as an area for research funding. Subsequently, agricultural scientists at Washington State University proposed the term "low-input sustainable agriculture" as an appropriate focus. No doubt partly as a result of the ease with which it can be turned into the acronym LISA, this became a standardized term in the 1989 National Academy of Sciences report on *Alternative Agriculture* and related discussions.[26] Then under Subtitle B of Title XVI of the Food, Agriculture, Conservation, and Trade Act of 1990, LISA was broadened and replaced by the Sustainable Agriculture Research and Education (SARE) Program, with four regional bases and a diverse membership, including representatives ranging from family farms and agribusiness to nonprofit, governmental, and academic organizations. Politicians never want to leave out possible constituents, but in the process more economically powerful players tend to co-opt small-scale interests.

Concepts of sustainable energy use and sustainable agricultural production are both more limited concepts than that of sustainable development, although in the hands of Lovins and Rodale a clear if implicit expansion takes place. Another term, which broadens rather than concentrates the focus, is that of "sustainable society." Here the work of Lester Brown and the Worldwatch Institute was the most influential. For Lester Brown the issue was not sustainable economic growth or sustainable energy production—but sustainable society. Brown's tactic was to take the notion of sustainability and enlarge its context in such a way as to reform its meaning.[27]

It was in 1978 that Denis Haynes, a cofounder with Senator Gaylord Nelson of Earth Day 1970 (which became an annual Earth Day celebration), wrote the first Worldwatch tract employing the term "sustainable society."[28] Then in 1981 Lester Brown's book *Building a Sustainable Society* made a sustained and comprehensive case for sustainability broadly construed as an ongoing societal project. As Brown wrote in his preface, "This book is not an isolated effort to discuss the sustainable society, but rather part of the continuing research program of the Worldwatch Institute." The picture he sought to draw of a sustainable society "has of necessity been painted with a broad brush."[29] But taking a political approach, it too tried to enroll corporations, religion, universities, public interest groups, the communications media—and even such alternative or countercultural ideals as "voluntary simplicity" and "conspicuous frugality." In 1984 Worldwatch began issuing semiannual *State of the World* reports on "progress toward a sustainable society" promoting a variety of related themes.

The symptomatic character of this transition from sustainable development to sustainable society is highlighted by another book, one not directly associated with the Worldwatch Institute. Although it draws on Worldwatch publications and parallel sources, Lester W. Milbrath's *Envisioning a Sustainable Society: Learning Our Way Out* (1989) made its own case for why modern society is not sustainable and how the sustainable ideal responds to limits-to-growth ideas.[30] The postmodern society, insofar as it initiated a break with modernity, is, he suggested, better described as the sustainable society.

One further illustration of this expansive approach is Willem Vanderburg's effort to assess society not just in terms of sustainability with respect to the biosphere but in relationship to human life and culture, for which he coined the term "human sustainability." For Vanderburg, human sustainability, which he found lacking in advanced technological society, indicates "the ability of a community to create a way of life which is an expression of [its] values and aspirations and which is able to give meaning, direction and purpose to the lives of its members."[31]

This inflation of the notion of sustainability, when it moves from associations with development to associations with society, deserves to be stressed. As was pointed out earlier, the idea of progress originates in and presupposes a restriction of perspective. Once that restriction is set aside, the idea of progress is no longer nearly as obvious or as easy to defend. Discussions of sustainability tend to reemphasize certain cyclical patterns in nature and in history, and indeed discussions of sustainability themselves may exhibit some cyclical features. This tendency to break with previous and defining restrictions is a characteristic move in sustainability thinking. It is also part of what turns "sustainability" into a new name for the good broadly construed.

4. CRITICISMS OF SUSTAINABILITY

The greatest danger in this characteristic inflation of "sustainability" is that the term becomes what Uwe Pörksen calls a "plastic word."[32] Sustainability and sustainable development have become modular slogans everyone praises, but in their universal approbation people truly agree about little. It can mean almost anything. *Science* magazine editorializes about a "sustainable future for planet earth"; an update of *The Limits to Growth* appropriates "sustainable" in its subtitle; and a new journal *In Context* bills itself as "a quarterly of humane sustainable culture."[33] Hundreds of books and hundreds of articles on sustainability are published each year. This is form without substance, word without significance. Efforts to escape the slide into vacuousness have led to repeated analyses of the meaning, especially of the term "sustainable development."[34]

More substantive efforts to deal with the problem of conceptual inflation take the form of criticisms from both the right and the left of the political spectrum. From what might be called the Right comes an argument that "sustainability is not enough." From the Left—when it is not complaining that sustainable development is still development, which perforce contributes to environmental deterioration now on the scale of global climate change—comes the question "Sustainability for what?"

For sustainability-is-not-enough critics, sustainability denotes stasis or, at best, dynamic equilibrium. In relation to sustainable agriculture, one critic argues that "traditional agricultural systems that have met the test of sustainability have not been able to respond adequately to modern rates of growth in demand for agricultural commodities."[35] From this perspective, sustainable development—as slower and more careful development—will be overwhelmed by continuing increases in population and aggregate consumer demand; sustainable development is not itself sustainable.

Julian Simon is one of the most radical representatives of this school of thought. From *The Ultimate Resource* (1981) to *The State of Humanity* (1995), Simon argued that the limits-to-growth argument is so bad as to "not [even be] worth detailed discussion or criticism."[36] The dynamic of the modern techno-capitalist economy, as Arendt rightly observed, depends on the continuous discovery and creation of new conditions for the appropriation and consumption of resources. According to Simon, with nuclear fission and space travel, "we now have . . . the technology to feed, clothe, and supply energy to an ever-growing population . . . to go on increasing forever, improving our standard of living and our control over our environment."[37] Within such a framework, the ideal of sustainability, too, is scarcely worth mentioning. Sustainability exists, as it were, only through attacks on sustainability.

With the "Sustainable for what?" question, a different set of critics ask for clarification about exactly what it is that is supposed to be sustained. Surely it is not just sustainable development itself or, as the World Bank defines it, simply "development that lasts"[38]—that is, that which sustains only itself. Sustainable development must, in sustaining itself, sustain and not undermine something else to which it is related: the natural world, society, a way of life, a culture. The danger, according to those who ask this question, is that sustainable development is simply meant to sustain a way of life in the West that leads to the destruction of other ways of life throughout the world. The suspicion is that sustainability is proposed as a means to sustain, without questioning, a way of life that is not the good that it appears to be.

Migrations of sustainability across the cultural landscape can have two different although not mutually exclusive kinds of results: (a) it may dilute the concept, subtly transforming it in the process; and (b) it may insinuate its principles into new areas. One can notice both tendencies at work here. Sustainable development need not require growth, but it does imply input-output management, the core principles of which it thus introduces into aspects of society beyond engineering and economics. Perhaps it is because the term retains this character that someone like Rodale is inclined to favor other terms such as "organic" and "regeneration."

Wolfgang Sachs, in a provocative article originally titled "The Gospel of Global Efficiency," suggests that precisely insofar as "Worldwatch alerts us to the need for the efficient use of means [it also elevates] the rules of micro-economics to imperatives for national (and even global strategy)." Sustainable development theorists readily become what Sachs terms "eco-developers," who simply expand the perspective of traditional economics "by surveying the broad range of life-supporting factors in order to assure the sustainability of yields over the long term."

> Worldwatch implies the world-wide victory of [the] specifically modern outlook.... The Worldwatch utopia of a sustainable world appears to be peopled by a fairly recent version of *homo sapiens*: the efficiency-conscious individual.[39]

In pursuit of the ideal of sustainability there emerges an addiction to management that looks on the world, in the influential image of engineer Buckminster Fuller, as a space ship in need of an operating manual.[40] Such an outlook must be recognized as, at the very least, at odds with much of traditional culture. Perhaps there is a need to reconsider not just a course of development that may no longer be viable in the face of environmental constraints—and the resultant effort to replace development with sustainable development—but also the very aspirations that have led to such a course and the contemporary search for new engineering means to preserve it.

The ideal of development has roots not just in Rostow's theories of international economics but also in a typically modern fear of what is presented as its only alternative—scarcity. At the deepest level, criticism of sustainability must address this fear of scarcity and the modern proposal to flee scarcity by means of techno-engineering development. But in the face of the empirical evidence of the problem of scarcity in the world today, how can one possibly raise such a question?

Consider one of the most dramatic forms of scarcity, namely, shortages of food that lead to famine. The developmentalist view is to think that dire natural or agricultural shortages are to be addressed by agricultural engineering. But as economist Amartya Sen has shown, famines are not nearly so often a consequence of failures in agricultural production as they are of failures in distribution and sharing.[41] Although it is true that drought and flood often precede mass starvation, declines in food production alone do not explain most famines; even in desperate times there is typically enough food in the country to go around or enough money to import it. Disaster strikes, Sen's research argues, not simply when there is no more food, but when what food exists is no longer shared throughout a society. Such failures to share arise today when the poor can no longer afford to buy food following the loss of what Sen variously calls entitlements, capabilities, or functionings—that is, because of a sudden loss of employment or a surge in food prices.

What Sen points to as a loss of abilities to function within the society others have referred to as a loss of solidarity in an economy of subsistence. In the words of two commentators on this phenomenon,

> Ironically, only modern and affluent societies are convinced about the importance of scarcity as a determinant of social behavior, while traditional societies tend to rest on a different conviction. In traditional cultures, wishes and desires are not seen as endless or indefinite but as religiously and culturally constrained. Of course, traditional societies are used to shortages, but precisely because of a commitment to the constraint of desire such shortages are dealt with in ways quite different from those typical of modernity.[42]

The idea of scarcity—from which development and its modification, sustainable development, are proposed as the means of escape—has a history. It is a social construct of the early modern period that has become defining of modern economic life.[43] But at its most profound level, what the modern world calls scarcity might better be thought an economy of subsistence—and, indeed, that which is sustainable in the truest sense.

5. CONCLUSION

This essay began by suggesting that sustainability is an idea that a Socratic philosophy might want to question. It is appropriate to conclude by reiterating this point—and to add a word of caution. A danger of critical reflection on prereflective names for the good is the opening of a door to nihilism or despair. One must beware of calling sustainability a lie.[44] The limits to growth and the need for development are both real. The ideal of sustainable development remains a tremendously attractive name for trying to acknowledge and bridge both realities and their sometimes conflicting claims. But the precise parameters of that bridging is yet to be determined and will in fact only emerge in the course of our interdisciplinary and intercultural discussion of the meaning of sustainable development.

In 1977, when Paul VI became the first Pope to set foot on the South American continent, he declared that "the name of peace in our time is development." We now know that development may be the name for many things, but peace is not necessarily among them. Development destroys as it proceeds, undermines as it builds, attacks the environment, and makes war on indigenous cultures. In an attempt to save development from itself the ideal has been reformulated as sustainable development—one might even say, peaceful development.

Although this may to some extent be a contradiction in terms, in the process sustainable development has also become a synonym for that first, pre-reflective name of the good as that which is our own. Just as Socrates taught us that the good as our own calls for philosophical criticism to save it from itself, so now does the idea of sustainable development.

ADDENDA: ECONOMICS, PHILOSOPHY, ENGINEERING, AND ECOMODERNISM

1. An overlooked conceptual shift during the historical period analyzed in this essay was from sources to sinks. The 1972 *Limits to Growth* report, chapter two, "The Limits to Exponential Growth" (and note the limits are not to growth per se but to exponential growth), devoted almost twice as much analytic space to limitations in resources (physical mostly, i.e., land, food production, minerals but also social) as to limitations on how much waste or pollution the Earth (or society) could absorb. Although

> curves of various kinds of pollution can be extrapolated into the future, [unlike with resource depletion] no upper bounds have been indicated for the exponential growth curves of pollutants . . . because it is not known how much we can perturb the natural ecological balance of the earth without serious consequences. It is not known how much CO_2 or thermal pollution can be released without causing irreversible changes in the earth's climate, or how much radioactivity, lead, mercury, or pesticide can be absorbed by plants, fish, or human beings before the vital processes are severely interrupted.[45]

By contrast, the 30-year revision (Meadows et al., 2005) devoted a whole chapter on "Sources and Sinks" in which the question of anthropogenic climate change plays a significant role. Beyond the "limits-to-growth" discourse, the issue of greenhouse gas emissions has become the most widely discussed dimension of sustainable development. The issue of sinks has dwarfed that of sources.

2. Nicholas Georgescu-Roegen's important work on *The Entropy Law and the Economic Process* (1971) also deserved reference. According to one assessment,

> This book is a must for all engineers, an imperative for engineering economists and most desirable reading for those who are in any way connected with economic development. It is a philosophical treatise, an appeal to the author's fellow economists, to view the economic process from a biological and thermodynamic, rather than a mechanistic, perspective.[46]

Georgescu-Roegen argued for a fundamental distinction between agricultural and industrial processes in a way that poses ontological and epistemological questions for engineering. Drawing on Georgescu-Roegen, his student Herman Daley developed the discipline of ecological economics and the concept of a steady-state economy, which obviously implies a radical rejection of sustainable development as an oxymoron. According to Georgescu-Roegen, however, this remains insufficiently radical. The true engineering challenge is not to produce sustainable development or even a steady-state economy of zero growth. "The most desirable state is not a stationary, but a declining one.... growth must... be reversed".[47] Georgescu-Roegen is the inspiration for what French economist Serge Latouche calls *décroissance* or degrowth.[48] Is it possible to imagine a degrowth engineering?

3. Despite extensive public discussion of sustainability in economics, political science, environmental science and engineering, and environmental studies, including in numerous political documents—see, for example, the 17 UN Sustainable Development goals of 2015 that emerged from the United Nations Conference on Sustainable Development (UNCSD also known as Rio 2012, Rio +20, or Earth Summit 2012—the concept has been given surprisingly little sustained philosophical attention. The first article in the journal *Environmental Ethics* (founded 1979) to include "sustainability" in its title did not appear until the late 1990s; there is no article on sustainability in the *Routledge Encyclopedia of Philosophy* (1998), in the *Encyclopedia of Philosophy*, second edition (2006) or in the *Stanford Encyclopedia of Philosophy* (as of January 2019).

The best early philosophical prosecution of the sustainability discourse well beyond the present essay exists in work emerging from the Society for Philosophy and Technology, initiated in a 1993 conference organized by José Sanmartín at Peñiscola, Spain; two arguments directly complementary to the present essay, for example, can be found in the proceedings volume: Stanley R. Carpenter (1993) and César Cuello Nieto and Paul Durbin (1993). In 1996, Mexican colleagues organized a special, bilingual SPT conference in Puebla, Mexico, in which *sustentabilidad* and *medio ambiente* were major themes; the proceedings volume (Martínez Contreras et al., 1997) includes more than 20 papers exploring relevant ontological, epistemological, and ethical issues. The philosophical problematics of sustainability was further prominent in the first two numbers of the SPT electronic journal *Techné*. Threaded throughout, I would argue, is a persistent idea that for sustainability to be anything more than cliché or question begging will require engineering to break free from its current cultural captivity. Insofar as this is the case, it calls for searching inquiries into what it is about engineering that makes it so receptive to an unsustainable servitude.[49] Coming at things from a different angle, a few social scientists engaged in engineering education reform have been trying to establish linkages between sustainability as an ethical concept and engineering. Here the general argument has been that sustainability is more than a strictly engineering issue with social justice dimensions that engineers tend to overlook.[50] Here again it would be useful to at least ask whether there might be features inherent to engineering thought and practice that give rise to some of the resistance social justice issues.

4. In engineering itself a pivotal publication was Dutch environmental engineer Karel Mulder's *Sustainable Development for Engineers: A Handbook and Resource Guide* (2006). Mulder is important, as well, because his work has a philosophical bent. The first five (of ten) chapters address philosophically relevant issues, such as the need for and nature of sustainability, relationships between engineering and social political structures, how to measure sustainability, and more.

Another more radical but less influential effort to rethink engineering in light of the challenge of sustainability is by Canadian mechanical engineer Willem Vanderburg. His concern for "human sustainability" (referenced in the main body of the essay) and what he sees as an inability of the contemporary engineering way of life to adequately express human values and aspirations is significantly extended in *The Labyrinth of Technology* (2000) with the promotion of what he calls "preventive engineering": essentially a form of engineering that *plus respicere*, takes more into account (see also Vanderburg 2005 and 2016).

There is, however, a problem with Vanderburg's and related diagnoses of the engineering way of life as unable to express truly human values and aspirations, an argument that has been made repeatedly since the emergence of modern engineering. Vanderburg's argument was already well developed in early work by his mentor, Jacques Ellul (see, e.g., Ellul, 1962, for one clearly stated version). The problem is that many engineers do in fact find the engineering way of life and the creative advancement of engineered convenience both meaningful and rewarding (see Florman, 1976). So do many of those living in its mass production and consumer technofetish cultural penumbra. Otherwise, why would people so strongly resist changes to "our [engineered] way of life" and so many people on the periphery seek to move toward the center? What is to be done when a majority of people are so strongly committed to what Vanderburg and others think is an illusion? Those sympathetic to Vanerburg's diagnosis are faced with Kassandra's problem: What to do when nothing can be done?

5. Then there are the disruptions of deconstructive ecocriticism and ecomodernist engineering as challenges to conventional sustainability discourse. The

former is illustrated by Timothy Morton's manifesto for a new environmental romanticism that abandons any too-easy valorization of nature as other:

> *Ecology without Nature* argues that the very idea of "nature" which so many hold dear will have to wither away in an "ecological" state of human society. Strange as it may sound, the idea of nature is getting in the way of properly ecological forms of culture, philosophy, politics, and art. . . . Putting something called Nature on a pedestal and admiring it from afar does for the environment what patriarchy does for the figure of Woman.
>
> (2007, pp. 1 and 5)

By attacking the idealization of nature as other, Morton seeks to promote a wider environmental aesthetics, "from low to high, from pastoral kitsch to urban chic, from Thoreau to Sonic Youth."

Morton admits that his celebration of emotional connection with nature in all its forms, together with denial of difference between nature and artifice, runs the danger of being misread as a critique of deep ecology. In his own view, however, what he aims for is a "really deep ecology." "Ironically, to contemplate deep green ideas deeply is to let go of the idea of Nature, the one thing that maintains an aesthetic distance between us and them, us and it, us and 'over there'" (p. 204). As he argues elsewhere, the real practice of sustainability depends on accepting the fact that objects have become hyperobjects, and (as others might put it) we need to learn to love our monsters.

As counterfoil to this poetry of impurity and the engineering constructed is the sober practicality of "The Economodernist Manifesto" (2015, see http://www.ecomodernism.org) and its reenvisioning of Julian Simon's cornucopian futurism:

> To say that the Earth is a human planet becomes truer every day. Humans are made from the Earth, and the Earth is remade by human hands. . . . A good Anthropocene demands that humans use their growing social, economic, and technological powers to make life better for people, stabilize the climate, and protect the natural world. . . . In this, we affirm one long-standing environmental ideal, that humanity must shrink its impacts on the environment to make room for nature, while we reject another, that human societies must harmonize with nature to avoid economic and ecological collapse.

The sustainable way forward is not through some romantic emotional unity with nature but through greater separation. Humans must use their technology to become independent of nature and thus allow nature once again to thrive on its own. From the time when agriculture replaced hunting and gathering human technology and engineering have separated humans from and made them less dependent on nature. "Even as human environmental impacts continue to grow in the aggregate, a range of long-term trends are today driving significant decoupling of human well-being from environmental impacts." Through advancing artifice, from genetic engineering of crops to engineering design of materials for construction and nondestructive energy generation, humans are making themselves progressively less dependent on the natural environment. We must, as it were, create our Spaceship Earth on the Earth so that the Earth can go its own way without being exploited by humans.

DR. STRANGEGREEN OR HOW I LEARNED TO STOP WORRYING AND LOVE THE ANTHROPOCENE

Once it happened, what was the alternative? We had intentionally to accept what we had unintentionally engineered.

NOTES

1. *Memorabilia* I, i, 16.
2. The *Oxford English Dictionary*, second edition (New York: Oxford University Press, 1992), gives a first citation in Sowell (1972); Redclift (1994) includes remarks on the etymology of "sustainable" but fails to note the conversion from adjective to substantive. Ruse (1994) too facilely roots the idea in notions of a "balance of nature" and then simplistically identifies it with equilibrium. The concept of balance or stability in nature has been carefully critiqued by Shrader-Frechette and McCoy (1993), pp. 32–47.
3. For one summary review of such disagreements, see Worster (1993), pp. 142–55.
4. See, for example, Eliade (1959).
5. J. C. Squire, "In Continuation of Pope on Newton," *Poems* (1926).
6. Winston Churchill, Parliament Bill speech, November 11, 1947; in *Winston S. Churchill: His Complete Speeches, 1897–1963*, ed. Robert Rhodes James, vol. 7: 1943–1949 (New York: Chelsea, 1974), p. 7566.
7. For a complementary analysis of the theory of progress to that given in section one, see Mitcham (1989).
8. Meadows et al. (1972).
9. See Peccei (1977).
10. Mesarovic and Pestel (1974).
11. *World Conservation Strategy: Living Resource Conservation for Sustainable Development* (Gland, Switzerland: International Union for Conservation of Nature and Natural Resources, United Nations Environment Programme, and World Wildlife Fund, 1980).

12. The *World Conservation Strategy* argument is continued in Thibodeau and Field (1984).

13. World Commission on Environment and Development (1987). Page references in the text are to this volume.

14. Daniel Sitarz (1994), p. 4. One remarkable anomaly in this passage is the idea that environment and development were "seemingly unrelated" until *Our Common Future* made its case, since the Commission was explicitly set up to deal with this issue. Indeed, the relation of environment and development had been a major issue at a previous UN Conference on Human Development organized by Maurice Strong and held in Stockholm in 1972. There environmentalists had presented development as a threat to the environment, while Third World developmentalists, led by Indian prime minister Indira Gandhi, had responded by criticizing environmentalists as a threat to development. For publications representative of the official "anti-development" ideology see, for example, Friends of the Earth (1972); Jackson and Dubos (1972); and Strong (1973). The Rio Earth Summit of 1992, also organized by Maurice Strong, can be interpreted as a twentieth-anniversary attempt to bridge the gap that had opened at Stockholm.

15. See, for example, Carley and Christie (1993); Van den Bergh and Van der Straaten (1994); and Redclift and Sage (1994).

16. Rostow (1956).

17. For reference and analysis of the Truman speech, see Esteva (1992).

18. Rostow (1960), pp. 7–9. Rostow's theory grew out of an earlier study (1952; second edition, 1962) and was restated in an edited conference volume (Rostow, 1963) as well as placed in larger historical context in Rostow (1990). As an aside, the persistence of Rostow's view of sustainability is promoted by uses such as that provided by an article on "Sustainable Yield," which appeared in *Forbes* (December 14, 1987) the same year as the publication of *Our Common Future*. In the straightforward words of *Forbes*: "Unless you've got an awful lot of capital, living well off it without penalizing your heirs isn't easy" (p. 204). But, of course, for *Forbes* this means increase your capital investment—which is exactly what the transition to economic takeoff also requires. There is no recognition that this strategy could, in perhaps another sense, penalize one's heirs.

19. Arendt (1958), pp. 252–53.

20. Schumpeter (1942).

21. Lovins (1976 and 1977). See also Lovins (1975) and Nash (1979).

22. Among his many books, see especially Howard (1927 and 1940).

23. Rodale (1981), p. 14. This book was rewritten and revised from Rodale (1972). Although the 1972 version argues for an expanded idea of organic living, it is not as explicit as the 1981 volume.

24. See also in the same tradition Jackson et al. (1984).

25. Schumacher (1973). See also Wood (1984).

26. NRC (1989). For a less official but important complementary volume, see also Edwards et al. (1990).

27. The Worldwatch Institute was founded in 1974 by Brown, who had worked for ten years (1959–1969) as a Department of Agriculture international development

specialist and then for five years (1969–1974) for the Overseas Development Council. From the beginning Worldwatch was a no frills, savvy organization bucking some of the trends of traditional think tank culture. Worldwatch marketed its research as much as it went grant hunting for support. See Cohen (1986).

28. Hayes (1978).
29. Brown (1981), pp. xi and xii.
30. Milbrath (1989).
31. Vanderburg (1995).
32. Pörksen (1989).
33. Ableson (1991); Meadows et al. (1992), in which the "limits-to-growth" argument nevertheless remains essentially intact. *In Context* (1983–1995) was a publication of what became the Context Institute (context.org).
34. Relevant samples of this literature are Redclift (1987), Tisdell (1988), and Lélé (1991).
35. Ruttan (1988), p. 128.
36. Simon (1981), p. 286. See also Simon and Kahn (1984), pp. 34–37. Simon (1995) is an update of Simon (1981) that does not even bother to mention, much less discuss, the limits-to-growth thesis—or arguments for sustainability of any kind.
37. Meyers and Simon (1994), p. 65.
38. World Bank (1992), p. 34. Quoted from Sachs (1993).
39. The quotes are from an edited, privately circulated version of Sachs (1989), pp. 16–19. See also two edited volumes: Sachs (1992 and 1993). Of further relevance: Lemons and Brown (1995).
40. Fuller (1969). Explicit proposals for the high-engineering management of the environment are exemplified in the NASA "Mission to Planet Earth" as described by Malone and Corell (1989) and the special "Managing Planet Earth" issue of *Scientific American*, vol. 261, no. 3 (September 1989), which includes William D. Ruckelshaus's "Toward a Sustainable World," pp. 166–70, 172, and 174.
41. Sen (1981).
42. Tijmes and Luijf (1995), p. 328.
43. On this issue, a comprehensive study is Achterhuis (1988). For an English precis, see Achterhuis (1993). Compare Xenos (1989).
44. Achterhuis (1994).
45. Meadows et al. (1972), pp. 80–81.
46. English (1975), p. 226.
47. Georgescu-Roegen (1975), p. 369.
48. See, for example, Latouche (2010).
49. One important monograph on sustainability from outside the SPT community is Davison (2001).
50. See, for example, Lucena et al. (2010) and Bell (2011).

Essay 11

Engineering Ethics Education in the American Context

Retrospect and Prospect

In the United States, professional ethics has become a required topic in the engineering education curriculum. This practice has a history that invites reflection. The present essay is one modest contribution, oriented not as much toward historical knowledge or the advancement of pedagogical effectiveness as toward promoting critical reflection on engineering in a broad historico-ethical context. It aims for a retrospective on the recent emergence of engineering ethics, the place it now occupies in engineering curricula, coupled with a reflection on what place it might occupy in the future.

The argument is divided into four parts. Given the prominence of codes in engineering ethics education, part one briefly summarizes an interpretative model of code development leading to the contemporary consensus that engineers have a paramount responsibility to protect public safety, health, and welfare. Part two is a short interlude with some quantitative observations. Part three then considers the ways engineering ethics and its codes have been introduced into engineering curricula through textbooks. Against this background, part four speculates about possible transformations in engineering ethics, especially the idea of a policy turn and globalization, and in the process raises questions about the nature and future of engineering.

An implicit theme is that although the engineering curriculum has progressively enhanced the professional ethics component, it lacks the crucial intellectual dimension of a critical history of ideas about engineering and engineering ethics. It is remarkable, for instance, that engineers are said to be responsible for the protection of public safety, health, and welfare, but in fact seldom take—and are certainly not required to take—courses dealing with the historical and social character of public safety, public health, or societal welfare. It is apparently assumed that such notions are nonproblematic common knowledge.

Essay 11

A BRIEF HISTORY OF KEY IDEAS IN ENGINEERING ETHICS

The emergence of engineering ethics as a defined component of professional engineering may be framed as a simplified, three-phase trajectory:

- Phase one, late nineteenth and early twentieth centuries: Engineering ethics emphasized loyal obedience in an organizational context.
- Phase two, prominent mainly in the first half of the twentieth century: Engineering ethics briefly imagined as pursuit and promotion of technical efficiency.
- Phase three, from the mid-twentieth century to the present: Professional engineering holds paramount the protection of public safety, health, and welfare.

This analytic framework rejects a belief common among engineers that engineering history can be traced (as in Finch, 1951) back to the Romans and Egyptians or is (as in Koen, 2003) coeval with humanity. The conceptualization of engineering is a contested issue, with different approaches having been taken by engineers, historians, sociologists, philosophers, and others. In the present case, the historiographic position is that of Michael Davis (2002), who argues for an identification of engineering through social organization. In support, it may be noted how the wide adoption of the terms "engineer" and "engineering" are of post-1500 provenance and emerged to name a distinctively post-1500 activity that took shape when engineering as a type of military specialization was repositioned into the civilian sphere in conjunction with the Industrial Revolution. A key event in the historical construction of English-speaking engineering was the 1828 formal chartering of the British Institution of Civil Engineers (ICE), which dedicated itself to progressively "directing the great sources of power in nature [as revealed through science] for the use and convenience of [humans]."

In their early development, professional engineering associations—fragmented by technical (and social) boundaries into civil, mechanical, electrical, and other forms of engineering—included ethics only implicitly. Like their British counterparts, in their early years neither the American Society of Civil Engineers (ASCE, founded 1852), the American Society of Mechanical Engineers (ASME, founded 1880), nor the American Institute of Electrical Engineers (AIEE, founded 1884) informally promoted an ethos of professional solidarity as custodians of technical knowledge and subservience to organizational authority, taught primarily through the osmosis of apprenticeship and example. During the late 1800s and early 1900s, this implicit professional code of conduct began to be given explicit formulation.

Orthogonal to the articulation of codes stressing explicitly loyal obedience, there emerged an ideal of leadership in technological progress through pursuit of technical perfection or efficiency. As institutional economist Thorstein Veblen put it in *The Engineers and the Price System* (1921), there was an inherent tension between engineers, who aimed for efficiency in both industrial systems and consumer products, and business, which prioritized monetary profit. If engineers were freed from subservience to capitalist business interests, Veblen thought, their own standards would lead to the creation of a more sound economy and better consumer products.

Dissatisfactions with the ethical adequacies of both loyal obedience (too subservient to corporate interests) and technocratic efficiency (not consistent with democratic principles) prepared the way for a third phase. This phase took off after World War II, as engineers became aware of the social impacts of their work and of corresponding social responsibilities. The key idea from this period was the rise to codified prominence of a principle stressing the priority of protecting public safety, health, and welfare.

One locus of this development was the Engineers' Council for Professional Development (ECPD), founded in 1932 as an organization charged with bridging divides between multiple engineering specializations. Early on, the ECPD drafted a model ethics code for its constituent societies. Over successive iterations, from 1947 to 1977, this code eventually formulated the first of seven "fundamental canons" as: "Engineers shall hold paramount the safety, health and welfare of the public in the performance of their professional duties."

As the ECPD was developing its model ethics code, the American Society for Engineering Education (ASEE) was taking shape by way of the unification of two previous associations. In 1957 an ASEE Ethics Committee undertook to work with the ECPD and eventually the National Society for Professional Engineers (NSPE, founded 1934) to develop appropriate materials for teaching engineering ethics (Alger, 1958, offers a brief for the initiative). Across eight years of extensive consultation with practicing engineers, the committee worked "to describe our ethical problems and to show how thoughtful engineers think they should be solved" (Alger et al., 1965, p. vi). The published report, *Ethical Problems in Engineering* (1965), was a watershed in engineering ethics that collected 127 problems with practical commentary on possible responses by and for engineers working in different sectors (consulting, industry, government, and more). This was the first substantial casebook of engineering ethics problems designed specifically for pedagogical use.

In 1980 the education component of the ECPD was spun off as the Accreditation Board for Engineering and Technology (now simply named ABET) to certify engineering degree programs. Then at the turn of the century, as part of a broad overhaul of the accreditation process, a new set of accreditation criteria (ABET EC 2000) listed eleven outcomes for graduate engineers, one of which explicitly called for "an understanding of professional and ethical responsibility." At the same time ABET shied away from any explicit recommendations about the specific content or pedagogy of engineering ethics instruction.

ENGINEERING ETHICS: SOME QUANTITATIVE OBSERVATIONS

During the first two phases, engineering ethics was learned through apprenticeship and active membership in professional societies. There was little in the way of the explicit promotion of engineering ethics at the college level, certainly nothing by way of required courses, and virtually no reflection on the social dimensions of engineering practice or its philosophical assumptions. Mostly there was just implicit acceptance of the ideology of engineering as the underappreciated foundation of civilizational progress throughout history. Such books as Daniel Mead's *Contracts, Specifications and Engineering Relations* (1916) and Francis Harding and Donald Canfield's *Legal and Ethical Phases of Engineering* (1936) included chapters on personal ethics in the context of legally conducting engineering business. Law predominated over ethics, as in Mead, in which only one chapter out of twenty was devoted explicitly to ethics. And even there, as Mead candidly admitted, what was included was "inadequately treated. He did little more than advocate "the 'square deal' in the relations of the engineer and architect with the contractor and with all others with whom they have relations" (Mead, 1916, p. iii). North American developments up until the period after World War II were at best piecemeal and pragmatic, in accord with the larger character of U.S. intellectual culture, including its antiintellectualist streak (Hofstadter, 1963).

As ideas from the third phase percolated down into engineering education programs, it also become a field of sustained scholarly research and publication, as reflected in a collection by Michael Davis (2005) of 57 papers, mostly from the 1980s and 1990s. Davis's colleague, Vivian Weil (1984), associated the rise of engineering ethics with mid-1970s consumer and environmental protection movements and related social criticisms of science and technology but admitted difficulties in curricular institutionalization. One factor was uncertainties about precisely who were professional engineers. Was it only graduates of engineering colleges or licensed professional engineers, or maybe even those who acquired their knowledge and skill from apprenticeship should be included? Entanglements between engineering, business, and government created further ethical challenges.

An n-gram on the terms "engineering ethics" and "engineering and law" discloses two further aspects of the history. One is a persistence of discourse about engineering and law that goes back to the formation of engineering societies but with peaks in the second and early third phases of engineering ethics development. Another is rapid emergence of discourse about engineering ethics beginning, as Weil noted, in the late 1970s, but with an earlier more modest peak during the late-second and early-third phase. Engineering ethics first emerged as a widely discussed topic in English language textbooks around 1900, exhibited an initial spike in the 1920s and 1930s, then makes a dramatic jump in the late 1980s and after, corresponding roughly to the three phase model.

For a complementary retrospect, consider the *Journal of Engineering Education*, the journal with the highest impact factor in the field. In 1893 members of the American engineering community, in order to promote a shift from apprenticeship education to the formal teaching of applied scientific and technical knowledge, by sharing curricular experiences from different engineering programs, established the Society for the Promotion of Engineering Education (SPEE). For the next hundred years a series of SPEE serials under slightly different titles continued to reinforce the importance of formal technical education. During World War II the Engineering College Research Association (ECRA) was formed as a complementary organization (Anonymous, 1942), which after the war merged with SPEE to create a new combined American Society for Engineering Education (ASEE), which in 1993 transformed its less scholarly periodical *Engineering Education* into a research publication, *Journal of Engineering Education* (JEE). But it was not until ABET EC 2000 that engineering ethics began to become significant in JEE.

Another quantitative indicator of intensified engineering ethics discourse is the journal *Science and Engineering Ethics*, founded in 1995. During its first five years of publication, original papers were overwhelmingly oriented toward ethics in science (e.g., research ethics, conceptualization of responsible conduct of research, plagiarism, teaching scientific integrity). During the first 10 years (1995–2004), out of a total 356 papers, only 50 (or 14 percent) focused on engineering. During the next 10 years (2005–2014), engineering crept up slightly but not to any dramatic extent.

ETHICS INTO ENGINEERING EDUCATION

In the real-world context of professional engineering life in the United States, the new key idea of paramount obligation to protect public safety, health, and welfare could not help but give rise to a suite of particular problems. One way to bring ethics to bear on the analysis of these problems was to approach them piecemeal. A leading step in this direction was taken by the Center for the Study of Ethics in the Professions, the home base for both Michael Davis and Vivian Weil. With a grant from the Exon Education Foundation, Weil edited six pamphlets for classroom use on loyalty (Baron, 1984), responsibility for harms (Curd and May, 1984), technology assessment (Cameron and Milland, 1985), risk (Sagoff, 1985), whistle-blowing (Petersen and Farrell, 1986), and conflict of interest (Wells et al., 1986). Yet in only one instance (the Wells et al. volume) was there any direct collaboration with engineers; all other authors were philosophers or social scientists writing alone.

Already in the 1970s, however, individual engineers, especially professors of engineering, had begun to became involved, often working in tandem with academic philosophers. This was stimulated by a series of widely publicized cases perceived as examples of engineering negligence or improper subordination to economic interests as well as by federal funding for engineering ethics research. Among the leading cases for which textbooks were developed was a series of catastrophic DC-10 airliner disasters traceable to questionable engineering designs (Fielder and Birsch, 1992), an instance of whistle-blowing during construction of the San Francisco Bay Area Rapid Transit (BART) system (Anderson et al., 1980), and negligent design on the Ford Pinto automobile that contributed to a number of fatal accidents (Birsch and Fielder, 1994). Beyond the already mentioned environmental and consumer movements, protests against the Vietnam War (1960s), political corruption associated with Watergate (early 1970s), and the nuclear meltdown at Three Mile Island (1979) were other factors promoting the insertion of ethics into the classroom. Such factors contributed not just to the emergence of engineering ethics but also of a more general ethics of technology—including especially biomedical and environmental ethics as two major components of a broad applied ethics movement. To some extent, engineering ethics was just another manifestation of emergent applied, practical, and professional ethics activities.

The publication of three major engineering ethics textbooks between 1980 and 1983 marked out a new, more synthetic, interdisciplinary phase. All were involved in some way with federal government grant programs that, in turn, reflected a social sense of urgency and questioning of science, engineering, and technology.

The first was a two-volume collection coedited by Robert J. Baum and Albert Flores on *Ethical Problems in Engineering* (1978 and 1980). This publication was associated with a codirected National Endowment for the Humanities funded "National Project on Philosophy and Engineering Ethics" (see Flores, 1977; anonymous, 1977; and Baum, 1977) administered through the Center for the Study of Human Dimensions in Science and Technology of Rensselaer Polytechnic Institute (from which grew the RPI Department of

Science and Technology Studies). The NEH project was a three-year effort that sought to broaden the discussion of engineering ethics by bringing over a dozen two-person teams of philosophers and engineers to examine value issues-related engineering knowledge and practice. Team projects included the preparation of case studies on selected ethics problems, curricular development, and the drafting of recommendations for professional engineering societies. For the next decade, the Baum-Flores collection was the single best source of materials and remains an important historical document.

The second textbook was computer engineer Stephen H. Unger's *Controlling Technology: Ethics and the Responsible Engineer* (1982). Unger participated in the Baum project and received further funding from the new Ethics and Values in Science and Technology (EVIST) program at the National Science Foundation. (The creation of EVIST in 1972, with Baum as its first director, is further witness to the critical spirit of the times; see Hollander and Steneck, 1990.) Unger's involvement, on behalf of IEEE, in investigating the situation surrounding the BART whistle-blowers, and his association with the creation of an IEEE interest group on the Social Implications of Technology—subsequently the IEEE Society for the Social Implications of Technology, which publishes an important quarterly, *IEEE Technology and Society Magazine* (1982–present)—made him an influential presence in the profession. Unger's book surveys cases, argues the importance of professional ethics and for more vigorous ethics activities on the part of professional societies in support of practicing engineers, and puts forth his own model ethics code, appeared in a second edition in 1994 and remains in print. (For a much richer account of IEEE debates during this period, see Tang and Nieusma, 2017.)

Still a third important textbook creation from this period was produced by the philosopher-engineer team of Mike Martin and Roland Schinzinger: *Ethics in Engineering* (1983). The philosopher Martin and engineer Schinzinger volume, which has gone into multiple editions, takes an approach representative of the Anglo-American analytic style in philosophy; that is, it adopts a mixed utilitarian and rights-based ethical perspective and presents ethics as dependent on critical moral reasoning. At the same time, it makes a provocative argument for engineering as social experimentation, and seeks to draw out implications, although in a nonradical manner. It is one of the most widely used texts in the field.

Finally, a fourth important textbook that grew out of NSF engineering ethics support—this time, for the creation of a series of ethics education scenario cases—was written in 1995 by Charles E. Harris, Jr., Michael S. Pritchard, and Michael J. Rabins (the first two being philosophers, the third an engineer). The title, *Engineering Ethics: Concepts and Cases*, indicates its adoption of a pedagogical strategy that became standard. It also went into multiple editions.

Although there have been other additions to the engineering ethics textbook literature supplementing the work of Baum-Flores, Unger, Martin-Schinzinger, and Harris-Pritchard-Rabins, the last two of these have become the most widely used. (One primary complement is a reader anthology by Deborah Johnson, 1991.) Subsequent texts have to some degree institutionalized what may be described as a largely internalist and individualist focus—that is, a focus on individual professional responsibility to promote public safety, health, and welfare—using a mix of analytic ethics and case studies with some modest introduction of social implications, always with explicit reference to the ethical codes of various professional engineering societies. Textbooks consistently include some select set of professional engineering ethics codes.

In the early years of this federally funded engineering ethics research there occurred some historical events that became touchstone case studies for major pedagogical effectiveness. Following the 1984 industrial disaster in Bhopal, India, there was the space shuttle *Challenger* disaster of January 1986—to be followed three months later by the nuclear accident at Chernobyl, Ukraine. The *Challenger* disaster was personalized by mechanical engineer Roger Boisjoly, who—after being invited by Caroline Whitbeck's engineering ethics students at MIT in January 1987 to give an account his experiences—became a missionary for engineering ethics education. Having made this pitch first at MIT (see Boisjoly, 1991), he began to do so across the United States to anyone who would listen. His talk at the American Society for Engineering Education meeting in June 1988, for instance, was a moving testament that surely influenced eventual formulation of ABET EC 2000 standards. Rational arguments for taking engineering ethics into the classroom were being reinforced by historical circumstances and media attention.

CONTEMPORARY POSSIBILITIES: A POLICY TURN?

With the creation of ABET EC 2000 standards, however, phase three reached a kind of plateau, thus preparing the way for a possible phase four. It is always risky to attempt to characterize the present, but venturing such a risk, it might be called a "policy turn" in engineering ethics. The policy turn is defined by a dissatisfaction with individualist or personal professional ethics. Personal responsibility is necessary but not sufficient. Personal ethics—including personal professional ethics—must be complemented with, and on occasion action to transform, institutional arrangements and policy directives. Social institutions and policies are what set contexts for the pursuit and practice of engineering. Engineering ethics education is called upon to take these new dimensions into account.

A few nodal events in emerging support for such a claim would include the following:

- When engineering educators in Europe undertook to write their own first engineering ethics textbook (Goujon et al., 2001), they chose to distinguish their approach from that of their North American colleagues by including major sections on institutional ethics and public policy.
- Other publications pointing in this direction have been Richard Devon's (1991) argument for a "social ethics of technology" and Joseph Herkert's (2002) for a new focus on "macro-ethics" in engineering education, developing a concept from John Ladd (1980). Devons's social ethics and Herkert's macro-ethics share a lot in common with notions of policy.
- William Wulf's keynote address, as president of the National Academy of Engineering, at a 2003 workshop on *Emerging Technology and Ethical Issues in Engineering* (2004), gave official sanction to the interests of Devon and Herkert when he too called for complementing micro-ethics with macro-ethics.
- The third edition of the Harris–Pritchard–Rabins textbook on *Engineering Ethics: Concepts and Cases* (2005) included, for the first time, a discussion of policy issues.

These inflection points in engineering ethics can be associated with a general rise in the public profile of science policy, in research on science policy, and in the establishment of science policy centers such as the Consortium for Science, Policy, and Outcomes (at Arizona State University) and the Center for Science, and Technology Policy Research (at the University of Colorado, Boulder) that include philosophers. There is also an emerging discourse associated with the philosophy of science policy (Frodeman and Mitcham, 2002; Mitcham and Frodeman, 2006) and ethics policy (Mitcham and Fisher, 2012), which asks questions that could just as well be directed to engineering. Finally, there now exists a nascent movement in philosophy, especially in certain fields of applied ethics, that has been referred to as a policy turn in philosophy (Briggle et al., 2006). As in engineering ethics, the policy turn in applied philosophy argues that individualist ethics is not enough—and that philosophy has an obligation to become more involved with and willing to learn from public policy concerns and activities.

What might the implications of this nascent policy turn be for engineering ethics and engineering education in general? This is a difficult question that will only be speculated on by way of three comments:

First, the ideals of public safety, health, and welfare—the paramount values of professional English-speaking engineering—have histories in multiple senses. Safety, health, and welfare have been and will continue to be conceived differently by different peoples. To know something about these differences is necessary active intelligence in trying to live up to the core values. In this sense ethics always benefits from the history of ideas, since any ethical good has a history. But in the present case, one might state as a lemma to Melvin Kranzberg's fifth law: "All history of technology is important, but the social history of engineering ethics and its ideals is more important—at least to engineering ethics." (Kranzberg's fifth law: "All history is relevant, but the history of technology is the most relevant"; see Kranzberg, 1986.) It might even be added that there is very little in the way of such a history of ideas, which thus marks out a challenge and responsibility that could take science, technology, and society (STS) studies into modestly new territory.

Second, policy and policy initiatives also have intellectual histories and social contexts. Like "engineering," the very word "policy" has a conceptual and social history that is not without relevance to the theory and practice of a policy turn. The origins of the policy concept even have some parallels with that of engineering, and a case can be made that policy is peculiarly allied with the theory and practice of social engineering. Additionally, over the course of time and in multiple human situations, it is arguable that more policies run themselves out to failure than turn into successes. For anyone contributing to a policy turn in STS studies, in ethics, or in philosophy, to become acquainted with how policies have been formulated and worked out would appear crucial. Thus we could propose a second lemma on Kranzberg's fifth law: "All policy history is important, but the intellectual and social history of science, technology, and engineering policy is more important."

Third, any assumption in regard to engineering ethics that the policy of affirming responsibility for the protection of safety, health, and welfare is nonproblematic and is almost guaranteed to mislead. Appreciating the contingencies involved with the enactment and pursuit of various interpretations of such ideals could promote rethinking the character of engineering itself. Ultimately it might even be worth considering the possibility that the age of engineering as we have known it is coming to a close.

CODA: POST-ENGINEERING

The idea that the age of engineering as we have known it may be coming to an end has cropped up in a number of discussions related to engineering ethics broadly construed. Two examples are the previously mentioned reports from the U.S. National Academy of Engineering (NAE). In late 2001 the NAE undertook to envision the likely character of engineering in 2020 and then to assess engineering education in light of this vision. The first step was released in 2004; the second appeared in 2005.

The 2004 report began with two chapters outlining the technical and societal changes of the next decade as disclosed through reflection on four possible scenarios—the future as dominated by scientific revolution, or biotechnological revolution and social reaction, or natural disasters, or global conflict between civilizations. It then argued, to quote from the executive summary, for developing a new cadre of "engineers who are broadly educated, who see themselves as global citizens, who can be leaders in business and public service, and who are ethically grounded." This would entail enhancing "analytical skills, creativity, ingenuity, professionalism, and leadership" (NAE, 2004, p. 5).

An assessment of this assessment may begin with four observations. First, engineers are among the most self-reflective of professionals. Seldom do academic disciplines or professions analyze their possible futures with such zeal. Only rarely do chemists or physicists or biologists examine the future of these disciplinary professions. Slightly more common is such discussion among physicians and lawyers. But as Steven Goldman (1988) has documented, between 1918 and 1987 the engineering community issued no less than 21 significant self-assessments (many of which made passing reference to engineering ethics). Although salutary, such determined self-reflection no doubt reflects a measure of professional insecurity.

Second, engineering increasingly plays an ambivalent social role. Not only engineers but also economists sometimes present engineering as an engine of progress and necessary tool for international competitiveness. Such is argued at length by Norman Augustine and the Committee on Science, Engineering, and Public Policy (2007) in *Rising Above the Gathering Storm*. In support it cited the fact that the political leaders in China—with its rapidly expanding economy and rapid economic growth—have often been educated as engineers. But this evocation of the ideal of technocracy also points up how engineering has readily been captured by various authoritarian social, economic, or political interests—and the historical insensitivity of engineers to issues of social justice or human rights. Insofar as such insensitivity is inherent to engineering methodologies and stance, if it can be altered, will engineering still be engineering?

Third, the methodological base of the engineering of 2020 report is scenario planning. Scenario planning is creative and useful, within limits. In the present case there is a basic assumption that is insufficiently questioned: that engineering is an unquestionable good and should be advanced. The attempt to identify challenges was done only to adapt engineering to these challenges, not to consider the possibility that the age of engineering may be coming to a close—that the engineering profession does not in fact have the unquestionable social value that engineers would like. Critical reflection on the core analysis of this study points toward the following possibility: that engineering is just not (or will not continue as) what it used to be, that the received learned profession is properly waning and must relearn what it is. To give this prospect positive form, consider that possible futures of 2020 point, as much as for the need of a reformed conception of engineering, toward something we might call "post-engineering."

Finally, fourth, no one has provided a more penetrating sketch for the possibility of a post-engineering than MIT Dean of Undergraduate Education, Rosalind Williams (2002). In a provocative memoir focused around her five years as dean, Williams explores the self-contradictions in engineering, especially how engineering has undermined if not destroyed the ways of life on which engineering and engineers have from the 1500s depended. The phenomenon of "post-engineering" (a term Williams does not use) is the paradox of an "expansive disintegration":

> There is no "end of engineering" in the sense that it is disappearing. If anything, engineering-like activities are expanding. What is disappearing is engineering as a coherent and independent profession that is defined by well-understood relationships with industrial and other social organizations, with the material world, and with guiding principles such as functionality. . . . Engineering emerged in a world in which its mission was the control of non-human nature and in which that mission was defined by strong institutional authorities. Now it exists in a hybrid world in which there is no longer a clear boundary between autonomous, non-human nature and human-generated processes.
> (Williams, 2002, p. 31)

Surely this is a phenomenon on which a new engineering ethics, enhanced by an alliance with policy, should take into account.

The argument for such a post-engineering ethics and policy may be outlined as follows: A world transformed by the possibilities of continuing scientific revolution, or biotechnological revolution and social reaction, or new and emerging technologies, or concatenations of natural disasters, or global conflict between civilizations is indeed one in which—to quote from the main body of *The Engineer of 2020* report—"engineering schools may have to create new engineering degree programs to attract a new pool of students interested in a less rigorous engineering program as a 'liberal' education" and engineers must be "educated to understand and appreciate history, philosophy, culture, and the arts" (NAE, 2004, pp. 39 and 52). But the obvious question is: Why call such an expanded, liberal arts-based program or a program focused on engineering, history, social science, philosophy, culture, and the arts—why continue to call this engineering? Is this not like insisting that when alchemy was transformed into chemistry it should still be called alchemy? Or when natural philosophy was transformed into science, that it should have continued to be referred to as natural philosophy? What is the justification for such a rhetorical ploy? New realities call for new names. *Quo vadis* engineering?

Essay 12

Notes on Engineering Ethics in Global Perspective

(with Gary Lee Downey and Juan Lucena)

Globalization can refer to a multidimensional (economic, political, cultural, environmental) intensification of international connections and to taking a multidimensional, interdisciplinary perspective on particular problems or issues. In the case of engineering ethics and engineering education, both meanings are relevant. It is increasingly important for engineering ethics in any particular national context to take account of the globalization of engineering practice and to attempt interdisciplinary ethical responses to engineering challenges—two approaches that are not easily disaggregated. In the present essay, however, the focus will be on the former.

Professional ethics has become a well-established dimension of engineering education and practice in a number of countries—and may even be described as undergoing a process of globalization. Since the mid-1980s the U.S. Accreditation Board for Engineering and Technology (ABET) has required engineering programs to include the teaching of professional ethics—a requirement given specificity in 2000, with the stipulation that one of eleven demonstrable outcomes should be "an understanding of professional and ethical responsibility." That same year the Japan Accreditation Board for Engineering Education (JAABE) likewise began to require accredited programs to have an "understanding of . . . engineers' social responsibilities (engineering ethics)." Other contemporaneous expressions of commitment to engineering ethics can be found, as in the *Charte d'Ethique de l'Ingénieur* (Charter of Ethics of the Engineer) published by the *Conseil National des Ingénieurs et Scientifiques de France* (CNISF or National Council of Engineers and Scientists of France) in 2001 and the *Ethische Grundsätze des Ingenieurberufs* (Ethical Foundations of the Engineering Profession) issued by the *Verein Deustcher Ingenieure* (VDI or Association of German Engineers) in 2002.

Despite the obvious similarities of these initiatives, however, emergent interests in engineering ethics are the products of distinct historical trajectories. For example, in issuing its new criteria, ABET, which traces its history back to the 1930s, was concluding a decade-long modification of its accreditation system. By contrast, JABEE, which was only created in 1999, was introducing the concept and practice of accreditation for the first time. In the French case, the "Charter of Ethics of the Engineer" was a new document issued by an engineering alumni organization, with no direct implications for any educational curriculum. The "Ethical Foundations of the Engineering Profession" likewise had only indirect educational import yet derives from a history that goes back to a post–World War II revival of German engineering professionalism.

Such contrasts suggest that any robust understanding of engineering ethics calls for national comparative studies and reflection globally on the ethics of a profession that is itself transforming the globe. In conjunction with modest elaborations on the three cases just mentioned, the following notes are offered as contributions to this ongoing process. At the same time, it must be admitted up front that the interpretations here are limited and may well be distorted by the American cultural context even as they grow out of a desire to practice intercultural dialogue.

JAPAN: ENGINEERING AND PROFESSION AS HOUSEHOLD

In Japan engineering and science have not been treated as separate enterprises to the same extent as in the United States or Europe. For example, an early influential code-like document was the "Statement on Atomic Research in Japan" issued by the Japanese Science Council (which includes both scientists and engineers) in 1954. This statement set forth what became known as "The Three Principles for the Peaceful Use of Atomic Energy": All research should be conducted with full openness to the public, should be democratically administered, and should be carried out under the autonomous control of the Japanese themselves.

These principles reflected a desire among Japanese elites during the 1950s to find ways subtly to distance themselves from American interests and policies (recall that the Allied occupation ended in 1952). Immediately after World War II, the United States had broken up the *zaibatsu* (財閥) or family-controlled vertical monopolies that had been the basis of Japanese industrial modernization since the Meiji period, then prohibited all Japanese research in aviation, atomic energy, and any other war-related area. By 1951, however, following the communist victory in China and outbreak of the Korean War,

U.S. policy began to shift toward encouraging certain kinds of military-related science and engineering and the incorporation of Japan into the Western alliance. Yet the Three Principles were in opposition to, for example, the U.S. policy of secrecy in atomic research, and in order to avoid publicity and the possible development of opposition, the JSC statement was not initially translated into English. It is also a policy that, although formulated by scientists and engineers, was adopted by the government, reflecting significant sociopolitical connections between the Japanese technical community and the state. During this period the Japanese also began to create *keiretsu* (系列) of corporations with interlocking business relationships (Sumitomo, Mitsubishi, and Mitsui being major examples) to replace the suppressed *zaibatsu*. Many engineers worked and found their homes in these organizations in ways that naturally influenced their professional lives.

Beginning in the 1980s scientists and engineers developed an interest in ethics reflective of but with continuing distinctions from what was taking place in the United States. This is illustrated, for instance, by the JSC declarations on "The Basic Principles of International Scientific Exchange" (1988) and "The New Science Scheme: Science for Society and the Fusion of Humanity and Natural Sciences" (2003), both of which emphasized a responsibility on the part of scientists and engineers to promote sound scientific development and to help educate the public about important issues related to scientific and engineering developments. In 1999 the Japan Accreditation Board for Engineering Education (JABEE) was also established, the agency that has given special attention to engineering ethics education.

The promotion of professional engineering ethics instruction in the late 1990s was stimulated in part by some high-technology accidents such as a sodium leak at the Monju fast-breeder reactor (1995) and a disastrous criticality accident at the Tokaimura reactor (1999). In this respect, there are similarities to the situation in the United States, where DC-10 crashes and the *Challenger* accident contributed to the rise of ethics education to defend professional prestige and autonomy. But instead of quickly importing American individualism and an ideal of the heroic engineer into an interpretation of the Japanese engineering community, it might be better to test the relevance of anthropological studies such as Dorinne Kondo's *Crafting Selves: Power, Gender, and Discourses of Identity in a Japanese Workplace* (1990). Although Kondo's research draws primarily on her time working in a small artisan shop and does not mention engineers, her analysis nevertheless suggests a need to recognize factors in the engineering profession other than those operative in an American context.

One distinctive influence on engineering ethics in Japan is the role harmony plays in Japanese culture in general and for Japanese engineers specifically, in both their domestic and international professional relations.

A key term here is the kanji 家 that, depending on context, is pronounced *ie* (roughly house) or *uchi* (home); Kondo generally interprets *ie* as "household" and *uchi* as "belonging." In both cases, her point is that working in an *ie*, which includes more than a strictly biological family (in her case, a shop or business) is to be involved in more than an economic relationship. The term *ie* can refer to households as task performance units based on work and as sites of identity formation, which enjoin them members' loyalty and love, "company as family" (Kondo, 1990, p. 121). "The *ie* is not simply a kinship unit based on blood relationship, but a corporate group based on social and economic ties." Moreover, a person's identity does not so much exist prior to such a household as it is defined in terms of one's position within an *ie*. "Subordinating one's individual desires to that of the household enterprises takes on the character of moral virtue," Kondo observes. "Pursuing one's own plans and disregarding the duties toward the household smacks of selfish immaturity" (Kondo, 1990, p. 131). The household serves as a center for attachment or *uchi*.

In preparation for such belonging, students begin competing to demonstrate their qualifications for corporate household attachments long before entering higher education, in kindergarten or even preschool. The country is widely known for the extended preparation for entrance examinations to higher education institutions that determine life career paths (see Vogel, 1971). But these examinations are not just individual achievements so much as demonstrations of mature other-directedness developed through the disciplined acceptance of hardship. It is in this sense that preparation for the exam is about "polishing the heart" [*kokoro*]. As Kondo puts it, "In Japanese society generally, hardship is considered one pathway to mature selfhood.... [E]ndurance and perseverance are among the most frequently cited virtues in Japanese society.... Learning to stick to a task, no matter how difficult or unpleasant, thus strengthens the *kokoro*" (Kondo, 1990, p. 109).

Those who achieve the highest entrance examination scores are able to enter engineering programs at prestigious national universities. Yet in contrast with engineering students in the United States, once matriculated they have little left to do to warrant good employment. Japanese students typically regard their university years in more relaxed terms. Although those in engineering have more work than students in other degree programs, even for them university life constitutes something of a time-out from household duties. Having departed from the family household of origin they are transitioning to the corporate household that will serve as the basis of identity and obligation for the balance of their working lives.

This distinctive approach to reckoning identity and responsibility through household-like social groups has a long history, one nodal point of which was establishment of the Japanese nation-state under what is known in the West as the "Meiji Restoration" (1866–1869). Stimulated by the challenge of the West—as manifested, for instance, by U.S. Commodore Matthew Perry's forced opening of Japan to world trade in the 1850s—the new imperial government explicitly restructured Japan as a "family state" (Shibata, 2004, p. 76). Survival could best be assured through the fulfillment of obligations to a family state that made possible an unusual openness to adaptations from the West, such as industrialization, provided such imports can be given a Japanese form.

One example of an import followed by a process of Japanization was technical education. In 1868, Yozo Yamao, who had been studying abroad in Glasgow, returned to become vice minister of education with the goal of establishing an engineering school. The Imperial College of Engineering was founded in 1873 with Scotsman Henry Dyer recruited to serve as its head. The government then systematically replaced British professors with Japanese graduates until finally, in 1886, the college was merged into what is now the University of Tokyo (becoming the first engineering college in the world to be located within a university).

The University of Tokyo engineering program in turn has been a major source for managers and directors of the most powerful technology-based corporations. One of the first replacement faculty was Fujioka Ichisuke, a founder of Toshiba. Another corporation, Hitachi, had eleven directors prior to 1941, all but one a University of Tokyo engineering graduate. Other graduates founded Toyota and Nissan (Odagiri, 1998, pp. 143–46). Although the post–World War II occupation authorities dismantled the militaristic hierarchies of the Japanese *ie*, new forms of household formation emerged. As Prime Minister Suzuki Kantaro already proclaimed shortly after Emperor Hirohito announced Japan's surrender in 1945: "It is essential that the people should cultivate a new life spirit of self-reliance, creativity and diligence in order to begin the building of a new Japan, and in particular should strive for the progress of science and technology, which were our greatest deficiency in this war" (Morris-Suzuki, 1994, p. 161).

During the postwar period, according to Kondo, the "company as family" became the basis of the Japanese employment system and was characterized by welfare paternalism, seniority promotions, lifetime employment, and worker identification with the firm. The decades between 1945 and the early twenty-first century nevertheless presented numerous challenges to this social order, especially in corporations and among the engineers on whom so much of the Japanese economic success depended. Many studies have noted changes in and the weakening of corporate *ie* culture in the context of globalization. As Hideo Ohashi, of the Japan Federation of Engineering Societies, has argued, "Globalization forces Japanese companies to shift to [a] more globally competitive management system" (Ohashi, 2001, p. 2).

In this context (which deserves more elaboration than given here) it is possible to interpret late twentieth-century actions promoting professionalization as responses. As two leading interpreters of Japanese engineering have pointed out, until recently "*engineering as a profession* [was] unequivocally absent in Japan" (Luegenbiehl and Fudano, 2014, p. 616, their italics). When the corporation weakened as a household for engineers, the profession emerged as an alternative. In the words of Ohashi again, speaking to his fellow engineers, "We need a revolution of our consciousness, ... to respecting [ourselves as] professionals" (Ohashi, 2001, p. 3). When working engineers in the early 2000s began to take continuing education classes in engineering ethics, they also received information documenting engineering accomplishments. There was little emphasis on becoming whistle-blowers who risk job and career in the name of individual honesty and autonomous judgment. In the course of promoting ethical decision-making, professional engineering societies were offering themselves as new *uchi* or centers of belonging for engineers in order to help them struggle with change. In this sense, ethics is a means "to secure international acceptance of engineers' qualifications" (Kawashima et al., 2004, p. 101).

The new JABEE criteria for engineering education that include explicit mention of engineering ethics further fits this interpretation. In first place among the eight new standards for assessing engineering education programs is teaching "the ability and intellectual foundation for considering issues from a global and multilateral viewpoint." Second place goes to an ethics-related standard of learning to appreciate "the effects and impact of technology on society and nature, and of engineers' social responsibilities." In the Japanese engineering curriculum, what is primary is not the development of abilities in mathematics and science or in engineering analysis—the first in rank outcomes of ABET accreditation—but learning to consider issues from a point of view that rises above self-interest, overcomes selfish immaturity, and locates personal concerns and interests in relation to those of others engaged in the general pursuit of harmony.

Further evidence for the importance of *ie* over individualism in the new engineering professionalism can be found in a 1999 action by the Japanese Society of Civil Engineers to replace an anomalous statement of "Beliefs and Principles of Practice for Civil Engineers" from 1938 with a new "Code of Ethics for Civil Engineers" (Luegenbiehl and Fudano, 2014). The new "Code of Ethics" bypasses corporations and admonishes engineers to "adhere to the ethical principles of self-disciplined moral obligation when applying advanced technology" while repeatedly stressing responsibilities to society at large. The first provision, for example, states that the civil engineer shall "apply his/her technical skills to create, improve, and maintain 'beautiful national land,' 'safe and comfortable livelihood,' and 'prosperous society,' thus contributing to society through his/her knowledge and virtue with an emphasis upon his/her dignity and honor." A sense emerges of the professional engineering society as a household through which obligations can legitimately be formulated and fulfilled for the common good. The national movement to promote professionalization and ethics among engineers may thus be read as a move to establish the viability of a new household, the engineering profession, that functions as a replacement for corporate households while retaining primary obligation to the national household.

ENGINEERING ETHICS AS INSTITUTIONAL PROTECTION IN HONG KONG

Another distinctive case in Asia that can be briefly mentioned is that of the Hong Kong Institution of Engineers (HKIE, founded 1947). As a British Crown Colony, the professional organization of engineers in Hong Kong originally developed not just on the British model but as a branch of British institutions. During the 1970s, with the dawning realization that Hong Kong would in the near future (in 1997) be returned to Chinese sovereignty, local engineers undertook to provide Hong Kong with an independent engineering association. Part of this activity involved some intensive discussion of professional ethics, with a special conference being organized in 1980 on "Professional Ethics in the Modern World."

In 1984 the HKIE formally adopted a set of "Rules of Conduct" that differed in a few key respects from the parent organization. Although the primary obligation remained the responsibility to the profession, this was modified by the following statement: "When working in a country other than Hong Kong [the Hong Kong engineer should] order his [or her] conduct according to the existing recognized standards of conduct in that country, except that he should abide by these rules as applicable in the absence of local standards" (paragraph 1.9).

The basis of this modification had been clearly spelled out in a subsequent discussion. At an interprofessional symposium in December 1985, F. Y. Kan of the Hong Kong Institute of Surveyors identified the role of his professional association as the promotion of the status of surveyors and the usefulness of the profession. "So far," he is reported to have said,

> The role [has] not changed but, with the Sino-British agreement in operation [to return Hong Kong to Chinese sovereignty], there might be a tendency to a far-reaching effect on the professions. There was, therefore, a need to break away from U.K. qualifications. However, professional competence must be maintained and this could bring institutions into the political field.
> (Luscher, 1986, p. 39)

In a world where engineering readily comes into contact with the political field—something that is the case not only in Hong Kong—it can be useful for engineers to consider how such issues may have been negotiated by their technical colleagues elsewhere.

FRENCH ENGINEERS, PROGRESS, AND THE RATIONAL STATE

As noted, the French "Charter of Ethics of the Engineer" was not meant to become part of standard engineering curricula. Its creator, the CNISF, merely coordinates the activities of alumni associations for engineering schools and is not even particularly well known. At the same time, it is notable how this charter explicitly links engineering with the concept of progress, describing engineers as the source of innovation and the engine of progress: *"L'ingénieur est source d'innovation et moteur de progress"*—a view undoubtedly held by many U.S. engineers but is not a statement included in any American ethics code.

A key to understanding the disinterest in ethics in French engineering education rests with their elite status. As the French journalist Jean-Louis Barsoux (1989) explains, "In France, engineering education does not play second fiddle to medicine, law, or architecture—it is *the* recognized way to the top, both socially and professionally." Barsoux is referring to "state engineers," that is, those who work as high-level civil servants in the national government. Although these engineers have been in the minority at least since 1900, their high social prestige spreads to all others.

The ethics of French state engineers is the result of a multidimensional socialization process resting on a distinctive educational program. Students who want to become engineers have to undergo many years of rigid preparatory education and then compete for positions in elite schools, the so-called *grandes écoles,* by sitting for the *concours,* a combined written and oral exam whose scores are published in local newspapers and which determine who will be granted admission to which schools. In this respect, there are similarities with the Japanese system (although the postsecondary educational experience is quite different). The ethical character of this process is suggested by the fact that entry into a *grande école* is not called "admission" but "promotion," and that eventual graduates will forever identify themselves as cohorts based on their year of matriculation. Furthermore, the rankings continue throughout their studies, at which point the highest-ranked students remain on pathways leading to senior positions in government ministries. By entering an engineering school, prospective state engineers join a system in which they serve as both leaders and embodiments of French society.

In contrast with the theory of progress prominent in the United States—that its end is free-market individualism maximized by means of the low cost mass production of goods and services—since the Enlightenment the French view has been that the goal is rational unity achieved through sound mathematical principles. Such rational unification takes place best in government, protected from the special interests and restricted economic perspectives of private enterprises. Examples of the French commitment to rational planning are legion. As historian Cecil Smith writes, "Ever since the birth of the Corps des Ponts et Chaussees in the eighteenth century, French state engineers have promoted the complementary notions of rational public administration in the general interest and planning on a national scale" (Smith, 1990, p. 659). In another deep background historical account of the distinctive character of French engineering, Ken Alder concludes that these "engineers have been *designed* to serve" (Alder, 1997, p. 86).

During the early twentieth century, a group of graduates from the most elite of the technical schools, the *École Polytechnique* (aka *L'X*), established the think tank *X-Crise* to promote a philosophical alternative to capitalism, communism, and fascism. They called it "planism." Among them was Jean Coutrot, an engineer-intellectual and founder of the *Centre d'Etudes des Problemes Humaines* [Center for the study of human problems]. According to Coutrot, the leadership of engineers was rooted in engineering analysis: "It is to the engineers, today, that it falls to construct better societies because it is them and not the legalists or politicians who hold onto the necessary methods" (Clarke, 2001, p. 81). For Coutrot and other engineers who were concerned about the dehumanizing effects of mass production, communist collectivism, and fascist centrism, "The central problem of their time was the question of how to organize a society that was both rational and human" (Clark, 2001, p. 84). In many respects, what French engineers achieved in this instance was the imaginary ideal of the technocracy movement that emerged in the United States during the same period.

After World War II, state engineers secured complete jurisdiction over electricity, rail transport, and atomic energy, all in the name of rational national planning in the general or public interest. As Smith explains, "They acted as planners, economists, urbanists—'inter-ministerial generalists,' drafting legislation and then the decrees to implement it" (Smith, 1990, p. 692). The influence of state engineers spread through a greatly enlarged "para-public" sector that included electric power, gas, coal, banks, airlines, telecommunications, Renault, and the French national rail system, SNCF. "As true as it is that public engineers acted as an elite all too confident in the power of 'superior light' [*lumieres superieures*] to determine the 'general interest,'" Smith concludes, "it is no less true that for 250 years they sustained an ethos of public service rarely found elsewhere" (p. 693). This is an ethos acquired at the *grandes écoles*.

Since their eighteenth-century founding, engineering educators in the most elite *grandes écoles*—that is, the *École des Ponts et Chaussées* (1747), *École des Mines* (1783), and *École Polytechnique* (1794)—have placed the highest value on mathematical knowledge. For French engineers, demonstrating the ability, commitment, and discipline to become proficient in the mathematical foundations of engineering confirms that one has the moral character and reliability to warrant the trust of the Republic. Students who have been promoted into the national system of rational deliberation and action geared toward increasing social order have already demonstrated everything necessary to warrant a position of national leadership. They have mastered the principles and values that constitute engineering ethics in France; indeed, one might reasonably claim that engineering constitutes the dominant ethic of France. Further testifying to their status, the annual military parade on Bastille Day, which publicly celebrates the accomplishments of the Republic, is led by second-year students from the *École Polytechnique*. For engineers who have already demonstrated their character through competence, to be required to take a separate course in ethics would be superfluous if not insulting.

Why then did the collective organization of alumni associations feel any need to formulate and disseminate a code or charter? The move may be understood as one of many efforts in and around French engineering education to adapt to the increasing value accorded the private sector as a measure of national worth after the end of the Cold War. An American-led shift in the dominant image of international relations from grand conflict between two philosophies of political economy to a model of economic competitiveness based in nation-states has forced other countries to adapt to a North American model of progress oriented toward the free-market production of low-cost goods for mass consumption. In response, the *grandes écoles* reluctantly initiated international student exchange programs and new career pathways oriented more toward private industry. In particular, expecting engineers to participate increasingly in international workplaces beyond Europe, schools have begun expanding the nontechnical dimensions of engineering education.

It is in this context that "ethical reflection on the engineering profession" has gained a modest foothold. In 1995, the *Commission des Titres d'Ingenieurs* [Engineering titles commission], established in 1934 to protect the formal title "graduate engineer," updated its nontechnical requirements to include "foreign languages, economic, social and human sciences and a concrete approach to communication problems as well as providing openings to ethical reflection on the engineering profession" (*Comité d'Etudes sur la Formation des Ingenieurs*, 2000). This modest effort nevertheless has yet to generate much concrete activity.

GERMANY: ENGINEERING AND *BILDUNG*

The VDI "Ethical Foundations of the Engineering Profession" (which the VDI website translates less literally as "Fundamentals of Engineering Ethics") stresses in unique ways that engineers should "know the relevant laws and regulations of their countries" but should honor them only "insofar as they do not contradict universal ethical principles" (paragraph 1.3). Moreover, in cases of value conflict, engineers are admonished to choose "the values of humanity over the dynamics of nature," "human rights over technological implementation and exploitation," and "public welfare over private interests" (paragraph 2.4). How did universal ethical principles become such a major commitment—one much stronger than the U.S. commitment to protecting public safety, health, and welfare—and what does it mean for German engineers themselves?

The immediate interest among German engineers in the impacts and effects of technologies on humanity can be traced to the post–World War II period. Having been co-opted by the National Socialists during the 1930s, the VDI was revived in 1947 with an international engineering education conference on *"Technik als ethische und kulturelle Aufgabe"* [Technology as ethical and cultural task]. The problem for VDI members was precisely that they had accepted the ideal of what engineer-historian Thomas Hughes (1980) calls "culture-determined technology," in which they failed to challenge Nazi cultural leadership. A major postwar task was thus to break free from such a cultural captivity, a project that began with this conference and continued through an active collaboration with anti-Nazi German philosophers in a series of four additional meetings between 1950 and 1955 on the general theme of technology and humanity. Indeed, the strong collaboration between engineers and philosophers is itself a distinctive feature of German engineering culture.

As its contribution to this effort, the 1950 conference drafted an "Engineer's Confession" that employed a religious rhetoric to offer a vision of engineering as a spiritual vocation. According to this statement, the engineer "should place professional work at the service of humanity . . . [and] should work with respect for the dignity of human life and so as to fulfill his service to his fellow men without regard for distinctions of origin, social rank, and worldview." To include an explicit commitment to humanity as a whole constituted a self-criticism by German engineers, who previously had understood themselves as advancing civilization by serving Germany. At the same time, a significant continuity was the idea of technology as a vocation, the understanding of which points toward the distinctive German notion of *Bildung*, formal education oriented toward spiritual growth and perfection.

As a result of the suppression of political revolutionary movements in the mid-1800s, German culture and education became a major vehicle for the expression of German aspirations for unification. Already in 1807, philosopher Johann Gottlieb Fichte had argued in his *Reden an die deutsche Nation* [Lectures to the German nation] the significance of *Bildung* as a means to unify and develop Germany (First Lecture, five paragraphs from the end). Germany could become great and contribute to human development through a *Bildung* that was, however, conceived as grounded in and an extension of Greek and Latin culture.

Throughout most of the nineteenth century, the professions of law, medicine, philosophy, and theology monopolized *Bildung* in this strong sense because of the preparatory curriculum taught in the elite secondary schools or gymnasia. Only those students who had mastered classical studies in Greek and Latin philology were thought able to manifest the *Geist* or spirit that perfected human nature. The significance of this *Bildung* derived in part from its contrast with technical training and work. *Techniker* or technicians, who actually functioned like what in other countries were called engineers, underwent an educational program separate from that of the gymnasium, with the gymnasium degree or *Abitur* being the only path into the university.

Early attempts to enhance the cultural prestige of technical learning and work included creation of the Association for the Promotion of Technical Activity in Prussia (1821) by Prussian finance minister Christian Peter Beuth. Understanding *Bildung* and cognizant of the negative effects of industrialization on English workers and landscapes, Beuth sought to promote a distinctively German industrialism that imbued technology with art and emphasized aesthetics as an evaluative criterion (Brose, 1992). According to Beuth, industrialization would be acceptable in Germany as a site for the emancipation of *Geist* by means of a new form of *Bildung*. He thus unsuccessfully stipulated that art and aesthetics be included in the curricula of nascent technical schools serving the working classes of society.

An educational movement that proved more immediately successful involved establishing *Technische Hochschulen*, technical postsecondary schools or institutes, which included among their responsibilities fundamental research on *Technik*, a concept that covered both technical products and the technological processes for their production. First set up during the middle part of the century, the new institutes gained greater visibility and status after the 1870s during the unification of Germany under the Prussian-led Second Reich. Advocates for the technical institutes also established a new form of quasi-academic secondary education in *Oberrealschulen*, whose "realism" included teaching modern rather than classical languages.

In 1885 a commission of the VDI (which had been founded in 1856) concluded a review of the structure of German education and its implications for engineers by demanding that the courses students followed into and through the technical institutes have the same legal standing as those through the gymnasia to university. According to the Commission,

> The engineer in the eyes of many was—and partly still is—an advanced artisan, neither requiring nor deserving the higher *Bildung* offered by the Gymnasium. We declare that German engineers have the same needs with respect to their general *Bildung* and wish to be subject to the same standards as the other higher professions
>
> (Gispen, 1990, p. 146).

Kaiser William II approved this request in 1892 by giving *Oberrealschulen* graduates the right of admission to the engineering corps, in 1899 accepting them as eligible for employment in the civil service, and in 1900 granting them equal status to graduates of the classical gymnasium.

In the early twentieth century, members of this new professional engineering community defended the thesis that the emancipation of the human spirit included not just classical culture but also *Technik*. In *Lebentige Krafte* (1904), for instance, the German engineer Max von Eyth even argued contra Hegelian philosophers and Prussian lawyers that technology rather than reason should be seen as the vehicle for the unfolding of *Geist* or mind/spirit. As historian Jeffrey Herf summarized Eyth's view,

> There was more Geist in a beautiful locomotive or electric motor than in the most elegant phrases of Cicero or Virgil. Technology, like poetry, dominates matter rather than serves it. . . . [T]echnology was actually more cultural than culture itself.
>
> (Herf, 1986, p. 159).

Feeling empowered by an increasing national commitment to industry, engineers openly challenged the value of the universities and "praised their own achievements as 'national' ones and engineers as 'pioneers of German value and culture'" (Herf, 1986, p. 156). A German historical analysis focused on engineering during the Third Reich described leaders in the professional technical community as promoting an "anticapitalism of engineers" (Ludwig, 1974, p. 52). Such an engineering ideology easily adapted to National Socialism as political movement oriented toward a new German culture that promised to promote a technology free of purely utilitarian interests and a chaotic free-market capitalism.

During the Third Reich, engineers were enrolled in the designing and manufacturing of war technologies, including death camps. Post–World War II, when engineers were struggling to reconstitute and free the VDI of Nazi contamination and reposition their own sociocultural roles, they naturally wanted to extend their vision beyond any hypothetical German-ness to include humanity as a whole. In the 1970s this new sense of social responsibility led the German engineering profession to engage with the emerging discipline of technology assessment. German engineers undertook to conceptualize broad ethical responsibilities for assessing technologies according to eight metrics of value in three categories: functionality, economy, and material standard of living; safety, health, and environmental quality; and development of individual personality and quality of social life. The German term *Technikbewertung* [Technology evaluation] as a translation of "technology assessment" stresses going beyond the kind of restricted cost-benefit analysis that became a norm in the United States. Moreover, individual engineers were not left alone to evaluate technologies on the basis of personal conscience but were presented with guidelines that had been authorized by the engineering community as a whole in dialogue with philosophy.

The subsequent reworking of these guidelines in the 2002 "Ethical Foundations of the Engineering Profession," as with engineering ethics discussions in Japan and France, can be read as an effort to adapt to a world dominated by economic competitiveness, with its emphasis on low-cost production for mass use. On the one hand, German engineers were struggling to build a system in which technology evaluation is not only an ideal but also reduces costs. On the other hand, a reaffirmation of a responsibility to engage in technology assessment offered evidence that *Technik* is still about *Bildung*, which is the website tab under which its ethics code is found. Simplifying and reaffirming universal ethical principles is one way to achieve both ends.

ENGINEERING ETHICS AS SOCIAL REFORM IN SWEDEN

Efforts to critically assess relationships between professional engineering and national interests can be found in other European countries as well. In this regard Sweden provides the instructive case of a neutral country that used its engineering prowess to develop a strong military by relying on a well-developed domestic weapons industry. One of the leading weapons-producing corporations has been Bofors, a primary supplier of advanced field artillery, anti-aircraft artillery, and ship artillery to the Swedish armed forces. Known not only domestically but internationally for such technologies, in the 1960s Bofors increased its exports. In principle, exports of military weapons were prohibited. But the government could legally waive this restriction for special cases, which nevertheless became increasingly questionable.

In 1969 a young engineer named Ingvar Bratt, who had been educated at the Royal Institute of Technology in Stockholm, was hired by Bofors to work on the electronics of weapons-related projects including a missile and anti-aircraft gun, which were exported to Malaysia in 1977. During the 1970s and 1980s, however, Bratt became politically active and by 1982 was publicly opposed to all weapons exports, even approved ones.

Rumors arose that the unapproved countries of Dubai and Bahrain had acquired Bofors's missile technology. When Bratt discovered evidence that Bofors missiles had been exported to a company in Singapore, he shared this information with journalist Cecilia Zadig, who identified the company as an arms dealer. This suggested Singapore as a possible approved country through which Dubai and Bahrain were receiving arms. In 1984 Bratt resigned from Bofors and promoted a public debate leading eventually to passage of a Swedish whistle-blower protection law.

This exposé contributed to development of a new code of engineering ethics, one that downplayed company loyalty, which had been a focus in the previous code, and emphasized responsibility to "humanity, the environment, and society." In response to the view that engineers were often those who contributed to social or environmental problems, the new ethics code stressed the social and ecological responsibility of engineers. (This note relies heavily on Welin, 1992.)

THE DOMINICAN REPUBLIC:
AN ENGINEERING ETHICS FAILURE

Engineering ethics codes in developing countries provide still another point of comparison. Engineering, architecture, and surveying had been introduced to the Dominican Republic by the Spanish in the early 1500s, but engineering did not become a formal course of study until the 1900s, and there was little difference between engineers and architects until 1945 when the first professional engineering society, the *Asociación Dominicano de Ingenieros y Arquitectos* (AIDA) was formed. In 1963 AIDA was transformed to the *Colegio Dominicano de Ingenieros, Arquitectos y Agrimensores* (CODIA or Dominican Association of Engineers, Architects, and Surveyors) and was required by the government to draft a code of ethics. This code focused on relationships within the profession, with other professions, with clients, and the promotion of national interest. According to an analysis by STS scholar César Cuello Nieto (1992), there was neither mention of public safety, health, or welfare nor any reference to responsibility or concern for the potential negative effects of engineering on society or the environment.

When during the 1980s numerous engineering failures and catastrophes occurred as a result of professional negligence and corruption, there were calls for government regulation. But civil engineer Orlando Franco Battle argued that part of the problem rested with a weak tradition in professional engineering ethics and promoted new guidelines for the ethical and responsible exercise of the civil engineering profession, inspired by the code of the American Society of Civil Engineers. Yet Franco Battle was unable to bring about a change in this code.

Indeed, whether a reform of the professional ethics code would have substantially impacted the problem of substandard work was questionable. A survey by Cuello Nieto in 1990 revealed that most CODIA members had not read the existing code and even if they had, did not take it seriously. Engineering was thought to be simply the best-paying job in the country, with medicine is the most prestigious. This implied that engineers had chosen their profession simply for economic benefit—and, in fact, one survey revealed that two-thirds of the engineering professors thought societal interest was secondary to self-interest. According to Cuello Nieto, the situation cannot be understood apart from "the cultural context of Dominican society and its history of more than thirty years of rule by the Trujillo family dictatorship" (Cuello Nieto, 1992, p. I113).

ENGINEERING ETHICS AS ALTERNATIVE DEVELOPMENT IN CHILE

A second comparison of engineering ethics in a developing country can be found in Chile. In Chile, as in many countries other than the United States, professional codes such as the "Code of Professional Ethics of the Engineers of the Colegio de Ingenieros de Chile [Association of Engineers of Chile]," actually have the force of law, as a result of having been formulated, in this case, in response to general legislation calling for such codes in all professional organizations. Although the *Colegio* was founded in 1958, its code was not formulated until it was required by the authoritarian regime of Augusto Pinochet (1973–1990). At the same time, as Marcos García de la Huerta (1991) has argued, Pinochet's two-decade dictatorship severely compromised almost all professional practices. This is a degradation that García de la Huerta has himself worked to overcome by publishing one of the first studies specifically on engineering ethics in Latin America (García de la Huerta and Mitcham, 2001). It should be noted, however, that 15 years earlier a Spanish language engineering ethics textbook reflecting strong North American influence was published in Puerto Rico (Lugo, 1985).

The Chilean code, like many others, includes little by way of positive guidance. There is, for instance, no mention of any responsibility to public safety, health, and welfare. Instead, the code consists primarily of an extended list of actions that are contrary to sound professional conduct and that are thus punishable by professional censure. Among many unremarkable canons against conflict of interest, graft, and more, however, is one rejecting "actions or failures to act that favor or permit the unnecessary use of foreign engineering for objectives and work for which Chilean engineering is sufficient and adequate." Such a canon, emphasizing national interests, can also be found in other codes throughout Asia and Latin America, from India to Venezuela.

It is important to note that such a canon need not have simply nationalistic implications. Judith Sutz, for example, a computer scientist in Uruguay, in an essay raising important questions about the directions of information technology research in Latin America, argues,

> The basic question is, What do Latin American engineers want? Do they want to seek original solutions to indigenous problems? Or do they only want to identify with that which is more modern, more sophisticated, more powerful—disregarding real usefulness—in order to feel like they "live" in the developed world?
>
> (Sutz, 1993, p. 304)

Many countries experience a serious difficulty in addressing their own real problems. Driven by what René Girard (1965) calls mimetic desire, engineers and scientists often devote themselves to high-tech research that brings international prestige rather than to less glamorous but more useful tasks. One serious challenge to professional engineering in the age of globalization will be the extent to which various national and cultural differences can be maintained in the face of such pressures.

CONCLUSION: GLOBALIZED DIVERSITY

As these examples suggest, the progressive concern for engineering ethics in different countries may well be another manifestation of globalization. Because of their situation in the largest economy in the world, leadership in engineering ethics development in the United States undoubtedly influences other nations. But as with other aspects of globalization, border crossing often leads to subtle transformations. The fact that engineering ethics has been pursued in the United States to promote professional unity and autonomy does not mean that it will necessarily function in the same way in other sites.

In Japan, the early twenty-first-century interest in engineering ethics among professional societies and the promotion of ethics education by a new Japanese engineering accreditation organization offers a case of consciously imported influence, in part to achieve international recognition of domestic engineering programs. But engineering ethics in Japan can also be interpreted in terms of its relationships to the uniquely Japanese social institution of the *ie* household and efforts to develop the engineering profession as a household center of belonging alongside existing corporate households.

In Hong Kong, in anticipation of reunification with the People's Republic of China, engineers sought to create professional associations as what as sites for maintaining differentiation in the future. This has obviously been to some degree successful, insofar as (as of February 2019) the HKIE website avoids any explicit reference to the PRC. A milestones chronology does not mention the reunification of 1997 and a set of relations with professional organizations outside Hong Kong lists, in alphabetical order, "Australia, Canada, Ireland, the Mainland, New Zealand and the United Kingdom."

In France, formal education in engineering ethics has attracted little interest. Explicit courses in engineering ethics are seen as unnecessary if not insulting to those elite engineers who have already become civil servants dedicated to rational national progress. Indeed, in such a context, for nonelite schools to adopt education in engineering ethics might even be interpreted as an open admission and acceptance of subordinate status—although global competitive pressures as well as new pan-European efforts may be leading in this direction.

In Germany, a post–World War II reassessment of the relation between engineering and the traditional ideals of humanistic *Bildung* has led to a new commitment of engineers to the good of humanity as a whole. A longtime commitment to social responsibility through the production of high-quality technology further promoted adaption of technology assessment as a major feature of engineering ethics.

In an advanced country such as Sweden as well as in developing countries such as the Dominican Republic and Chile, engineering ethics has played a role of opposition to corruption, although in some situations more successfully than others. In Chile, the effort to distinguish technical development from too much imitation of rich countries could also be compared to Hong Kong's efforts to retain some independence from what are conceived as exogenous political forces.

Of further relevance to globalization of engineering ethics is the work of three transnational professional engineering associations: the *Unión Panamericana de Asociaciones de Ingenieros/União Panamericana de Associações de Engenheiros* (UPADI or Pan-American Federation of Engineering Societies), founded in 1949; the *Fédération Européenne d'Associations Nationales d'Ingénieurs* (FEANI or European Federation of National Engineering Associations), founded in 1951; and World Federation of Engineering Organizations (WFEO), founded in 1968. WFEO includes UPADI, FEANI, and a number of other regional engineering associations from the Arab world, Africa, South and Central America, Asia and the Pacific, and so on.

WFEO was midwifed by the United National Educational, Scientific, and Cultural Organization (UNESCO) and describes itself as "the internationally recognized and chosen leader of the engineering profession and cooperates with national and other international professional institutions in being the lead profession in developing and applying engineering to constructively resolve international and national issues for the benefit of humanity." As the most global of engineering associations, its model code of ethics has a globalizing influence. In the preamble to its model code, however, it recognizes that seldom does one size fits all. In WFEO words,

> The inherent nature of "professionalism" is that as engineers we always have a duty to others and an obligation to "do the right thing". Exactly who the "others are", and what the "right thing" is, will be a matter of continual balance. . . . Imposition of duties upon members which they cannot realistically satisfy and the inclusion of provisions which restrain commercial activity and have a negative effect is not the function of a code of ethics.

The explicit attempt to reject the idea that ethics should unduly restrain commercial activity may nevertheless reflect a rather specific idea about both "others" and "the right thing."

Recognition of how engineering ethics follows different trajectories with distinctive implications has implications for how to think about engineering ethics from within any one country or culture. Who openly advocates instruction in engineering ethics, who passively ignores such initiatives, and who openly resists may indicate something about both the function of ethics and those who are content with their current situation or even those who might be seeking change. In many instances, the real force of globalization depends on how its challenges are internalized—and the global dialogue that deserves to emerge from such differential internalizations.

ACKNOWLEDGMENT

The discussion of engineering ethics in the Japanese, French, and German contexts, along with some comments on globalization, draws text from Downey, Gary Lee, Juan Lucena, and Carl Mitcham, "Engineering Ethics and Engineering Identities: Crossing National Borders," in Steen Hyldgaard Christensen, Christelle Didier, Andrew Jamison, Martin Meganck Carl Mitcham, and Byron Newberry, eds., *Engineering Identities, Epistemologies, and Values: Engineering Education and Practice in Context*, vol. 2 (Dordrecht: Springer, 2015), pp. 81–98. I am grateful to my colleagues Downey and Lucena for their permission to use this material.

ADDENDUM: A FURTHER NOTE

During the second decade of the twenty-first century, the relationships between engineering ethics and globalization are slowly attracting increased scholarly attention. Two books deserving special notice:

- Colleen Murphy, Paolo Gardoni, Hassan Bashir, Charles E. Harris, Jr., and Eyad Masad, eds., *Engineering Ethics for a Globalized World* (2015), a collection of 15 articles identifying key challenges and pedagogical implications; and
- Heinz Luegenbiehl and Rockwell Clancy's *Global Engineering Ethics* (2017), a monograph exploring key principles through case studies.

An insightful analysis referencing work by contributors to these two volumes that proposes a framework for further discussion can be found in Qin Zhu and Brent Jesiek (2017). After noting different ideas of the global, they argue for distinguishing and considering the strengths and weaknesses of four approaches to thinking about engineering ethics in relation to globalization:

- arguments for constructing an ethics code based on universal human values and thus applicable across societies and cultures;
- arguments for a code articulating shared functional characteristics internal to the engineering profession that can thus apply globally;
- cultural studies, emphasizing the importance of cultural differences in ethical decision-making that therefore question the possibility of any universal ethical code; and
- ethics and social justice studies that challenge students and professionals to reconceptualize engineering in terms of global justice.

To explore a potential synthesis of these diverse approaches, Zhu and Jesiek analyze how engineering might profit from the histories and experiences of other professions such as business and medicine with regard to the challenges of professional ethics and ethics education in the context of globalization. Qin Zhu, it might also be mentioned, has done leading work on engineering ethics in China, one of the more important contexts otherwise ignored in this essay.

Essay 13

Humanitarian Engineering

(with David Muñoz)

More obviously than science, engineering is strongly influenced by context. Unlike scientists, who claim to produce knowledge cut free from particular contexts and cases, engineers engage particular settings by designing, constructing, and operating structures, machines, and diverse products, processes, and systems. Engineering can only be engineering in context.

One way engineers acknowledge contextual dependency is to describe their work as designing under constraints. Such constraints may be physical, economic, political, cultural, ethical, legal, environmental, and more—but are not usually constraints that engineers themselves have chosen. They are simply presented to engineers who must incorporate them into designs.

However, to some extent and at deeper levels engineers do choose constraints. They choose or accede to those disciplinary constraints that make them engineers—and to subtle and not-so-subtle ideological constraints of the sociohistorical settings in which they go to work. Could engineering also adopt the context and constraints of humanitarianism?

1. SHIFTING CONTEXTS AND CONSTRAINTS

Engineers began to detach themselves from a military context during the Industrial Revolution in Great Britain. In 1828 the Institution of Civil Engineers (ICE) described the distinguishing activity of civil (nonmilitary) engineers as "the art of directing the great sources of power in nature for the use and convenience of man." The appeal to "use and convenience" especially implicated attachment to a sociohistorical setting with distinctive ideological constraints, those of a new bourgeois capitalist economic culture.

Shifting interpretations of how best to operationalize "use and convenience" in the profession have characterized engineering ethics ever since. For roughly the first hundred years, until the early 1900s, an operating assumption was that use and convenience was realized through industrial capitalism. The engineering ability to redesign the world by intensifying human exploitation of nature through industrial economic expansion was the most effective way to achieve use and convenience as an ideal of the newly moneyed middle class, sandwiched between the old aristocracy of inherited wealth and lower-class wage dependency.

New wealth is conveniently produced by engineering the mass production of goods that are themselves designed to be easy and convenient to use. For the owners, at least, it is use and convenience all the way down: use and convenience characterizing both process and product, means and ends, and undergirding lives less dependent on society than ever before in human history (lives of self-made individuals). John Galt, the hero of individualist libertarian philosopher Ayn Rand's *Atlas Shrugged* (1957), was an engineer. So was Wernher Von Braun, the German-American hero of the space race (but before that of Nazi rocket engineering).

Engineering has continued to be closely involved with both bourgeois conspicuous consumption and nationalist techno-economic projects. Many people—from engineers themselves to politicians—have argued that economic development if not civilization is synonymous with engineering achievement.

This was the vision of U.S. president Harry S. Truman, for instance, in the famous "point four" of his inaugural address of January 20, 1949. After committing the United States to supporting the United Nations, economic recovery, and opposing (Communist) aggression, Truman announced,

> Fourth, we must embark on a bold new program for making the benefits of our scientific advances and industrial progress available for the improvement and growth of underdeveloped areas. . . . Our aim should be to help the free peoples of the world, through their own efforts, to produce more food, more clothing, more materials for housing, and more mechanical power to lighten their burdens.

Although the word is not in Truman's speech, engineering clearly constituted an unstated foundation for what he had in mind. Truman injected engineering into an American nationalist agenda of what became the Cold War context. Engineers had won World War II (with the atom bomb) and now should win the peace through exporting technical development. Indeed, well before Truman gave the notion American expression, the United Kingdom had been enrolling engineering prowess in its predecessor colonialist enterprises.

Truman's point-four commitment was subsequently generalized by policy adviser Walt W. Rostow in *The Stages of Economic Growth: A Noncommunist Manifesto* (1960). Again, although he did not explicitly reference the term, engineering plays a key role in Rostow's sequence from traditional societies and the "preconditions for take-off" to "take-off," "drive to maturity," and "high mass consumption." Then, as part of their rebellions against imperialist control by countries in Europe, peoples in the colonies began initiating their own engineering take-offs.

Yet late in the same century in which engineering emerged as a civilian profession, there arose a number of movements critical of different aspects of capitalist industrialization. The unfairly disparaged "utopian socialism" of Robert Owen (1771–1858), the labor movement, and the revolutionary socialism of Karl Marx (1818–1883) are all cases in point. Humanitarianism is another instance, and as a criticism of some of the implications of nationalism and imperialism, it invites English-speaking engineers to consider the possibility of a contextual alternative to that with which their professional lives have been so entangled.

In fact, as questions began to be raised about the use and convenience of unregulated capitalist production, professional subservience was qualified. By the early twenty-first century a new interpretation of use and convenience had emerged as a norm: protecting public safety, health, and welfare. Within constraints, engineers were argued to be those professionals best able to help design and construct a world to realize public desires for health and welfare (use) with safety (convenience). The focus here will be on humanitarianism as a sociohistorical movement that presents still another possible context of relevance to engineering understood.

2. HUMANITARIANISM IN HISTORY

> Theory cannot equip the mind with formulas for solving problems nor can it mark the narrow path on which the sole solution is supposed to lie by planting a hedge of principles on either side. But it can give the mind insight into the great mass of phenomena and their relationships, then leave it free to rise into the higher realms of action.
>
> —Carl von Clausewitz, *On War*, book VIII, chapter 6

Paradoxically, it is not out of place to begin a review of humanitarianism with a quotation from one of the great modern war theorists. Not only has modern warfare given impetus to the humanitarian movement, but the point made by Clausewitz applies as much to humanitarian action as to warfare. There are no formulas for solving all problems of protecting human safety, health, and welfare in either humanitarianism or engineering. In fact, it was shortly after the crystallization of English-speaking engineering that a humanitarian challenge emerged to some of its local, unstated contextual assumptions. But this challenge calls for some conceptual differentiations.

Humanitarianism, Humanism, and Human Rights

Like engineering, humanitarianism is a complex phenomenon with a variegated history. First, philosophical humanitarianism must be distinguished from theological humanitarianism. Theological humanitarianism designates a Christian doctrine that affirms the humanity and denies the divinity of Jesus Christ. This was the original usage of the word, but not one of concern here.

Yet philosophical humanitarianism can take on religious connotations. In the mid-1800s, partisans of what was called the "religion of humanity" were often termed "humanitarians." Humanitarianism in this sense professes, absent revelation, a commitment to or faith in the advancement or perfection of the human race, as with the German philosopher Johann Gottfried Herder (1744–1803).

Philosophical humanitarianism also differs from simple humanism. Humanism can involve one or more of three claims. Claim one is that there is something about being human, which transcends all particular forms of humanity. In William Shakespeare's *The Merchant of Venice* (from the late 1500s) the Jewish banker Shylock responds to those who see him as less than human by saying, "Hath not a Jew eyes? Hath not a Jew hands, organs, dimensions, senses, affections, passions! . . . If you prick us do we not bleed?" (act III, scene 1). This Renaissance humanism challenged a natural tendency to limit moral sympathy by religion, race, or class.

Claim two is that human beings are of unique or special significance in the world, with some versions going so far as to conceive humans as the only beings of significance. This is the metaphysical theory of existential philosopher Jean-Paul Sartre (1905–1980) when he declares, "There is no other universe than a human universe, the universe of human subjectivity" (*L'Existentialisme et un humanisme*, anti-penultimate paragraph).

Still a third claim is that of secular or scientific humanism, which may well affirm the special reality of humans but also sees this reality as part of nature. In the words that open "Humanism and Its Aspirations" (2003, and sometimes referred to as "Humanist Manifesto III"), "Humanism is a progressive philosophy of life that, without supernaturalism, affirms our ability and responsibility to lead ethical lives of personal fulfillment that aspire to the greater good of humanity."

Although not incompatible with humanism in the second and third senses, humanitarianism as an ethical and political orientation need not involve either a metaphysical elevation of human beings or a scientific naturalism. Its minimal foundation is something closer to the first claim, with criticism of narrow forms of human community, especially nationalism. One succinct definition describes humanitarianism as active sympathy or compassion for all humans in need. Efforts to aid people in crisis situations, whether caused by other humans or nature, are commonly seen as exemplars of humanitarian action.

A commitment to humanitarian action should also be distinguished from a commitment to human rights, although the two are often associated. Humanitarianism typically involves an effort to alleviate human suffering by responding to human needs, but not necessarily on the basis of respect for human rights.

Humanitarian Universalism

There are many roots for the humanitarian criticism of restricted forms of community and the promotion of equality among humans. One is the cosmopolitanism of classical Greek and Roman philosophy. In contrast to the idea that our primary social bonds are to our fellow citizens in a *polis* such as Athens or Syracuse, some philosophers argued that the whole *cosmos* constitutes a kind of *polis*, making humans (or at least philosophers) members of a single community. Diogenes of Sinope (c.400s BCE), when asked his citizenship, is reported to have answered, "I am a citizen of the world" (*kosmopolitês* in Greek).

Another root is Christian missionary theology, which, from St. Paul to the German theologian and medical missionary Albert Schweitzer (1875–1965), has argued a supernatural version of universalism. Insofar as all human beings are created by and equal in the sight of God, they are individual members of a common community with obligations to care for one another. Louis Dumont (1986) has traced the religious to secular heritage and radical implications for the social order of this distinctly Western anthropology.

Still other roots are in the moral principles of the Enlightenment in both the empiricist and rationalist traditions. The Scottish empiricist philosopher David Hume (1711–1776) argued both the central importance of use and convenience and sympathy as the foundational moral sentiment. Use and convenience are pleasing because of our benevolent sympathy for others. The rationalist German philosopher Immanuel Kant (1724–1804) argued for recognition of a categorical obligation to treat all humans as ends in themselves.

Efforts to create a transnational community of scientists who share a common pursuit of knowledge and the labor movement organization of workers who are equally victims of economic oppression have been further influences. It would be interesting to consider the extent to which there is or is not a corresponding transnational community of engineers.

Yet none of these historical phenomena adopted the term "humanitarianism." Indeed, in its initial secular uses in the early 1800s the term was largely derogatory, as denoting excess in the promotion of humane principles over more realistic or patriotic ones. People more concerned about the poor in a foreign country than the welfare of their own families were disparaged as humanitarians. In the late 1800s, however, the term begin to take on positive connotations, as when the American sociologist Lester F. Ward described humanitarianism as aiming "at the reorganization of society, so that all shall possess equal advantages for gaining a livelihood and contributing to the [common] welfare" (1883, p. 450).

3. FIVE PHASES IN MODERN HUMANITARIANISM

For present purposes, humanitarianism will be restricted to a particular movement that has used the term to define itself. The history of a self-identified humanitarian movement can be outlined (drawing on Smyser, 2003) in five overlapping phases in the development of an active compassion for the basic needs of all persons irrespective of national or other distinctions.

Phase I (1800s): Rise of the Humanitarian Movement Proper

The humanitarian movement as such originated in the mid- to late 1800s in conjunction with the rise of the profession of nursing through the work of Mary Seacole (1805–1881) and Florence Nightingale (1820–1910) in the Crimean War (1854–1856) and Clara Barton (1821–1912) in the U.S. Civil War (1861–1865). But the key event was the reaction of Swiss businessman Henri Dunant (1828–1910) to the Battle of Solferino (1859), which ended the Second Italian War of Independence.

In nine hours of fighting, the Battle of Solferino resulted in approximately 30,000 Austrian, Italian, and French casualties. When Dunant, on a business trip, accidentally witnessed the industrial carnage of the Solferino battlefield and the way medical personnel from each army attended only to their own injured, he was stimulated to imagine a new kind of medical care that would address need irrespective of national identity. This vision led in 1863 to creation of the International Committee of the Red Cross/Red Crescent (ICRC), which now defines itself as "an impartial, neutral and independent organization whose exclusively humanitarian mission is to protect the lives and dignity of victims of armed conflict and other situations of violence and to provide them with assistance." Additionally, as it states on its website (icrc.

org) the ICRC "endeavours to prevent suffering by promoting and strengthening humanitarian law and universal humanitarian principles."

Phase II (Early 1900s): Humanitarianism beyond the Battlefield

In a second phase, the first half of the twentieth century witnessed development of new forms of humanitarianism expanding beyond medical care for military casualities in warfare. The ICRC became concerned with the plight of civilian noncombatants and for persons caught in natural disasters. New models of humanitarianism are exemplified in the work of Norwegian scientist and explorer Fridtjof Nansen (1861–1930) and of U.S. mining and civil engineer Herbert Hoover (1874–1962): Nansen in post–World War I work resettling refugees under the auspices of the League of Nations, and Hoover in relief work during and after the war as well as during the Great Mississippi Flood of 1927.

This period also witnessed the emergence of humanitarian NGOs other than the ICRC: for example, Baptist World Aid (1905), American Friends Service Committee (1917), Catholic Medical Mission Board (1928), Save the Children (1932), OXFAM (1942), and CARE (Cooperative Action for American Relief Everywhere, 1945). At the same time, the ICRC's knowledge of the Holocaust, war crimes, and crimes against humanity during World War II—along with its principled refusal to reveal these to the world because of its respect for national sovereignty—raised fundamental questions. Creation of the United Nations (1945) and international adoption of the Universal Declaration of Human Rights (1948) provided a further basis for questioning the primacy of national sovereignty.

Phase III (1950s–1960s): Humanitarianism as Free World Ideology

In a third phase, something like humanitarian development became a kind of free-world ideological alternative to communism. This was the explicit proposal of Truman and was embodied as well in the European Recovery Program or Marshall Plan (1947–1951). The creation of international agencies such as the United Nations Educational, Scientific and Cultural Organization (UNESCO, 1945), the UN International Children's Emergency Fund (UNICEF, 1946), the UN High Commissioner for Refugees (UNHCR, 1950), the Organization for Economic Cooperation and Development (OECD, 1961), the U.S. Peace Corps (1961), the World Food Programme (WPG, 1963)— together with a series of UN peacekeeping actions (India–Pakistan, 1949; Suez, 1956; Congo, 1960; et al.)—combined to give humanitarianism the character of an international anticommunist program.

Phase IV (1970s–1990s): Alternative Humanitarianisms

Beginning in the late 1960s, a fourth-phase humanitarianism began to separate itself from close association with anti-communism. A key event was the Nigerian Civil War in the breakaway province of Biafra (1969), which became the first televised international humanitarian crisis. The experience within the disaster relief community was one of gut-wrenching paradox: providing relief only enabled killing to continue and become more murderous.

Under such conditions, humanitarian aid workers began to challenge even more strongly the principle of respect for national sovereignty. Aid workers began to openly criticize governments on both sides of the civil war and governments outside the conflict supporting one side or the other. This crisis of conscience in the humanitarian community catalyzed the founding by the French physician Bernard Kouchner of *Médecins sans Frontieres* (MSF or Doctors without Borders) in 1971. MSF, which rapidly became the largest nongovernmental relief agency in the world, grew out of dissatisfaction with the inability of the ICRC to react independently of national government controls and its tendency to remain within safe boundaries; MSF refused to be limited by state sovereignty.

Phase V (2000–present): Humanitarianism Globalized and Questioned

Finally, in the context of the end of the Cold War (early 1990s) there arose two different trajectories in humanitarianism. The first, positive view can be called globalized humanitarianism. This trajectory is best represented by the "United Nations Millennium Declaration" (2000), in which memberstates recognized, "in addition to separate responsibilities to [their] individual societies, . . . a collective responsibility to uphold the principles of human dignity" and a duty "to all the world's people, especially the most vulnerable" (section I, paragraph 2). The national signatories affirmed a belief that

> the central challenge we face today is to ensure that globalization becomes a positive force for all the world's people. . . . [O]nly through broad and sustained efforts to create a shared future, based upon our common humanity in all its diversity, can globalization be made fully inclusive and equitable. These efforts must include policies and measures, at the global level, which correspond to the needs of developing countries and economies in transition and are formulated and implemented with their effective participation.
>
> (section I, paragraph 5)

The "Millennium Declaration" was extended into the Millennium Project, commissioned by UN Secretary General Kofi Annan in 2002 to develop a concrete action plan to eradicate the most extreme poverty by 2015.

A second, skeptical trajectory focuses on the persistence and even intensification of strident forms of nationalism and the rise of non-state-actor terrorism. From this perspective, the crisis of humanitarianism has become only more acute. Worry that ethnic-based nationalisms could lead to indefinite warfare was part of a European resistance to military action led by the United States in the Balkans. Additionally, the "humanitarian war" against Yugoslavia during the Kosovo campaign (1999) called into question the whole meaning of humanitarianism, as has the so-called war on terror that became a central feature of international relations post 2001.

Perspicacious articulations of the early twenty-first-century crisis in humanitarianism can be found in David Rieff and David Kennedy, two persons who have been deeply involved with the humanitarian movement. As Rieff observes,

> Humanitarianism is an impossible enterprise. Here is a saving idea that, in the end, cannot save but can only alleviate. . . . For there are, as Sadako Orgata, the former head of UNHCR, put it, "no humanitarian solutions to humanitarian problems." More than that, the pressures on humanitarian workers . . . have become all but intolerable.
>
> (Rieff, 2002, p. 86)

Or, in the words of Kennedy,

> We promise more than can be delivered—and come to believe our own promises. We enchant our tools. . . . At worst, . . . our own work [contributes] to the very problems we hoped to solve. Humanitarianism tempts us to hubris, to an idolatry about our intentions and routines, to the conviction that we know more than we do about what justice can be.
>
> (Kennedy, 2004, p. xviii)

At the same time, in spite of these well-recognized difficulties and failings, it is important to acknowledge many successes. As other observers have argued, humanitarian action has made important contributions to the lives of many of the powerless and poor (see, e.g., DiPrizio, 2002; Minear, 2002; and Terry, 2002). To recall again the counsel of Clausewitz, "Theory cannot equip the mind with formulas for solving problems nor can it mark the narrow path on which the sole solution is supposed to lie [although] it can give the mind insight into the great mass of phenomena and their relationships."

Essay 13

4. HUMANITARIAN ENGINEERING

> Technology influences disaster aid in two ways: First, it enables agencies to devise quick, although not necessarily appropriate, solutions to needs. Second, it influences the way in which needs are perceived and thus indirectly shapes many approaches.
>
> —Frederick Cuny, *Disasters and Development*
> (1983), pp. 138–139

It is against the dual background of engineering as context-dependent and the new context of humanitarianism that "humanitarian engineering" emerged. A seminal figure in its emergence was the civil engineer Frederick (Fred) Cuny (1944–1995).

The Fred Cuny Story

Cuny's story has been told a number of times, most fully in Scott Anderson's *The Man Who Tried to Save the World* (1999). Cuny studied civil engineering at Texas A&M University and on graduation took a job with constructing the new Dallas-Fort Worth international airport (begun in 1966). But he quickly went in search of a more fulfilling work and wound up flying relief supplies for CARE into the civil war between Nigeria and the breakaway region of Biafra (1967–1970). After the war he returned to Dallas, founded the Intertect Relief and Reconstruction Corporation, and became involved in a series of disaster relief operations.

One of Cuny's beliefs was that there was a distinct place for engineering in humanitarian disaster relief. More than physicians and nurses, engineers were also needed. Even in the field of development, which was dominated by agricultural specialists and agronomists, engineers were useful. In some ways he picked up and carried forward (mostly without knowing it) alternative technology ideas from the 1970s, such as those associated with E. F. Schumacher's *Small Is Beautiful* (1973).

Cuny's subsequent work in the Nicaraguan and Guatemalan earthquakes (of 1971 and 1976, respectively) led him to formulate what became known as the "Cuny approach" to using disasters as a catalyst to improve people's lives. Further disaster and development work undertaken during the Sudan–Ethiopia famine (1985) and with the Kurds in Iraq (1991); during the Somalia relief operation (of 1992); and to repair a water system during the siege of Sarajevo (1993–1994) extended Cuny's influence. Awarded a MacArthur Foundation Fellowship for 1995 to recognize "hard-working experts who often push the boundaries of their fields in ways that others will follow," Cuny was assassinated in Chechnya before he could be notified (Arenson, 1995).

Stimulated by the ideals of MSF and Cuny, the early 1990s witnessed creation of a humanitarian engineering NGO network. Using some permutation of the name "Engineers without Borders" (EWB), engineering students and their professors began independently to explore possibilities for engineering in the context of humanitarianism in diverse localities: *Ingénieurs Sans Frontires* (France), *Ingénieurs Assistance Internationale* (Belgium), *Ingeniería Sin Fronteras* (Spain), *Ingeniører uden grænser* (Denmark), and *Ingenjörer och Naturvetare utan Gräner*—Sverige (Sweden), *Ingegneria Senza Frontiere* (Italy), and others. At the same time, engineering and technology came increasingly to be recognized as able to play a crucial role in humanitarian action (see Cahill, 2005).

Other Precursors and Influences

Cuny's life and work may have exercised a strong influence on the humanitarian engineering ideal, but there were other precursors and influences as well. Two engineers who became U.S. presidents are cases in point.

Herbert Hoover (1874–1964), a mining engineer from the first class of Stanford University, became the 31st U.S. president (1929–1933) primarily on the basis of his success as the organizer of European relief work during and immediately after World War I (1914–1918) and then again as the first secretary of commerce (under Presidents Warren Harding and Calvin Coolidge) especially during the Great Mississippi Flood of 1927. Although disaster relief was a remit of the secretary of commerce, the governors of affected states requested that Hoover lead an effort that became a model for mobilizing and coordinating volunteers, local authorities, and national agencies. With a grant from the Rockefeller Foundation, Hoover also set up health units that helped address problems of malaria, pellagra, and typhoid fever throughout the region. From Hoover's own perspective, his success was grounded in his training as an engineer and his commitment to the promotion of efficiency through the use of analytic experts to identify problems and propose solutions.

Jimmy Carter (born 1924), the 39th president (1977–1981), graduated from the U.S. Naval Academy at Annapolis with a BS in engineering. After his term as president, he became a well-known humanitarian working, for instance, with Habit for Humanity and home construction that could be presumed to draw on his engineering skills. Carter was awarded the Nobel Peace Prize (2002) "for his decades of untiring effort to find peaceful solutions to international conflicts, to advance democracy and human rights, and to promote economic and social development."

Maurice Albertson and the U.S. Peace Corps

More directly relevant than Hoover or Carter, however, was the life and work of civil engineer Maurice (Maury) Albertson (1918–2009). After earning his doctorate in water resource engineering from the University of Iowa, in 1947 Albertson joined the faculty at Colorado State University, where he helped found the department of civil engineering and develop its focus on hydrology.

Then in 1960 the U.S. State Department hired Albertson to assess the feasibility of creating what was called a "point-four youth corps," referencing Truman's 1949 inaugural address. Democratic senator Hubert Humphrey had been promoting the idea of a volunteer youth corps to provide technical assistance in developing countries. Albertson's coauthored report *New Frontiers for American Youth: Perspective on the Peace Corps* (1961) described the Peace Corps as extending the reach of volunteer international service organizations into the promotion of American political ideals and listed among its principal project needs, "engineering (irrigation, community water supply, flood control, roads, surveying, bridges)" (p. 39).

When John F. Kennedy became president, Albertson was enrolled to help get a new Peace Corps organized. In short order it had over 10,000 volunteers serving in some 50 countries. Albertson subsequently became a consultant to such agencies as the World Bank, the United Nations Development Program (UNDP), and the UNESCO. In all cases, Albertson persistently questioned a too easy alliance with capitalism, as he said when awarded an honorary degree in 2006:

> We need to be motivated by service as well as by profit. We serve best by finding out what people want and helping them work to realize their dreams, not by going into a country and telling villagers what they need.
> (Press Release, 2006.)

Médecins sans Frontiers and Engineers without Borders

Just as influential as individuals has been the model of the NGO *Médecins sans Frontieres*. But there is a basic difference. Hoover, Carter, and Albertson all fundamentally accepted, even when they were frustrated by, the notion of national sovereignty. The U.S. Peace Corps, with which Albertson was so much involved, is actually an agency of a sovereign country and thus tends to reinforce the whole concept of sovereignty or the idea that national governments have the final say over what goes on within state boundaries.

Yet from its beginnings, humanitarianism involved questioning the concept of sovereignty and associated ideas such as national patriotism. And one of the fundamental tenets of MSF is to criticize national sovereignty as a final arbiter of boundaries for humanitarian action. MSF activists are committed to going where the problems are, even without the permissions of national governments, and to exposing the misbehaviors of governments toward their own peoples, insofar as these involve mistreating their citizens or depriving them of protection and care.

EWB, as well as humanitarian engineering in general, whether for practical reasons on reflecting the engineering tradition, has remained largely respectful of national sovereignty, thus to some degree compromising the humanitarian ideal. Engineering has been described as design within a context or under constraints—constraints largely imposed by physical, political, cultural, ethical, legal, environmental, and economic phenomena. Insofar as this is the case, humanitarian engineering is an effort to escape the "social captivity of engineering" by capitalism and nationalism or some other form of wealth and power (Goldman, 1991; see also Johnston et al., 1996). In doing so, however, humanitarian engineering seeks to work within a new self-imposed constraint of seeking to help meet the basic needs of underserved populations. Humanitarian engineering in the most general terms is the artful drawing on science to direct the resources of nature with active compassion to meet the basic needs of all—especially the powerless, poor, or otherwise marginalized.

5. CHALLENGES

> For many [humanitarianism] is best identified with the provision of relief to victims of human-made and natural disasters. For others, though, humanitarianism does not end with the termination of the emergency.... No longer satisfied with saving individuals today only to have them be in jeopordy tomorrow . . . many organizations now aspire to transform the structural conditions that endanger populations. Their work includes development, democracy promotion, establishing the rule of law and respect for human rights, and postconflict peacebuilding. These more ambitious projects expand [humanitarian work]—but, for better and worse, they coincide with and sometimes become part of the grand strategies of many powerful states.
>
> —Michael Barrett and Thomas G. Weiss,
> *Humanitarianism in Question* (2008), p. 3

Humanitarianism has become a progressively expansive project, now encompassing engineering. But like any expansion, this one brings not only benefits.

Engineers have long been focused on meeting human needs as these have been historically and socially understood. Ironically, in the midst of mid-twentieth-century American affluence the very concept of need became problematic. One response was the theory of a hierarchy of human needs proposed by psychologist Abraham Maslow (1954)—from basic needs for air, water, and food through needs for safety, friendship, and achievement to self-actualization—a hierarchy with implications for conceptualizing humanitarian engineering. Since the 1950s we have also become increasingly aware of failures by engineering practice and education in Europe and North America to address even basic needs. On our planet roughly one-fifth of the people lack access to clean water and millions of children under die every year from malnutrition and disease (Kandachar and Halme, 2008).

It is increasingly evident that the existing engineered infrastructure of the developed world will need to change in order to secure a sustainable future. Our water and energy use, food production, and patterns of expected material convenience will need to dramatically alter in order to accommodate a habitable world for our descendants. Engineering graduates must understand the global constraints they face—physical, economic, environmental, social, political, and cultural, all together—when attempting work toward socially just and sustainable solutions. This is a challenge to which humanitarian engineering might make a meaningful contribution. In pursuing the goals of humanitarian engineering and humanitarian engineering education, it is thus appropriate to consider two kinds of challenges.

Practical Challenges

Practical challenges for any humanitarian engineering education program run the gamut from what to include in a curriculum and how to include it to the difficulties of interdisciplinarity, travel, communication, and budgets. Following are some necessarily brief comments on each of these areas, related mostly to undergraduate program development.

The engineering curriculum is already full—indeed, overfull. It takes most students more than the standard four years to earn a bachelor's degree. Yet somehow more courses must be added or substitutions made, both of which faculty will resist. This is unlikely to happen without external stimuli.

Civil engineers may readily work with mechanical and electrical engineers, but interdisciplinary work that bridges technical, historical, social, and cultural knowledge is much more problematic and crucial. Interested humanities and social science faculty will need to develop distinctively appropriate courses in topics such as cultural anthropology, sociology and economics of development, histories of colonialism and postcolonialism, theories of ethics and social justice, and more. Interdisciplinarity is hard. The contemporary academic world is heavily disciplined, in many senses of the term.

Travel to and practical engagement in humanitarian engineering project sites are necessary components of humanitarian engineering programs. Yet foreign travel projects can be expensive. Might the same amount of money do more for those being served if it were provided to them directly? But then how would the students learn humanitarian engineering? Local sites are in fact abundant but seldom as exciting to students.

The need for robust communication skills means that students should be required to learn the basics about any country they plan to visit: government structure and current leaders, major geographical features, climate, history, religion, foods, sports, customs, and cultural characteristics. There will be greater challenges in non-English-speaking countries. At least some faculty and students should be fluent in the local language. A further aspect of communication is a need to maintain relationships with the same local people over years or decades in order to develop a sense of trust among all involved. In the fast turnaround cycles of academic life, not to mention undergraduate education, this is a formidable problem.

Theoretical Challenges

Theoretical issues are equally daunting. Before or after or in the middle of practice, how should we think about humanitarian engineering? How should we think about helping others? This essay has presented humanitarian engineering as a potentially important cross-fertilization between historical developments in engineering and in humanitarianism. But there are other, counter ways of thinking that challenge engineers and those of us who endorse the idea to ask questions about the fundamental viability of any such ideal. In many ways they are akin to some of the basic questions that have been asked within the humanitarian movement itself.

With regard specifically to humanitarian engineering, one question can be formulated quite simply as follows: What if the people who are offered humanitarian engineering assistance rejected it—especially when those who offer the help are genuinely convinced it will benefit those in need? Potential rejections can take many forms, from ignoring proffered assistance to passive aggressive resistance and sophisticated intellectual criticism. Rejections can also come from many different persons within a community. Some forms of rejection are unconscious, others conscious. Humanitarians will often have to struggle to appreciate and assess these rejections in their full complexity. They must be sensitive to, but need not necessarily accept at face value, all criticism.

Consider the example of Ivan Illich (1926–2002). In 1968 Illich, a Catholic priest and theologian who had worked in Latin America for a decade, gave a talk to a gathering of humanitarian aid workers in Mexico that concluded with the following words:

> Use your money, your status, and your education to travel to Latin America. Come to look, come to climb our mountains, to enjoy our flowers. Come to study. But do not come to help.

A transcript of this talk, subsequently titled "To Hell with Good Intentions," is widely available on the Internet (and included in Illich, 2018). It reiterates in even more biting terms a criticism of what Illich (1967) termed Christian "dogooding" the year before in the Jesuit weekly *America*.

The most profound theoretical challenge is thus to come to terms with our own interests and intentions and to consider carefully our understandings of what we are about. In this regard some questions for self-examination formulated by Jen Schneider, Juan Lucena, and Jon Leydens (2009) in their critique of what they term "engineering to help" are useful. Humanitarian engineers would do well, they argue, to ask themselves at least six questions:

1. What are your motivations?
2. What is the history and context of development projects in the area?
3. Who benefits and who suffers from the project?
4. Who is held accountable?
5. What are the possible unintended consequences?
6. Do we view communities as "less-than"?

Intellectuals and elites such as ourselves must undertake examinations of conscience. A sense of righteousness should no more be allowed to inhibit ignoring the good that others do than to rationalize failing to critically examine oneself.

CONCLUSION: HUMANIZING TECHNOLOGY

> How naturally inhumanity combines with technology. . . . We have experienced technology in the service of the destructive side of human psychology. Something needs to be done about this fatal combination. The means for expressing cruelty and carrying out mass killing have been fully developed. It is too late to stop the technology. It is too the psychology that we should now turn.
>
> —Jonathan Glover, *Humanity* (1999), p. 414

The argument of this essay may be restated as follows. First, engineering is a profession classically defined as a practical drawing on science to direct the resources of nature for human use and the convenience. As such, engineering involves systematic, context-dependent design work, operating under constraints. Historical and social determinations of the meaning of use and convenience are key external constraints.

Second, since the emergence of the engineering profession, social history—particularly in Europe but increasingly on a global scale—has given rise to what is called a humanitarian movement. This movement constitutes in part a critical reassessment of some common assumptions about use and convenience.

Third, the humanitarian ideal has a potential to sponsor new understandings and historically distinct practices in the engineering profession. Insofar as this is the case, something called humanitarian engineering may be described in general terms as the practical drawing on modern science to direct the resources of nature with active compassion to meet the basic needs of all persons irrespective of national or other differentiations—an active compassion that, in effect, is directed toward the needs of the poor, powerless, or otherwise marginalized persons.

In this regard the standard professional engineering ethical commitment to the protection of public safety, health, and welfare can be linked to humanitarianism. The fact is that significant groups of people throughout the world are lacking in provisions related to safety, health, and welfare. This obliges engineers to approach with some skepticism any career path that leads to engagement with nothing more than the design and manufacture of high-tech products for an advanced consumerist society.

Fourth, an educational curriculum in support of the practice of humanitarian engineering is in fact emerging in some quarters. But the needs question offers a special opportunity to develop understandings of those we see as peoples in need, as well as ourselves as need-seeing persons. The relation between needs and goods—between need and the good—is a salient entry point for humanitarian engineering reflection.

As an effort to promote discussion of humanitarian engineering, this modest essay aspires to be another effort to promote a new form of engineering and to help address some of its challenges—not so much at the practical level as at the level of philosophical reflection. The assumption is that if we think and clarify the meaning of humanitarian engineering, it is more likely to become an important feature in the spectrum of engineering actions that increasingly constitute the high-tech, advanced technoscientific world in which we live.

In a scholarly reflection on the historico-philosophical career of humanity in the twentieth century, the British philosopher Jonathan Glover reviewed the decline of traditional institutions of morality, especially religion and belief in the transcendent; examined what we know about the moral psychology of warfare and how the great wars of the century came about; noted the persistence and evil consequences of tribalism; summarized how terror functioned to debase human beings under Stalin and Mao, and how the Nazi experiment to create a master race led to holocaust and ruin. His goal, he wrote, was "an attempt to give ethics an empirical dimension [by using] ethics to pose questions to history and . . . history to give a picture of the parts of human potentiality which are relevant to ethics" (Glover, 1999, p. x). Accepting the absence of what he termed external moral law, he argued that "morality needs to be humanized: to be rooted in human needs and human values" (p. 406). Conspicuous by its absence in his narrative, however, is any discussion of engineering or of humanitarianism. There may be an absence of philosophical reflection in engineering, but there is perhaps even more of an absence of reflection on engineering in philosophy.

Essay 14

A Philosophical Inadequacy in Engineering

Engineering is a philosophically inadequate profession. This is not to claim that engineering is inadequate insofar as engineers fail to do philosophy. Such a claim might be true but trivial. Why should engineers be philosophers? Instead, the argument is that engineering is caught in a fundamental difficulty that is revealed by philosophical inquiry and thus may be described as philosophical in character. Reflective or critical analysis of engineering reveals that the profession is committed to an end that is not in fact integral to it. This philosophical inadequacy or deficiency leads to misunderstandings and false expectations both within and without the profession.

The flaw in engineering as a profession can be succinctly stated. Engineering is commonly defined as the art or science of "directing the great sources of power in nature for the use and the convenience of humans" (*McGraw-Hill Encyclopedia of Science and Technology*, 2008). This definition is but little changed from an 1828 formulation by Thomas Tredgold used in the Royal Charter of the Institution of Civil Engineers: "Engineering is the art of directing the great sources of power in nature for the use and convenience of man." But there is nothing in engineering education or knowledge that contributes to any distinct competence in making judgments about what constitutes "human use and convenience." Engineering as a profession is analogous to what medicine might be if physicians had no expert knowledge of health or to law if attorneys knew nothing special about justice.

The following argument for the philosophical inadequacy of engineering will proceed in three steps. It will review the conceptualization of engineering as a profession defined by two key features: technical knowledge and commitment to a service ideal. It then explicates in more detail the engineering service ideal. Finally, it argues there is something fundamentally deficient with regard to how this service ideal is or could be enacted, and conclude briefly with some implications.

Before proceeding, two qualifications: First, the argument, as is often the case in philosophy, takes place at a certain level of abstraction if not naiveté. It will, for instance, ignore an increasing body of literature, both sociological and philosophical, on the professions *tout court*. Indicative of a growing interest in a philosophical engagement with the professions is the fact that the first edition of the *Encyclopedia of Philosophy* (1967) had no entry on or index mention of the professions or professionalism, whereas the *Encyclopedia of Philosophy Supplement* (1996) and *Routledge Encyclopedia of Philosophy* (1998) both include relevant articles and cross-references. Most philosophical debates regarding the concept and meaning of professions and professionalism will nevertheless be ignored here in order to concentrate more narrowly on a single issue with regard to engineering.

Second, in the present instance, all references to engineering will be to the profession as it has developed and is practiced in the English-speaking world—especially in the United Kingdom and the United States. Despite this limitation, there are good reasons for believing the argument applies in some form to engineering as a profession in other sociocultural contexts. The profession has not emerged in the United Kingdom or the United States in isolation from developments elsewhere.

ENGINEERING DEFINED

Prescinding from the definition originated by Tredgold, the most comprehensive and careful analysis of engineering as a profession is found in the work of Michael Davis. According to Davis, who is summarizing standard views,

> An engineer is a person who has at least one of the following qualifications: (1) a college or university B.S. from an accredited engineering program or an advanced degree from such a program, (2) membership in a recognized engineering society at a professional level, (3) registration or licensure as an engineer by a government agency, or (4) current or recent employment in a job classification requiring engineering work at a professional level.
>
> (Davis, 1998, p. 32)

As Davis rightly notes, each of these four qualifications or criteria is philosophically deficient because each includes some form of the *definiendum* (engineer) in the *definiens* (engineering or engineer).

One effort to escape this inadequacy would be to take criterion (1) and consider the requirements for accrediting an engineering program. Do these requirements include reference to the *definiendum*? According to

ABET—"ABET" is the acronym of a former name, U.S. Accreditation Board for Engineering and Technology—all engineering programs are required to demonstrate the following outcomes:

a. an ability to apply knowledge of mathematics, science, and engineering;
b. an ability to design and conduct experiments as well as to analyze and interpret data;
c. an ability to design a system, component, or process to meet desired needs within realistic constraints such as economic, environmental, social, political, ethical, health and safety, manufacturability, and sustainability;
d. an ability to function on multidisciplinary teams;
e. an ability to identify, formulate, and solve engineering problems;
f. an understanding of professional and ethical responsibility;
g. an ability to communicate effectively;
h. the broad education necessary to understand the impact of engineering solutions in a global, economic, environmental, and societal context;
i. a recognition of the need for, and an ability to engage in lifelong learning;
j. a knowledge of contemporary issues; and
k. an ability to use the techniques, skills, and modern engineering tools necessary for engineering practice.

(ABET, 2007, p. 2)

As four of these eleven outcomes—(a), (e), (h), and (k)—again reference the *definiendum*, this approach fails to escape definitional inadequacy.

Similar efforts with respect to the other three criteria likewise fail. With regard to criterion (2), the self-description of recognized professional engineering organizations such as the British Institution of Civil Engineers or the U.S. National Society of Professional Engineers (founded 1934) include references to and assume knowledge of what constitutes engineering. With regard to criterion (3), registration or licensure as an engineer again rests on assumptions about or reference to engineering. Finally, with regard to criterion (4), it is unlikely that any job classification title of "engineer" would not also include an assumption about or reference to engineering.

Despite such philosophical weaknesses, Davis argues the common conception has pragmatic value insofar as the four criteria indicate the parameters within which the engineering profession exercises a certain degree of self-determining autonomy, similar to that found in what are generally accepted as two classic professions. Just as, practically speaking, physicians are those whom other physicians agree to recognize as physicians, and lawyers are those whom other lawyers accept as members of the bar, so engineers are those whom other engineers agree to recognize as such. Engineering is what people who call themselves engineers think and do—or attempt to think and do.

As a conclusion of his analysis, Davis thus reduces the four criteria to two, one concerned with thinking and another with doing: "(1) specific [specialized or technical] knowledge and (2) commitment to use that knowledge in certain ways (that is, according to engineering's code of ethics)" (Davis, 1998, p. 37).

Davis's practical definition is confirmed by the ABET accreditation criteria, which can themselves be grouped in two sorts: those having to do with a specialized knowledge or skill and those having to do with ethics. (The former is what engineer philosopher Taft Broome, 2010, means when he designates engineering as a "learned profession.) The knowledge and skills criteria are specified by outcomes (a), (b), (c) at least in part, (d), (e), and (k). The ethics criteria are present in (c) designing to "meet desired needs within realistic constraints such as economic, environmental, social, political, ethical, health and safety, manufacturability, and sustainability"; (f) "understanding . . . professional and ethical responsibility"; and (h) ability "to understand the impact of engineering solutions in a global, economic, environmental, and societal context." Outcomes (g), (i), and (j) are more generic and characterize any profession or occupation. They are necessary for well-educated citizens, never mind the roles they might occupy in society.

On another occasion, Davis summarizes his position by rejecting the attempt to formulate a philosophical genus-and-species definition of engineering. According to Davis,

> There can be no philosophical definition . . . that captures the "essence" of engineering (because engineering no more has an essence than you or I do). All attempts at philosophical definition will: a) be circular (that is, use "engineering" or a synonym or equally troublesome term); b) be open to serious counter-examples (whether because they exclude from engineering activities clearly belonging or because they include activities clearly not belonging); c) be too abstract to be informative; or d) suffer a combination of these errors.

Instead, as with other professions such as medicine or law, engineering should be understood as a self-perpetuating community of practitioners.

> There is a core, more or less fixed by history at any given time, which determines what is engineering and what is not. This historical core, a set of living practitioners who—by discipline, occupation, and profession—undoubtedly are engineers, constitutes the profession. They decide what is within their joint competence and what is not.
>
> (Davis, 2010, p. 17)

In other words, it is engineers as a historically constituted group who determine what counts as engineering, in terms of both thinking and doing or of specialized knowledge and commitment to use.

Davis's conclusion, it may be noted, accords with generally accepted accounts of the constituent features of any profession. According to Michael Bayles (1989), for instance, the working concept of a profession includes extensive training with a significant intellectual component and the provision of an important service to society. Davis elsewhere also proposes: "A profession is a number of individuals in the same occupation voluntarily organized to earn a living by openly serving a certain moral ideal in a morally permissible way beyond what law, market, and morality would otherwise require" (Davis, 2002, p. 3).

HISTORICAL EMERGENCE

In the present instance the concern is not so much with technical knowledge as commitment to use—and the relationship between the two. This is not to deny that engineering knowledge deserves more by way of historical and critical epistemological analysis than it has yet received. It would certainly be worthwhile to explore, among other issues, the emergence of engineering knowledge as a cognitive achievement and the relations between science and the engineering sciences, distinctions between craft and engineering skill, the methodologies of engineering design, and the epistemic structures of engineering knowledge about how the engineered world works. At the same time, it is relatively easy to find in the engineering educational curriculum more or less adequately delineated bodies of knowledge and skill sets that engineers themselves have decided are necessary for professional practice. All accredited engineering programs include in their curricula substantial courses in advanced mathematics, statistics, physics and chemistry (and sometimes biology), mechanics, thermodynamics, strength of materials, and design (or what Mitcham, 1994, has described as miniature making). In combination such studies can be argued to provide a relatively sound basis for technical competence and the exercise of informed judgment in the systematic design, manufacture, and operation of engineered structures, products, and processes. Otherwise engineering would suffer pragmatic failure as a profession—just as would medicine if physicians did not in practice possess more or less adequate knowledge and skill to treat illness and disease in order to achieve health or law if attorneys did not know how to legislate and to operate within existing legal structures to secure (at least procedural) justice.

But what is the commitment in engineering that indicates how its specialized or technical knowledge is to be used? What is the purpose, end, or service ideal of technical engineering knowledge and practice? What corresponds to the commitment in medicine to the promotion of health and in law to the pursuit of justice? (That the professional-ethical ideals have been contested with regard to both medicine and law does not obviate an effort to identify what might be the functional equivalent in engineering—which, once identified, might then also be contested.)

Historically, engineering arose from specialization within another profession, that of the military. The English word "engineer" and its cognates originally referenced a special group within the army—the corps of engineers—the members of which designed, built, and managed various types of fortifications along with "engines of war" such as battering rams and catapults. Indeed, during the same period in which societies of professional engineers were being established—for example, the British Institution of Civil Engineers in 1818 and the American Society of Civil Engineers in 1852—Samuel Johnson's *Dictionary of the English Language* (1755) defined the engineer as "one who directs the artillery of an army" and Noah Webster's *American Dictionary of the English Language* (1828) described the engineer as "a person skilled in mathematics and mechanics, who forms plans of works for offense or defense, and marks out the ground for fortifications." (As members of the military, engineers can be presumed to have studied the art of employing armed force to secure national policy and to have used that learning in obedience to authority.)

John Smeaton in the 1760s was among the first to attempt explicitly to detach engineering from its military origins by referring to himself as a "civil engineer"—meaning simply an engineer not in military service. In part this coinage seems to have been stimulated by an effort to carve out a kind of architectural work more closely allied with and based in natural science than was architecture itself. One contemporary commentator, for instance, argued that Dutch "hydraulic architecture" was really a form of engineering, precisely because it used and developed the new science of hydraulics. In part it reflected and no doubt contributed to a delimiting of the architectural profession—as had been classically reviewed, for instance, by Vitruvius's *De architectura*—and to a progressive reassigning of responsibility to a new profession for designing, constructing, and maintaining of roads, bridges, water supply and sanitation systems, railroads, and such. This reassignment is clearly reflected in the full statement of Tredgold's definition of civil engineering as

> the art of directing the great sources of power in nature for the use and convenience of man, as the means of production and of traffic in states, both for

external and internal trade, as applied in the construction of roads, bridges, aqueducts, canals, river navigation, and docks, for internal intercourse and exchange; and in the construction of ports, harbors, moles, breakwaters, and light-houses, and in the art of navigation by artificial power, for the purposes of commerce; and in the construction and adaptation of machinery, and in the drainage of cities and towns.

(Institution of Civil Engineers, 1828)

Human use and convenience is simply assumed to be synonymous with the advancement of commercial interests, a view that over the next hundred years will be subject to increasing social and philosophical criticism.

It is against this background that there have emerged a number of different contemporary formulations of engineering ethics in terms of a commitment to public safety, health, and welfare. The trajectory from Tredgold to the present is complex and made more so by the distention of engineering across numerous professional associations and the lack of unifying organizations, such as those manifested in medicine and law. (The basic history of engineering as a profession in the United States as narrated by Layton, 1971, is especially revealing in this regard; for relevant complementary studies, see Vesilind, 1995; Mitcham, 1997; and Davis, 1998.) Yet such complexity need not affect the present argument. Here it is sufficient to note that the "Code of Professional Conduct" of the Institution of Civil Engineers states in its third rule that "all members shall have full regard for the public interest, particularly in relation to matters of health and safety, and in relation to the well-being of future generations" (Institution of Civil Engineers, 2008). In parallel, the "Code of Ethics for Engineers" of the National Society for Professional Engineers even more forcefully declares in the first of six "fundamental canons" that "engineers, in the fulfillment of their professional duties, shall hold paramount the safety, health, and welfare of the public" (NSPE, 2007). Whereas in the early 1800s it was more or less assumed that the practice of engineering knowledge and skill would automatically rebound to human benefit, two centuries later an increasing need to assert the same and to ask engineers to be explicit in pursuing such a commitment has given rise to what is known as the paramountcy clause.

THE PROBLEM

Now comes the problem. The engineering curriculum in its technical component provides a solid basis for making sound technical decisions. But what about the ethical component? Although the professional ethical component of engineering education has become increasingly prominent, there are good reasons to doubt that it comes any way near close to being as sufficient as the technical component—or that it even in principle could become so. It is true that accredited engineering programs commonly include some limited course work in professional engineering ethics. But such courses are almost wholly limited to discussions of professional engineering ethics codes, their explication, and/or social difficulties related to enacting them, mostly in the form of case studies or hypothetical scenarios. A typical engineering ethics course may provide some general ethical theory as background to the codes of specific professional engineering organizations and then explore the difficulties in, for example, practicing loyalty and honesty, avoiding conflicts of interest, and protecting public safety, given the real-world context of countervailing economic, social, and political pressures and the limitations of professional autonomy within which engineers work (for illustration, see the two most widely used U.S. engineering ethics textbooks, Harris et al., 2005; and Martin and Schinzinger, 2005). On what basis does this qualify engineers to make informed judgments about what constitutes public safety, health, and welfare?

The initial version of ethical commitment was, as noted, to human use and convenience. Human use and convenience has never been a formal element of any engineering ethics code, in part because explicit ethics codes were not formulated until a hundred years after Tredgold but also because there would have been no doubt about the end of engineering knowledge and skill. The connection between technical engineering and the service ideal of human use and convenience would have been argued in the form of an enthymeme or syllogism with a suppressed premise like this:

Major premise: Technical engineering knowledge increases human power.
Minor (unstated or suppressed) premise: Human power is always used to enhance human use and convenience.
Therefore, engineering is for human use and convenience.

The articulation of an explicit service ideal for engineering and the substitution of public safety, health, and welfare for human use and convenience together can be interpreted to reflect an emergent awareness that the minor premise can no longer be assumed or go unquestioned.

To reiterate: There are two necessary (if not sufficient) elements of the engineering profession: possession of a body of technical knowledge and commitment to a goal of public service. Insofar as this is the case, engineering may be compared with the two classic professions of medicine and law, as indicated in Table 14.1.

In medicine, the specialized body of knowledge concerns human anatomy and physiology or some distinctive portion thereof and a commitment to the

Table 14.1. A Comparison of Three Professions

	Medicine	Law	Engineering
Specialized knowledge	of human anatomy and physiology	of the legal system	of the applied sciences, design methods, and artifact functionings
Public service	Human health	Justice procedural	Public safety, health, and welfare

promotion of human health. In the case of law, the specialized body of knowledge is the laws and legal procedures of a state and a commitment to justice—at least procedural justice. For engineering, analogously, the specialized body of knowledge includes the applied or engineering sciences and design methods along with a commitment to public safety, health, and welfare.

In medical school, not only do students take courses explicitly devoted to understanding the nature and meaning of health (and a host of opposing pathologies) but virtually all courses necessarily include some learning about health. In order to study how the human body is structured and functions, one must develop not just a descriptive but also a prescriptive appreciation of structure and function. What Ruth Millikan (1984) and some philosophers of biology call proper functions are an integral part of medical knowledge. Only from the perspective of proper functions can one learn to diagnose illness or disease and prescribe therapeutic responses. In the course of their studies physicians thereby develop a robust understanding of health, one that enables them genuinely to understand the anatomies and physiologies of their patients better than the patients could understand their own bodily functionings. Likewise in the case of law: The typical law school curriculum includes substantial doses of jurisprudence, and attorneys are genuinely able to understand what constitutes justice—at least procedural justice—better than their clients. Physicians and attorneys thus are reasonably committed to and charged with assisting those whom they serve and the public at large in understanding and pursuing health and justice.

On what possible basis, however, are engineers more qualified than anyone else to understand or determine the safety, health, and welfare that should be associated with engineered structures, products, or processes? In Walter Vincenti's lucid analysis of *What Engineers Know and How They Know It* (1990), there is no indication that engineers qua engineers know anything about safety, health, or welfare. These are simply not things engineers know, and there seems to be no way specific to engineering to acquire such knowledge. With regard to welfare, engineering education requires no courses in the subject, and surely welfare economists and psychologists are more expert in the nature and meaning of human welfare than engineers. With regard to health, physicians are obviously more expert than engineers. With regard to safety, it may be that engineers possess a modicum of relevant expertise, insofar as safety factor analysis is included in some technical courses. But what constitutes safety (or its obverse, risk) is a much more socially constructed notion than that of health. There is no such thing as "proper safety" as an objective feature of the engineered world, the way proper functions are understood as objective features of the biological world. Indeed, there is no such thing as a proper function for any artifact. In artifacts there are only what Pieter Vermaas (2007) terms "fiat functions," that is, ascribed functions, which can change or be added to when people come up with new creative uses. Although law and justice have strong socially constructed dimensions, the social construction of the legal system is one in which negotiation has reached at least provisional conclusion in constitutional, statutory or administrative law. The law is composed of what might be called well-entrenched fiat functions. With safety, however, except insofar as it is legally specified, social construction is seldom manifest in a clearly defined consensus and as a result remains continuously up for renegotiation.

With regard to safety, one engineer-philosopher, Samuel Florman, made the following objection: It would be crazy for engineers to determine "what criteria of safety should be observed in each problem" encountered. Things can always "be made safer at greater cost, but absolute freedom from risk is an illusion." Levels of safety are "properly established not by well-intentioned engineers, but by legislators, bureaucrats, judges, and juries [and it] would be a poor policy indeed that relied upon the impulses of individual engineers" (Florman, 1981, pp. 171 and 174). The most engineers can do is help clients and the public understand the relevant degrees of safety and then invite them to decide how safe is safe enough. Engineers qua engineers are no more qualified to make such determinations than anyone else; they legitimately participate in making such determinations, but only as consumers and citizens.

As the quotation from Florman suggests, this problem of the lack of any specifically engineering competence or expertise with regard to an implied professional commitment to public safety, health, and welfare has not gone completely unrecognized. In fact, one can identify at least three efforts to respond to the problem. One is that represented by Florman himself, who rejects an internal service ideal of public safety, health, and welfare in favor of external, democratically determined and governmentally enforced regulations for engineering practice. Florman even imagines that a special cadre of public service engineers could well contribute to the intelligent creation and administration of such regulations. But the consequences of imposing the protection of the public good on the consciences of all engineers would be chaos: "Ties of loyalty and discipline would dissolve ... organizations would shatter [and whistle blowing] would become the norm, instead of a last and desperate resort" (Florman, 1981, p. 171).

There are, however, two weaknesses with such a rejectionist stance. On the one hand, insofar as engineers are to have no special ethical obligations as engineers, the rejection of a service ideal seems to deprive engineering of professional status. On the other, by raising the possibility of a special cadre of guardian engineers in public service, Florman seems to presume the possibility of such status in a way that simultaneously promotes a fissure within the engineering community. Additionally, the extent to which the public service cadre possess any knowledge that would warrant its contribution to the establishment of regulations remains questionable.

An alternative to Florman's notion of engineers as observing minimal and in fact common ethical commitments to competence and integrity while working within externally determined frameworks provided by law or regulation is that of creating an alternative ethical ideal. In fact, Florman's very notion of a cadre of engineers who "do the research, write the codes, and make the inspections that keep technology in check" (Florman, 1981, p. 175) would seem to require some notion of an internal technical ideal. Thorstein Veblen in *The Engineers and the Price System* (1921), as a result of his economic analysis of the public welfare, argued that a major cause for the diminished quality of life in techno-capitalist society was the capture of engineering by captains of industry who sacrificed the making of good products to the making of money. For Veblen the "instinct of workmanship" and commitment to efficiency inherent to engineering should be liberated from the distorting shackles of commercial interests. Steven Goldman (1991), without endorsing Veblen's response, has made a similar argument with regard to the social captivity of engineering.

Veblen's proposal to turn decision-making over to engineers nevertheless flounders on two problems. One is that the resulting technocracy would be opposed to both democracy and the free market. Another is that efficiency as a concept is as context-dependent as safety. Efficiency as a ratio of output divided by input is dependent on what are counted as outputs and inputs. Output and input determinations are specified by contexts external to engineering—in the most rigidly quantitative terms by the economy but in softer forms by politics, culture, and fashion. It is difficult to see how the ideal of efficiency could escape the problem of the relation between economic internalities and externalities, as these are manifest by problems related, for instance, to environmentalist debates.

Still a third response is to subordinate public safety, health, and welfare to free and informed consent. This is the argument advanced initially by Mike Martin and Roland Schinzinger (1983) and then in a different form by Taft

Broome (1989). In the restatement of their argument in a widely adopted engineering ethics textbook, philosopher Martin and engineer Schinzinger argue for viewing "engineering as social experimentation." Engineering projects are experiments insofar as they are undertaken in partial ignorance, outcomes are uncertain, and future engineering practice is modified by knowledge gained as a result. More crucially, these experiments impact users, consumers, and those societies in which the engineered structures, products, and processes are created and deployed.

> Viewing engineering as an experiment on a societal scale place the focus where it should be: on the human beings affects by technology, for the experiment is performed on persons, not on inanimate objects. In this respect, albeit on a much larger scale, engineering closely parallels medical testing of new drugs or procedures on human subjects.

In consequence, "the problem of informed consent . . . should be the keystone in the interaction between engineers and the public" (Martin and Schinzinger, 2005, p. 92). Just as medical research with human subjects or participants is moral only to the extent it respects the free and informed consent of the persons involved, so must engineering undertake to respect the autonomy of those it affects. Commitment to public safety, health, and welfare as a substantive ideal is replaced by the ideal understood in proceduralist terms; the basic form of safety, health, and welfare is not to be subjected to risks, deprivations, or harms to which one has not knowingly acceded so that the practice of free and informed consent becomes the basic form of engineering ethics. Public safety, health, and welfare are then publicly determined by those affected through their free and informed involvement.

Engineer-philosopher Broome advances a slightly different but related argument. For him, the fundamental misconception about engineering is to think of it as an application of science partakes of its reputed cognitive certainties. In fact, engineering is based not on scientific knowledge but depends crucially on practical conjecture so that engineering takes the form of a heroic quest. Engineers have the obligation to include the societies in which they work in their heroic explorations and adventures. As Broome reformulates the public paramountcy principle:

> The engineer shall not claim that engineering necessarily assures good public health and welfare. Instead, [the engineer] shall advise the public of the risks associated with [engineering] and seek to obtain public acceptance of these risks.
>
> (Broome, 1989, p. 9)

But it is one thing for physicians, who work with individual patients, to practice free and informed consent—even undertake heroic medical efforts to save lives or discover new cures for disease. It is quite another for engineers to try to imitate such practices in projects that implicate whole groups, populations, and even future generations. Broome's notion of a heroic quest could easily morph from a techno-odyssey of adventure to a techno-campaign of conquest—from Magellan to Captain Ahab.

Again, the difficulties of establishing appropriate protocols and practicing free and informed consent with regard to individual research programs and human subject participants are legion; they can only be enlarged when the practice rises to the level that Martin and Schinzinger suggest is necessary. Although they admit the problematic character of informed consent, for which they seek to substitute "valid consent" grounded in a simple responsibility of engineers to provide users and consumers "information about the practical risks and benefits of the process or product in terms they can understand" (Martin and Schinzinger, 2005, p. 93), this still leaves protocols open to the point of vacuousness. (One option for operationalization they neglect to consider is the practice of consensus conferences, as has become something of a standard practice in Denmark.)

But would it not be more honest to adopt Florman's straightforward but tragic affirmation of "the existential pleasures of engineering" (Florman, 1976)? Grounded in creative self-realization, engineers and their supporters, "aware of the dangers and without foolish illusions," would nevertheless "press ahead in the name of the human adventure," believing that "without experimentation and change our existence would be a dull business" (Florman, 1981, p. 193). The ultimate commitment to safety, health, and welfare—indeed, to use and convenience—would then be conceived as the creative practice of engineering itself, safe in affirmation of ourselves as technological beings whose health and welfare is as much found in identification with the engineering spirit as with any more mundane forms of well-being.

CONCLUSION

The thesis argued here is that a philosophical analysis of engineering reveals a substantive inadequacy, not to say incoherence or contradiction, in the profession: a commitment to public safety, health, and welfare that is incapable of enactment. Although efforts have been made to respond to this fundamental problem, all have weaknesses—some of which fail to recognize the depth of the difficulty at issue, others of which may be worse than the original problem itself. Still another simply winds up affirming engineering creativity as an end in itself with spill over benefits.

The fundamental problem can be restated as follows. When a physician correctly diagnoses and treats an ill or diseased patient, a return to health on the part of the patient directly confirms diagnosis and treatment. When an attorney counsels a client, the counsel can be judged by its benefits or otherwise at court. It is true that physicians and lawyers may have a deeper appreciation of the therapy or legal brief than the public, but the bottom-line assessments by the public will readily agree with those of the experts. In both cases the professions are relatively able to achieve their ends—health or justice, respectively—with similar assessments regarding the degree to which this occurs made by professionals and those they serve.

With engineers, however, the situation is different. Engineers can come up with good structural, product, or process designs that go unappreciated, are criticized, or even rejected by the firms or clients for which they work—not to mention the marketplace into which engineering designs may be introduced. Good engineering is mostly recognizable as such only by other engineers. Technical efficiency and engineering elegance are often confirmed only in the breach by some happenstance of corporate decision-making or market success. To say the same thing in a different way: The first-order ends of health and justice operative in the professions of medicine and law, respectively, are not enclosed within some second-order end of public good; they are the public good. In engineering, by contrast, the first-order technical end, however defined, which was once assumed to be itself a public good, is now conceived as subordinate to a second-order end that is not operative in the profession itself. This is a contrast that would seem to stand even if one rejects as naive or qualifies the comparisons with medicine and law.

The situation with engineers may be compared to that of scientists. Seldom will the public have any basis for appreciating and accepting a new scientific discovery. Indeed, in many instances the public will resist or reject a scientific discovery as a problematic imposition on common sense or accepted beliefs. Only after a long period of accommodation are many scientific observations and theories allowed to replace the folk truths or falsities accepted by the general public because the first-order end of knowledge in science is not identical with a second-order end of the public good—or has to be argued to be so. In like manner, only over extended periods of time do some instances of efficiency and technical elegance slowly insinuate themselves into publicly accepted artifice. Quite common are the public adoptions of less-than-adequate engineering designs along with their false efficiencies and profit-making potentials enhanced by superficial glamor and the push of marketing. Such would seem to be at least one cautionary moral of the only engineering ethics textbook to take the paramountcy clause as its central theme, a novel with commentary by Alastair Gunn and Aarne Vesilind (2003).

The gap between first- and second-order ends in engineering can account for both student disinterest and professional resistance to the study and discussion of engineering ethics, especially when this study and discussion becomes focused in terms of public safety, health, and welfare. For despite the claims of some engineers and engineering ethics educators, there is considerable evidence that students and practicing professional engineers are not very interested in ethics—except perhaps to protect them from societal criticism. Because commitment to public safety, health, and welfare is simply not integral to engineering, to try to impose it on the profession intuitively strikes many students and professionals as artificial and inappropriate. Additionally, it is failure to recognize this gap that accounts for many misleading expectations among the public with regard to the responsibilities of engineers for technical failures and disasters. The public tends to expect more from engineering than is possible or appropriate. Only recognition of the philosophical inadequacy of the contemporary conception of engineering will enable both the profession and the public to develop a more sound understanding of a form of knowledge and skill that has nevertheless become increasingly important to the public.

ADDENDUM: THE SOCIOLOGICAL INADEQUACY OF ENGINEERING—RESPONSE TO DAVID GOLDBERG

An earlier version of this essay, "The Philosophical Weakness of Engineering as a Profession," was read at a Royal Academy of Engineering Workshop on Philosophy and Engineering (WPE 2008). At a Society for Philosophy of Technology meeting the next year at Twente University (SPT 2009) David Goldberg made an exceptionally thoughtful response, which was subsequently published in a proceedings volume (Michelfelder et al., 2013) as "Is Engineering Philosophical Weak?" (Goldberg, 2013).

My original paper argued for a distinction between two kinds of professions. Philosophically strong professions, such as medicine and law, rest on ideal ethical goals that are well embedded in the professional curriculum and practice. Weak professions, such as the military and business, either lack such ideal goals or only weakly include them in the specialized knowledge of a professional curriculum and practice. Ethical ideals instead are stuck on from outside or assumed from social context. The (somewhat intentionally provocative) argument was that engineering has more in common with weak than with strong professions.

My paper referenced five occupations—medicine, law, engineering, the military, and business—instead of just the first three. With some qualifications, Goldberg begins by accepting the proposal to consider which might be more philosophically strong (PS) or philosophically weak (PW) as potentially informative. (One important qualification was to reference "occupations" instead of "professions," which I agree is better.) However, he expected the scope of my presentation would be broader than ethics, since ethics is only one of a number of major branches of philosophy: "When I first heard the title of the talk, I thought it was a much needed philosophical wake up call for engineering to become more broadly philosophically aware, and I was surprised that the talk focused more narrowly on ethical or aspirational concerns" (p. 394).

Thus, before explicitly criticizing my argument, Goldberg ventured a broader comparison of his own in terms of whether or not an occupation is metaphysically reflective (if it considers its history), epistemologically reflective (if it consciously transmits knowledge), and ethically reflective (if it has an explicit code of ethics). "As a matter of simplicity we will say that an occupation is philosophically reflective (or philosophically strong . . .) if it is reflective on two of the three categories" (p. 394). According to his admittedly rough cut ("in the mode of engineering reasoning") but nevertheless insightful comparison,

> Medicine, business, and engineering are metaphysically weak because they are not reflective on their nature. . . . Engineering is alone in its epistemological weakness, because it is blind to the ways in which its models are distinguished from those of math and science. . . . Business is alone in its ethical weakness because [it has] are no generally accepted ethical standards.
>
> (p. 395)

The result is that engineering and business, lacking reflectiveness in two areas, are the two occupations that would be classified as PW. He summarized his assessment in a table similar to that in Table 14.2.

As an aside, Goldberg noted that on his analysis, philosophy would be classified as PS. However, since the philosophical profession lacks a code of ethics, it would not be as strong as either law or the military (a problematic conclusion relevant to the second case). More importantly, "it seems a bit unfair to take engineering to task in the one area in which it is philosophically reflective" when more generally, as Goldberg sees it, engineering is PW only "because it doesn't pay sufficient attention to its nature or its knowledge."

> As engineers approach philosophy for the first time, they will find a rather extensive discussion of ethics but only a meager discussion of epistemology and ontology. This, of course, is changing, but the palin usage of the terms philosophical strength and weakness, seems better reserved for this larger distinction than the one made solely on the basis of aspirational ethics.
>
> (p. 395)

Having made an informative complement to my own argument, one equally provocative, he turned to consider the case I made for philosophical

Table 14.2. Philosophical Strengths and Weaknesses in Five Occupations

	Metaphysics	Epistemology	Ethics
Medicine	Weak	Strong	Strong
Law	Strong	Strong	Strong
Military	Strong	Strong	Strong
Business	Weak	Strong	Weak
Engineering	Weak	Weak	Strong

weakness in engineering with regard to ethics. Goldberg's criticism of my argument hinges on a new distinction between *ends-in-themselves* or *aspirational* occupations (such as medicine and law) and *instrumental* occupations (the other three). Again, qualifications were necessary. Neither medicine nor law consistently realize "their companion ideals" of health or justice. Moreover, when they do so, Goldberg points out, it is primarily because of support "from higher order institutional arrangements" (p. 396). "Doctors and lawyers work on behalf of their individual clients, doing their utmost to help the client [which Goldberg calls "local ethical alignment"], and this pursuit of the client's interest is assumed—in the larger institutional setting of medical and legal practice—to align with the good of society [global ethical alignment]." It is "this and largely only this" fortuitous conjunction of local and global ethical alignment "that makes law and medicine 'aspirational' professions." "Doctors and lawyers don't have to try very hard to be good in [the] aspirational sense" (p. 397).

This pointing up of the unfortunate circumstances that make it trying for engineers to be good, can encourage inquiry into "whether engineering can be made more like law or medicine" (p. 397) Goldberg thinks of two possibilities:

> First, can we imagine a simple aspirational ideal for engineers in general? Second, can we imagine a rearrangement of the institutional structure of engineering that permits local ethical alignment to lead to the presumption of global ethical alignment?
>
> (pp. 397–98)

Goldberg's answer to the first is that, because engineering designs and constructs so many different kinds of artifacts to meet so many different needs, no unifying aspirational ethical ideal is possible. To further the argument, he considered the possibility of whether the instrumental occupation of the military might be infused with an aspirational ideal. (That the military is considered here is, of course, significant, as engineering arose as military engineering.) The obvious candidate is peace. But according to Goldberg, peace as a unifying aspirational ideal founders on problems of predictability and the necessity for collective action. It is undermined not just by the fog of war but the fact that the only effective army is one in which individuals subordinate their own perceptions (and even personal survival) to a hierarchical structure. The idea "of individual soldiers pursing peace on their own terms is almost too absurd to contemplate, and thinking about engineers universally pursuing some larger societal goal on their own is strikingly analogous" (p. 399).

Goldberg's response to the second question imagined two social institutional redesign extremes: one in which "absolute control" is given to engineers in some form of technocracy and another "fail-safe action model" in which engineers are prohibited from doing anything dangerous. Both are *reductio ad absurdum* possibilities, suggesting the rationality of existing arrangements in which a modestly regulated market economy allows engineers to make things that a democratically constituted public governs through consumer choice. This may be the worst form of governance, except for all the others (to adapt a famous quip).

Considering this third or middle way, however, reminds us about the extent to which all "occupations are increasingly ethically complex as their institutional complexity increases," including in medicine and law. HMOs now make it increasingly difficult for physicians as individuals to act simply in their patients' interest. Large legal firms in a litigious society are likewise constrained with regard to any simple client advocacy. Both physicians and lawyers are members of large teams and networks and behave accordingly. The history of the engineering profession in the United States is just another manifestation of this social reality.

In conclusion, Goldberg considered the proposal of the U.S. National Academy of Engineering that engineering in the twenty-first century should assume a set of Grand Challenges for humanity as its aspirational ideal. Goldberg is skeptical. Although "aligned with the zeitgeist," this proposal is not "particularly institutionally or ethically sophisticated, and . . . likely to face difficulty unless accompanied by institutional rearrangement" (p. 402) that is quite unlikely. Still, "breaking out of the current limited array of institutional options would be desirable to effect practical change along the lines of Mitcham's urgings" (p. 403).

> When medicine and law achieve approximations to health and justice and when engineering, business, and the military struggle against their tendency to be instrumental to the purposes of the powerful, both types of case are shaped by the institutional form of the occupations themselves. The wider spread recognition of this institutional shaping is important to all occupations, and such recognition may encourage creativity and innovation to find new—as yet undiscovered—institutional forms.
>
> (p. 404)

With Goldberg's measured conclusion, I fully agree, and I am grateful for an argument that reached assessments by a different route complementary to my own. The philosophical inadequacy of engineering is grounded in a sociological inadequacy—or, to use Goldberg's more adequate term, a sociological weakness. The engineering profession is not as strongly bonded a society as, say, that of physicians or attorneys.

My only caveat is that the original argument was not meant to suggest the need for engineering to develop a substantive ideal. Instead, what I tried to do was consider the ideal that the professional engineering community does in fact endorse—that is, the paramountcy of protecting public safety, health, and welfare (which Goldberg did not reference)—and show that it is not substantively present in engineering, especially not in engineering education. There are two possible implications: either the substantive knowledge that is required to pursue these ideals should be infused into engineering education and practice or the ideals should be abandoned and a more realistic view of the occupation adopted. (On this point, Goldberg and I might continue to disagree, since he seems to accept that engineering has an ethics by virtue of having a code of ethics.)

Instead, I would suggest that the de facto ideal is use and convenience, which is more substantively present both in the profession and educational programs. Engineers are known as problem-solvers. As such, they solve problems for the users and make things more convenient. It is this ideal that is also the structural basis for the ease with which engineering can be captured by economic and political interests. However, the extent to which use and convenience is truly ethical is an open question.

Essay 15

The True Grand Challenge for Engineering

Self-Knowledge

In 2003 the U.S. National Academy of Engineering (NAE) published *A Century of Innovation* celebrating "20 engineering achievements that transformed our lives" across the twentieth century, from automobiles to the Internet. Five years later, it followed up with 14 Grand Challenges for engineering in the twenty-first century, including making solar energy affordable, providing energy from fusion, securing cyberspace, and enhancing virtual reality. But only the most indirect mention was made of the greatest challenge: cultivating deeper and more critical thinking, among engineers and nonengineers alike, about the ways engineering is reshaping how and why we live.

What Percy Bysshe Shelley said about poets two centuries earlier applies even more to engineers today: they are the unacknowledged legislators of the world. Mary Shelly made the case dramatically in her novel *Frankenstein* (1818). By designing and constructing new entities, structures, products, and processes, engineers are influencing how we live as much as any laws enacted by politicians. Indeed, much legislation today is in response to transformations wrought by engineering. Would we ever think it appropriate to enact laws that transform our lives without critically reflecting on and assessing them? Yet neither engineers nor politicians deliberate seriously on the role of engineering in transforming our world. Instead they too often limit themselves to celebratory clichés about economic benefit, national defense, and innovation.

Where might we begin to promote more critical reflection within lives that are increasingly engineered? One natural site would be engineering education. In this respect, it is again revealing to note the role of the NAE Grand Challenges. Not only in the United States, but globally, the technical community is concerned about the image of engineering in the public sphere and its problematic attractiveness to students. A 2010 United Nations Educational,

Scientific and Cultural Organization (UNESCO) study *Engineering: Issues, Challenges and Opportunities for Development* lamented that despite a "growing need for multi-talented engineers, the interest in engineering among young people is waning in so many countries." The Grand Challenges have thus been deployed in the Grand Challenges Scholars Program as a way to attract more students to the innovative life. The American, British, and Chinese national academies have further collaborated on a series of student-oriented Global Grant Challenges Summits in London (2013), Beijing (2015), and Washington (2017). But to adapt the title of American science policy adviser Vannevar Bush's *Science Is Not Enough*, a cultivated enthusiasm for engineering is insufficient. More pointedly, to paraphrase Socrates, "The unexamined engineering life is not worth living for humans."

AN AXIAL AGE

In a reflective examination of world history, the German philosopher Karl Jaspers (1949) observed how in the first millennium BCE, human cultures at multiple locations on the great Eurasian land mass independently underwent a profound psychological shift that he named the Axial Age. Thinkers as diverse as Confucius, Laozi, Buddha, Socrates, and the Hebrew prophets began to ask what it means to be human. They insinuated into cultures the possibility of such radical questions as "What kind of person should I try to become?" "What is the good human being?" Humans discovered it was not necessary always to accept whatever lifeways they were born into; they could begin to subject a cultural inheritance to critical assessment.

The original Axial Age shook up culture and the socialization process—not, of course, equally for everyone all the time but slowly, over a long historical sedimentation process. At least a few leaders in an increasing spectrum of cultures realized they could chose to lead their lives in different ways than what was socially expected. Along with this, the Axial Age sponsored the emergence of religions and philosophies to help us think about what it means to be human.

Today we are entering a Second Axial Age, one in which we no longer have simply to accept the material environment into which we are born. Just as two-and-a-half thousand years ago the second nature of culture became subject to questioning, now the first nature of the physical environment is no longer a given; it is subject to being engineered and reengineered. Human environmental niche construction has expanded beyond its traditional trial-and-error localizations. Yet engineering education makes little to no effort to give engineers the means to reflect on themselves and their world-transforming enterprise. The Second Axial Age is only slowly giving rise to philosophies that can help us think about the meaning of our planet and what kind of planet we should try to design and construct. Because engineers are the primary unacknowledged designers and makers of our physical world, it is incumbent on them to consider—and to help us all think about—what it means to exercise such powers.

Engineering programs like to promote innovation in product creation, and to some extent in pedagogy. But they have undertaken little innovative reflection on what it means to be an engineer. Surely the time has come for engineering schools to become more than glorified occupational trade schools whose graduates can make more money than hapless literature and philosophy majors. Today we need engineers who can think holistically and critically about their roles in designing and constructing the world, intentionally or not—and who can assist their nonengineering fellow citizens in thinking that goes beyond superficial promotions of the new.

Where might engineers acquire some tools with which to cultivate such capacities? One good place to start would be precisely through engagement with the traditions of thought and critical self-reflection that emerged from the original Axial Age: what are now called the humanities.

TWO CULTURES *RECIDIVUS*

To mention engineering and the humanities in the same sentence immediately calls to mind C. P. Snow's famous criticism (1959) of those "natural Luddites," who do not have the foggiest notion about such technical basics as the second law of thermodynamics. Do historians, literary scholars, and philosophers really know anything that can benefit engineers? Why should engineers read poetry, novels, or philosophy?

Snow's "two cultures" argument, along with many discussions since, conflates science and engineering. The powers often attributed to science—such as an ability to overcome poverty through the increased production of goods and to land humans on the Moon—belong more to engineering. As a result, there are actually two two-culture issues. The tension between two forms of knowledge production (sciences and humanities) is arguably less significant than another between designing and constructing the world versus reflecting on what it means (engineering versus humanities).

Indeed, although there is certainly room for improvement on the humanities side, I venture that a majority of humanities professors in engineering schools today could pass the test Snow proposed to the literary intellectuals he skewered. Yet in my experience relatively few engineers, when invited to reflect on their professions, can do much more than echo libertarian appeals to an alleged need for unfettered innovation to fuel endless growth. Even the more sophisticated apologists for engineering such as Samuel Florman (*The Existential Pleasures of Engineering*), Henry Petroski (*To Engineer Is Human*), and Billy Vaughn Koen (*Discussion of the Method: Conducting the Engineer's Approach to Problem Solving*) are largely absent from engineering curricula.

The two-cultures problem in engineering schools is distinctive. It concerns how to infuse into engineering curricula the progressive humanities and qualitative social sciences, as pursued by literary intellectuals who strive to make common cause with that minority of engineers who are themselves critical of the cultural captivity of techno-education. There are, for instance, increasing efforts to develop programs in humanitarian engineering, service learning, and social justice. Yet like many humanities professors who teach engineering students, I have experienced a continuing engineering/humanities tension across working in three different engineering schools, especially when employed in an increasingly corporatized environment at an institution that orients itself toward efficient throughput of students who can serve as handmaids of an aggressive oil and gas industry.

On the one side, engineering faculty (administrators even more so) have a tendency to look on humanities courses as justified only insofar as they provide communication skills. They want to know the cash value of the humanities for professional success. The engineering curriculum is so full that they feel compelled to limit humanities and social science requirements, commonly to little more than a semester's worth, spread over an eight-semester degree program crammed with math, science, and engineering—salted, perhaps, with a few economics and business management classes.

WHY HUMANITIES?

Unlike professional degrees in medicine or law, which (in the United States) typically require a bachelor's degree of some sort before professional focus, entry into engineering is via the BS degree alone. This has undoubtedly been one feature attracting many students who are the first in their families to attend college. It is an upward mobility degree, even when there may not be quite the economic demand for engineers that the engineering community often proclaims. On the other side, humanities faculty (there are seldom humanities administrators with any influence in engineering schools) struggle to justify their courses. These justifications are of three unequal types, taking an instrumental, enhanced instrumental, and intrinsic value approach.

The first, default appeal is to the instrumental value of communication skills. Engineers who cannot write or otherwise communicate their work are at a disadvantage, not only in abilities to garner respect from people outside the engineering world but even within technical work teams. The humanities role in teaching critical thinking is an expanded version of this appeal. All engineers need to be critical thinkers when analyzing and proposing design solutions to technical problems. But why no critical thinking about the continuous push for innovation itself? Too often, the humanities are simply marshaled to provide rhetorical skills for jumping aboard the more-is-better innovation bandwagon—or criticized for failing to do so.

A second, enhanced instrumental appeal stresses how humanities knowledge, broadly construed to include the qualitative social sciences, can help engineers manage alleged irrational resistances to technological innovation from the nonengineering public. This enhanced instrumental appeal argues that courses in history, political science, sociology, anthropology, psychology, and geography—perhaps even in literature, philosophy, and religion—can locate engineering work in its broader social context. Increasingly engineers recognize that their work takes place in diverse sociocultural situations that need to be negotiated if engineering projects are to succeed.

In similar ways, engineering practice can itself be conceived as a technoculture all its own. The interdisciplinary field of science, technology, and society (STS) studies deserves special recognition here. Many interdisciplinary STS programs arose inside engineering colleges, and even after their transformation to disciplinary science and technology studies, some departments have remained closely connected to engineering faculties.

In the United States, the enhanced instrumental appeal further satisfies ABET (an acronym for what used to be the Accreditation Board for Engineering and Technology) requirements. In order to be ABET accredited, engineering programs must be structured around a set of student outcomes. Central to these outcomes are appropriate mastery of technical knowledge in mathematics and the sciences, including the engineering sciences, and the practices of engineering design, including abilities "to identify, formulate, and solve engineering problems" and "to function on multidisciplinary teams." Engineers further need to learn how to design products, processes, and systems "to meet desired needs within realistic constraints such as economic, environmental, social, political, ethical, health and safety, manufacturability, and sustainability" and possess "the broad education necessary to understand the impact of engineering solutions in a global, economic, environmental, and societal context." Finally, engineering students should be taught "an ability to communicate effectively" and "professional and ethical responsibility." Clearly the humanities need to be enrolled in the process of delivering the more fuzzy of these outcomes.

The challenge of professional ethical responsibility deserves highlighting. It is remarkable how, although professional engineering codes of ethics regularly identify the protection of public safety, health, and welfare as a primary obligation, the engineering curriculum shortchanges these key topics. There exists a field termed safety engineering but none called health or welfare engineering. And even if there were, because the promotion of these values is an obligation for all engineers, their study would need to be infused across the curriculum. Safety engineering is a specialization, not a common core requirement like math and science. Physicians, who also have a professional commitment to the promotion of health, have to deal with its problematic presence in virtually every class they take in medical school. It is fair to say that public health administrators know more about public health and welfare economists probably know more about the complexities of public welfare than most engineers know about these values and their enactment, which they are nevertheless professionally required to protect in the public (and thus socially constructed) domain.

The 2004 NAE report on *The Engineer of 2020: Visions of Engineering in the New Century* emphasized that engineering education needs to cultivate not just analytic stills and technical creativity but also communication skills, management leadership, and ethical professionalism. Meeting almost any of the subsequent NAE list of Grand Challenges, many engineers admit, will require extensive social context knowledge from the humanities and social sciences. The humanities are accepted as providing legitimate if subordinate service to engineering professionalism even as they are regularly shortchanged in engineering schools.

But it is a third, less instrumental justification for the humanities in engineering education that will be most important for successfully engaging the ultimate grand challenge of self-knowledge, that is, of thinking reflectively and critically about the kind of world we wish to design, construct, and inhabit in and through engineering. The existential pleasures of engineering, not to mention its economic benefits, are insufficient to leading a distinctly human life. Human beings are not only geeks and consumers. People are also poets, artists, religious believers, citizens, friends, and lovers in various degrees all at the same time. The engineering curriculum should be more than an intensified vocational program that assumes students either are, or should become, one-dimensional in their lives. Engineers, like all of us, should be able to think about what it means to be human. Indeed, critical reflection on the meaning of life in a progressively engineered world is a new form of humanism appropriate to our time—a humanities activity in which engineers might lead the way.

RE-ENVISIONING ENGINEERING

Primarily aware of requirements for graduation, engineering students are seldom allowed or encouraged to pursue in any depth the kind of humanities that could assist them, and all of us, in thinking about the relationship between engineering and the good life. In American technological universities they often sign up for humanities classes on the basis of what fits their schedules—although they then sometimes discover courses that not only provide relief from the forced march of technical thinking but broaden their sense of themselves and stimulate reflection on who they really want to be. Recently a student in an introduction to philosophy class told me he was tired of engineering physics courses that always had problem sets to solve. He wanted to think about the nature of reality.

When such students drop out of engineering, as some of mine have, the humanities are likely to be blamed, rather than credited with expanding a sense of the world and life. The cost-benefit assessment model in colleges progressively coarsens higher education. As Clark University psychologist Jeffrey Arnett (2014) argues, emerging adulthood is a period of self-discovery during which students can explore different paths in love and work. It took me seven years and three universities to earn my own BA, years that were in no way cost-benefit negative. Bernie Machen, President of the University of Florida, has been quoted (in Berrett, 2014) as telling students that their "time in college remains the single-best opportunity . . . to explore who you are and your purpose in life." Engineering programs, because of their rigorously constrained curricula, tend to be the worst offenders at cutting intellectual exploration short. This situation needs to be reversed, in the service of both engineering education, and of our engineered world. If they really practiced what they preached about innovation, engineering schools would lead the way with expanded curricula and even BA degrees in engineering.

Physicist Mark Levinson's insightful documentary film *Particle Fever* (2013) explores a divide between experimentalists and theorists analogous to that between engineering and the humanities. But in the case of the Large Hadron Collider search for the Higgs' boson chronicled in the film, the experimentalists and theorists work together, insofar as theorists provide guidance for experimentation. Ultimately, something similar has to be the case for engineering. Engineering does not provide its own justification for transforming the world, except at the unthinking, capitalist controlled bottom-line level; by itself it fails to provide reflective guidance for what kind of world we should design and construct. We wouldn't think of allowing our legislators to make laws without our involvement and consent; why are we so complacent about the arguably much more powerful process of technical legislation?

What Jaspers in the mid-twentieth century identified as an Axial Age in human history—one in which humans began to think about what it means to be human—exists today in a new form: thinking about what it means to live in an engineered world. In this Second Axial Age, we are beginning to think not only about the human condition but about what has aptly been called the techno-human condition: our responsibility for a world, including ourselves, in which the boundaries fluctuate between the natural and the artificial, between the human and the technological. Just as a feature of the original Axial Age was learning to affirm limits to human action—not to murder, not to steal—so we can expect to learn not simply to affirm engineering prowess but to limit and steer our engineering actions.

As a theme of Second Axial Age thinking, the Anthropocene should be credited to a distinctive way of being human instead of humanity in general. What originated as an accidental Anthropocene must mutate into a conscious Anthropocene: consciousness that is connected with conscience and the practice of restrictions on human action into nature. Engineering, which grew out of the Enlightenment, is called on to grow into a new enlightenment about itself.

Amid the Grand Challenges articulated by the NAE, there must be another: the challenge of thinking about what we are doing as we turn the world into an artifact and the appropriate limitations to this engineered-engineering power. Such reflection need not be feared; it would add to the nobility of engineering in ways that little else could. It is also an innovation within engineering in which some are leading the way. For example, in the Netherlands—a country that, given its dependence on the *Deltawerken*, comes closest to modeling an anthropocenic planet—a collaboration of philosophers and engineers in 2007 founded what has become the 4TU Centre for Ethics and Technology based at the technological universities of Delft, Eindhoven, Twente, and Wageningen. That same year saw the NAE initiate a Center for Engineering Ethics and Society to address "the ethical and social dimensions of engineering." And at its 20th anniversary celebration in 2014, the Chinese Academy of Engineering sponsored sessions on the philosophy of engineering and technology. Is it not time for leaders in engineering organizations everywhere to take such innovative professional engineering actions to heart—and go further?

The real Grand Challenge of engineering is not simply to transform the world. It is to do so with critical, philosophical reflection on what it means to engineer the world. In the words of the great Spanish philosopher José Ortega y Gasset, in an early philosophical meditation on technology (1939), to be an engineer and only an engineer is to be potentially everything and actually nothing. Our increasing engineering prowess calls upon all of us—engineers and nonengineers alike—to reflect more deeply on who we are and what we want to engineer.

In examining those in his city who made claims to knowledge—especially, politicians, poets, and artisans—Socrates concluded that artisans have a stronger case than others. Artisans, a historical precursor of engineers, truly possess practical knowledge about making things. Yet in dialogue with fellow citizens who boasted of their prowess in making and doing things, Socrates also referenced the words inscribed on the Temple of Apollo at Delphi: Know yourself. This is a counsel that engineers—and all of us whose lives are informed by engineering—could well enlarge and apply anew to ourselves.

ADDENDUM: SIMONDON'S DREAM

In an argument that exhibits some affinities with this essay, Gilbert Simondon's *On the Mode of the Existence of Technical Objects* (1958) also calls on engineers to play a special role in contemporary culture through a new level of philosophical self-knowledge. After a half century of being largely ignored in English-speaking philosophy of technology (but not elsewhere: see Bontems, 2018), in the early 2000s Simondon became subject to increasing discussion; his book was finally published in an approved translation (2017), and a number of interpretative studies began to appear (see, e.g., Dumouchel, 1992; Schmidgen, 2012; Barthélémy, 2015).

Simondon's approach to engineering is through ontology, in which there are four distinct modes of existence: physical, biological, psychosocial, and technical. Although each mode is distinct, they are not opposed. The biological mode includes the physical. The existence of plants and animals includes physics and chemistry. Likewise, the technical is a new mode but not one opposed to the biological or psychosocial. Like Heidegger, Simondon seeks to disclose the technical mode through a phenomenology and history. Simondon's phenomenology, however, is a mechanology that focuses on technical evolution inside technology (not in its social impact) to reveal an essence of technicity that is not anything technical.

According to Simondon, the technoscientific world (not his term) manifests two false and conflicted attitudes toward technical objects. On the one hand, they are thought as no more than dumb tools or instruments for human use, with use being stuck onto them from outside, as it were. On the other, in the form of autonomous machines or robots, they are feared as something other than or opposed to humans.

The first attitude is a holdover from a culture constituted by craft making, which is a kind of childhood of the mode of existence of technical objects. Although there was a limited truth to this attitude in early history of the being of technical objects, it also rested on a failure to appreciate the deep reality of technicity, a reality that Simondon desires to disclose by looking back from the vantage point of engineering. The second attitude emerged as an uninformed cultural response to engineering and engineered objects. This again

Simondon desires to correct through a phenomenological analysis of the rich experience of the coming to be of machines.

In both attitudes, the technical object is thought to be a kind of slave: in the former, as a properly docile and subservient one; in the latter, as a slave in potential revolt against its masters. Philosophy previously transformed culture by teaching that slaves were human too, not an alien presence, and thus properly treated as companions it was proper to respect and with whom it was possible to cooperate to create a new kind of society. Now philosophy must do something analogous for technical objects: infuse into culture an appreciation of their truly human aspects so that humans and nonhuman technical objects can collaborate to engender a new, expanded sense of society (an anticipatory version of Latour's actor network theory). In Simondon's own words from the opening paragraph of his book,

> We would like to show that culture ignores a human reality within technical reality and that, in order to fully play its role, culture must incorporate technical beings in the form of knowledge and in the form of a sense of values. Awareness of the modes of existence of technical objects must be brought about through philosophical thought, which must fulfill a duty through this work analogous to the one it filled for the abolition of slavery and the affirmation of the value of the human person.
>
> (2017, p. 15)

As is common with phenomenology and efforts to call attention to overlooked or hidden aspects of experience, Simondon's analytic description of technical objects and engineering experience deploys a distinctive vocabulary. Simplifying his analysis, however, Simondon argues at least two distinctive ideas: One is the primacy of process over substance, that is, of the evolutionary emergence of technical individuals out of a preindividual background (the process of "individuation"). A special form of individuation is a movement described as from "abstract" (crudely formed) to "concrete" (well-formed) technical objects. At its origins a hammer will have been a crude rock tied to a stick in which neither part is particularly well formed and the two could easily break apart; through refinement over time, the head and handle will become integrated into a well-functioning unit (more individuated).

Another idea is that in the case of machines, evolution is not toward autonomy or independence but toward multipurposiveness and flexibility (Simondon's term is "indeterminacy"). A tool such as a hammer cannot be used for much more than hammering. By contrast, a computer can be used for many things, from crunching numbers to writing texts and running scientific simulations. The flexibility of advanced machines calls for humans to engage with their powers in acts of coordination and management of a complex human–machine assemblages. The liberation of slaves calls for a shift in governance from authoritarian dictatorship toward democratic leadership.

For Simondon, the engineer, when treating the world as what Heidegger calls *Bestand* does not so much challenge being in the form of *Gestell* as enter into the rich pluripotentialities of being and orchestrate the coming forth into existence of technical objects. This activity may be more apparent in the process of engineering than in craft making but is also true of craft making. The maker never simply imposes form on matter (as hylomorphic metaphysics would have it) but acts to cultivate the emergence of form, in ways that are more traditionally associated with agriculture. Heideggerian *poiesis* is present even, and at deeper levels, in engineering.

The arrival of a historical situation that (to venture an extended quotation, retaining the gendered language of the 1950s) allows

> man to see the technical relation functioning in an objective way is the prime condition for the incorporation of the knowledge of technical reality and of the values implied by its existence into culture. Now, these conditions are realized in the technical ensembles employing machines that have a sufficient degree of indeterminacy. For man, the action of having to intervene as a mediator in this relation between machines grants him a situation of independence in which he can acquire a cultural vision of technical realities. The engagement in the asymmetrical relation with a single machine cannot provide the necessary distance for what one might call technical wisdom. Only a situation in which there is a concrete link with machines and a responsibility toward them, but which is liberated vis-à-vis each one taken individually, can provide this serenity of having technical awareness. Just as literary culture, in order to constitute itself, needed the wise individuals who had lived and contemplated the inter-human relation with a certain distance that gave them a serenity and a depth of judgment while nevertheless maintaining an intense presence among human beings, technical culture cannot constitute itself without developing a certain sort of wisdom, which we will call technical wisdom, in men who feel their responsibility toward technical realities, but who remain disengaged from the immediate and exclusive relationship with a particular technical object. It is rather difficult for a worker to know technicity through the aspects and modalities of his daily work on a machine. It is also difficult for a man who is the owner of machines and who considers them productive capital to know their essential technicity. It is the mediator of the relation between machines alone who can discover this particular form of wisdom. Such a function, however, does not yet have a social place; it would be that of the production planning engineer if he was not preoccupied by immediate output, and governed by a finality external to the operating system of machines which is that of productivity. The function whose basic lines we are attempting to draw would be that of a psychologist of machines, or of a sociologist of machines—what we might call a mechanologist.
>
> (p. 159)

Neither artisan, nor industrial worker, nor capitalist owner—not even the specialized engineer—but only the "planning engineer," functioning in detachment from anything less than society or culture as a whole in order to orchestrate (to pick up a metaphor Simondon uses earlier, p. 4) humans and machines, can realize and bring into being a technosocial order as an end or good in itself. Such an engineer, a kind of Nietzschean engineer, will eschew thinking about decorating consumer products so as to attract more purchasers but focus instead at the levels of communication and power network infrastructures so as to inventively coordinate the manifold of technical objects—the technical object of technical objects, as it were.

There is much more to be said about Simondon's philosophy of engineering which, more than most, is grounded in a distinctive ontology. (Simondon's disagreement with Norbert Wiener and cybernetics as a philosophy of engineering is especially salient.) It is nevertheless possible to inquire about the extent to which Simondon's vision of engineering philosophy and philosophical engineers as founders for a cultural synthesis of humans and machines is more than a dream:

- What kind of engineering education might bring engineers out of the cave of capitalist industrialization into the sunlight of engineering as a good in itself? (Might it be systems engineering?)
- What is the relationship between Simondon's high-level engineering enlightenment and the more mundane experience of the existential pleasures in engineering (Florman, 1976)? (Might the more simple joys of making technical objects keep one tied to lower levels?)
- How will a culture free from the blockage to invention address the existential political needs for stability and ethnic belongingness or the gap between rich and poor? (Can invention resolve all social and political disharmonies?)
- In a cadre of enlightened engineer-philosophers, how is technical wisdom to escape manifesting its own class interest? *Quis custodiet ipsos custodes?*

In the light of such questions, it is difficult not to see Simondon's dream as merging on fantasy. Indeed, similar questions cannot help but cast a shadow over the argument of essay 15 as well.

Essay 16

From Engineering Ethics to Politics

(with Wang Nan)

Is engineering ethics a type of engineering or a type of ethics—and therefore of philosophy? Ideally it might be both. Yet prior to the 1950s it would have been difficult to find any discussion of engineering ethics that had much philosophical depth. It was just engineers trying to develop professional conduct guidelines for themselves rather than independent analysis and criticism of moral assumptions common in the profession. Prior to the second half of the twentieth century engineering ethics had about as much to do with philosophy as the work of a shade-tree mechanic has to do with automotive engineering. Then beginning in the 1950s and again in the 1970s, as a result of discussions that took place initially in Germany and independently in the United States, respectively, engineering ethics began to acquire new seriousness in engineering and eventually became a presence in the field of applied ethics.

During the second half of the twentieth century, engineering ethics became increasingly engaged with professionalized philosophy and expanded in a trajectory that moved from discussions based initially in two particular countries through discussions in many countries. The expansion has raised questions related to globalization and suggests the need for a movement from ethics to politics. This essay offers a summary sketch of such developments, with some reference to the contexts in which they took place. We make a point not just of describing different developments in engineering ethics but of trying to identify the problems that gave rise to distinct engineering ethics discussions. In conclusion we suggest the emergence of new problems and try to point toward the future.

PRE-PHILOSOPHICAL ORIGINS

In English the terms "engineering ethics" and "ethics in engineering" tend to be interchangeable. The later nevertheless more clearly declares that the subject matter concerns ethical questions in engineering. This means that different conceptions of what engineering is will have implications for engineering ethics.

For purposes of the present discussion, we provisionally adopt Michael Davis's definition of engineering in social and historical terms. For Davis, like all professions,

> engineering is self-defining (in something other than the classical sense of definition). There is a core, more or less fixed by history at any given time, which decides what is engineering and what is not. This historical core is not a concept but an organization of living practitioners who—by discipline, occupation, and profession—are undoubtedly engineers.
>
> (Davis, 2015, p. 67)

As constituted by discipline, by occupation, and by profession, engineering has undergone continuous emergence from the late 1600s to the present.

One aspect of this emergence of the engineering profession has been the development of engineering ethics. Initially ethics was submerged in occupation. As Davis notes, engineers were first denominated as such in France in 1676 with creation of the military *corps du génie* (engineering corps). The late 1700s and early 1800s witnessed formation in England of the original professional engineering societies as nonmilitary organizations. The Institution of Civil Engineers (ICE, officially founded in 1818 in London, but with roots that go back to the informal Society of Civil Engineers founded by John Smeaton in 1771) had no explicit code of ethics. At the same time, Thomas Tredgold's definition of engineering (formulated for the ICE Royal Charter of 1828) as "the art of directing the great sources of power in Nature for the use and convenience of man" implicitly associated engineering with the use value theories of classical political economists.

Yet it was not until the late 1800s and early 1900s in the United States that engineers began to discuss engineering ethics as such. The first appearance of the term "engineering ethics" in the title of an independent publication comes from a Carnegie Library of Pittsburgh *Monthly Bulletin* bibliography on the subject (McClellan, 1917). In his preface to the 17-page collection of modestly annotated references, the author notes how it was prepared in response to "numerous requests [which had] come to the Technology Department for material on Engineering Ethics" and explicitly identifies engineering ethics with "ethics for engineers." Ethics for engineers, however, is not the same as ethics in engineering; it subordinates ethics to what engineers do and aims to help them function more effectively as engineers. Ethics in engineering, by contrast, can sometimes make engineering more difficult if not oppose it altogether.

In the references themselves, the first use of "engineering ethics" in a title occurs in an 1893 news story about a discussion in the American Society of Civil Engineers (ASCE, founded in 1852 and the oldest professional engineering society in the United States) concerning the desirability of appointing a committee to explore drafting an ethics code. The resulting discussion in the later 1890s and first decade of the 1900s, together with related discussions in other engineering societies, led eventually to imitation of the professional associations of lawyers (American Bar Association) and of physicians (American Medical Association), which adopted codes of professional ethics in 1908 and 1912, respectively. The American Institute of Electrical Engineers (AIEE, later to become the Institute of Electrical and Electronic Engineers or IEEE, the largest engineering society in the world) adopted the first professional ethics code in 1912; the ASCE adopted another two years later in 1914. In both cases, the primary duty of the engineer was described as serving as a "faithful agent or trustee" of some employing company—a duty that has been argued to reflect the origins of engineering in the military, where obedience to authority is a primary obligation.

What was the perceived need that these codes aimed to address? According to Edwin Layton's historical narrative on the sociological development of the American engineering profession, the key element was what he termed *The Revolt of the Engineers* (1971) against subservience to corporate interests. The codes aimed to help engineers resist persistence efforts, both economic and political, to deprive them of their rightful authority over the design and construction of large-scale projects—deprivations that sometimes resulted in dam collapses and bridge failures—and undermined pursuit of the technical ideal of efficiency. Unfortunately the dominance of corporate interests even within the profession forced the codes to stress some version of company loyalty so that the original aim was often subverted. In David Noble's Marxist analysis, American engineering was actually "guided as much by the capitalist need to minimize both the cost and the autonomy of skilled labor as by the desire to harness most efficiently the potentials of matter and energy" (Noble, 1977, p. 34).

As significant as their functional features is that these early engineering ethics codes were formulated minus any consultation with professors of ethics or philosophy. Instead, despite the presence in American public life of philosophers such as William James, John Dewey, and others, engineering ethics was originally the product of what might be called folk philosophy.

INITIATING ENGINEERING-PHILOSOPHICAL DISCUSSIONS: GERMANY

The first clear engagements between engineers and philosophers as separate professional traditions took place in Germany. The background for this engagement was emergence in the late 1800s of the first efforts of general philosophical reflection on technology (in German, *Technik*, which can also mean engineering). As Carl Mitcham (1994) has summarized, this development of what he terms engineering philosophy of technology, ethics is subordinate to the articulation of an engineering worldview. He quotes, for instance, one apology for this worldview from the Russian engineer Peter K. Engelmeier (1855–c. 1941), writing in German:

> *Techniker* [technologists or engineers] generally believe they have fulfilled their social tasks when they have delivered good, cheap products. But this is only part of their professional task. The well-educated engineers of today are not found only in factories. Highway and water transport, urban and economic management, etc. are already under the direction of engineers. Our professional colleagues are climbing ever higher up the social ladder; the engineer is even occasionally becoming a statesman. . . . This extension of the engineering profession not only seems welcome, it is the necessary consequence of the enormous economic growth of modern society and augurs well for its future evolution.
>
> (Mitcham, 1994, p. 26, revised)

In the context of this societal ascent, in order to achieve the recognition they deserve, engineers should work to articulate their worldview as a general philosophy of technology.

According to Mitcham this was most fully realized in the German context in the philosophical efforts of the German inventor and engineer Friedrich Dessauer (1881–1963). Like Engelmeier, Dessauer not only argued his views with conscious reference to such key modern philosophers as G. W. F. Hegel and Immanuel Kant but also drew on and entered into dialogue with Plato and Aristotle from the premodern period of European philosophy and with contemporary Marxist, existentialist, and other thinkers. The result is a general engineering philosophy that sees engineers, especially through the act of invention, as coming in contact with Platonic forms or closing the gap between Kantian phenomena and noumena. Modern engineering involves "participating in creation" and constitutes *"the greatest earthly experience of mortals"* (quoted from Mitcham, 1994, p. 33, emphasis in the original). Dessauer's philosophical praise of engineering integrated ethics into epistemology, metaphysics, and even aesthetics.

Such a positive philosophical interpretation of engineering was dramatically questioned by World War II, which occasioned pessimism concerning highly developed technology and a tendency to condemn engineers as morally irresponsible contributors to destructive warfare. The actions of German engineers during the National Socialist regime (1933–1945) seemed to undercut engineering idealism. Because many of its members had been compromised by involvement with Nazism, the *Verein Deutscher Ingenieure* (VDI or Association of German Engineers) undertook to promote a new philosophical reflection among engineers. This led to a more sustained dialogue between engineers and philosophers than had previously taken place in Germany or in any other country.

Prior to World War II, German engineers had attempted to construct an ethics and philosophy of engineering on their own. They were autodidacts who read philosophical texts but did not ask philosophers themselves to think with them about engineering. After the war, they began to engage with philosophers and invited their help.

In the early 1950s, for instance, the VDI sponsored a series of conferences on "The Responsibility of Engineers," "Humanity and Work in the Technological Era," "Changes in Humanity through Technology," and "Humanity in the Force-Field of Technology." In all cases, professional philosophers were invited to discuss the issues with professional engineers. Out of the first conference came a *Bekenntnis des Ingenieurs* [Confession of Engineers], a Hippocratic-like oath for VDI members, and later the formation of a special *Mensch und Technik* [Humanity and Technology] special interest group composed of engineers and philosophers. Broken down into working committees on such themes as "Pedagogy and Technology," "Sociology and Technology," "Religion and Technology," and "Philosophy and Technology," the group produced by the mid-1970s a series of publications focused broadly on engineering, technology, and values.

This work in turn led to replacement of the now dated "Confession" and to further interdisciplinary engineering–philosophy research, especially on the theory of technology assessment. With regard to professional ethics, one *Mensch und Technik* working committee report in 1980 proposed simply that "the aim of all engineers is the improvement of the possibilities of life for all humanity by the development and appropriate utilization of technical means" (VDI, 1980, p. 41). With regard to the foundations of technology assessment, a second working committee in 1986 identified eight fields of value (environmental quality, health, safety, functionality, economics, living standards, personal development, and social quality), mapped out their interrelations, and developed recommendations for their implementation in the design of engineering products and projects. The practice of comprehensive, interdisciplinary technology assessment effectively became a recommended professional ethical obligation for German engineers. The best introduction to engineering ethics in Germany is a volume edited by philosophers Hans Lenk and Günter Ropohl (1987), which includes as an appendix the *Verein Deutscher Ingenieure, Ausschuß "Grundlagen der Technikbewertung": Vorentwurf für eine Richtlinie "Empfehlungen zur Technikbewertung"* [Association of German Engineers, "Foundations of Technology Assessment" Committee: Preliminary draft of a "Recommendations for Technology Assessment" guideline), parts 1–3 of five parts.

This distinctive interaction among engineers and philosophers for the first time created a philosophically rich engineering ethics within a national professional engineering community. At the same time, by the early 2000s some philosophers associated with the VDI began to question the adequacy of their work. Günter Ropohl (2002), for instance, described what he termed "the mixed prospects of engineering ethics." Although a new awareness had emerged among German engineers of endogenous ethical responsibilities, because their work was done mostly in teams under conditions highly influenced by economic and social pressures, they also acknowledged the extent to which they were subject to exogenous influences. Indeed, in a world undergoing globalization, "the world society—beyond the individual, the corporation and the national state—is appearing as the fourth level of responsibility in technology" (Ropohl, 2002, p. 154). Whether and to what extent this is a threat or an opportunity and how it is to be managed became an issue of continuing discussion. It also pointed toward the importance, beyond ethics, of politics and political philosophy.

INITIATING ENGINEERING-PHILOSOPHICAL DISCUSSIONS: UNITED STATES

In the United States the stimulus that brought philosophers and engineers together came more from outside than from within the profession. In the mid-1970s, policy initiating program officers at two federal agencies—National Endowment for the Humanities (NEH) and National Science Foundation (NSF)—awarded seminal grants to promote research on engineering ethics. In one case, Carl Mitcham's application for funding to support extension of his bibliographic research in the philosophy of technology (Mitcham, 1973) stimulated Richard Hedrich at NEH to propose a more synthetic project that funded Paul Durbin's editing of the *Guide to the Culture of Science, Technology, and Medicine* (1980), to which Mitcham contributed a chapter.

In a more immediately relevant case, NEH created a National Project on Philosophy and Engineering Ethics codirected by Robert Baum and Albert Flores that ran for five years, until 1982. Announcing the project in the journal of the National Society of Professional Engineers, one project leader observed, "Philosophers have not yet entered into constructive partnerships with engineers similar to their efforts in the field of medicine." This was not surprising, since "engineers have generally not been aware of the potential contributions philosophers might make [and] philosophers have on the whole failed to appreciate and understand the social and intellectual significance" of ethical problems in engineering. So the NEH, promoting what became known since an "applied turn" in American philosophy, took the initiative. Since the announcement went on to explain, "The National Project of Philosophy and Engineering Ethics has been designed to recruit 15 to 18 professional engineers from both the academic and non-academic engineering communities, who are interested in teaming up with professional philosophers to formulate, develop, and implement projects dealing with the ethical problems in engineering" (Flores, 1977, p. 28).

In still another influential initiative, NEH and NSF initially collaborated, and then the new Ethics and Values in Science and Technology (EVIST) program at NSF, under the program leadership of Rachelle Hollander (from 1975 to 2006), began to award a series of related grants (see Hollander and Steneck, 1990). One of the earliest and most influential went to Vivian Weil, director of a newly established Center for the Study of Ethics in the Professions at the Illinois Institute of Technology, in support of a series of philosopher and engineering workshops on ethical issues in engineering (see Weil, 1979 and 1983; and Hollander, 1983). Another EVIST grant supported compilation at the Center of a book-length bibliography on engineering ethics (Ladenson et al., 1980).

Together these external initiatives not only began to infuse philosophical ethics into professional engineering, but they also raised interest about engineering in American philosophy. In 1980, for instance, the biennial meeting of the Philosophy of Science Association for the first time hosted a symposium of invited papers on ethical and philosophical issues related to engineering. Three participants (Gravander, 1980; Hodges, 1980; Rogers, 1980) acknowledged debts to these philosophy and engineering projects.

These external initiatives occurred as American engineers themselves had been expanding their understandings of the profession. One pivotal event occurred in California in the 1930s when two engineers reported some illegal activities of their supervisors (who were subsequently tried and convicted), but the reporting engineers found themselves expelled from the ASCE for unethical conduct by failing to act as a "faithful agent or trustee" of their employer. One of the engineers continued unsuccessfully into the 1950s to seek vindication. Discussion of this and related cases led in the mid-1970s to a fundamental revision of the ASCE ethics code. The first principle of the new code was the following: "Engineers shall hold paramount the safety, health, and welfare of the public in the performance of their professional duties."

In parallel developments, leading AIEE engineers in the early 1900s had begun to challenge Tredgold's classic definition by conceiving of engineering as focused not just on exploiting the forces of nature for human benefit but of pursuing this end through the management of other human beings as well (see McMahon, 1984, chapter 4). Over the course of time, however, it was increasingly recognized that the implementation of any such expanded vision, especially in conjunction with increased recognition of the manifold societal impacts of technology, presented distinctive challenges. As historian Matthew Wisnioski (2012) has richly chronicled, the late 1960s and early 1970s witnessed a blossoming of dissent within the engineering community that both responded to and mirrored public concerns about nuclear weapons, product liability, environmental pollution, and the technological transformation of society. In 1974 the IEEE (created in 1963 by merger of the AIEE and the Institute of Radio Engineers or IRE), like the ASCE, affirmed in a revised "Code of Ethics for Engineers" an obligation to "protect the safety, health and welfare of the public." Initially relegated to the fourth of four articles, by 1990 "making decisions consistent with the safety, health and welfare of the public" had become the first of ten principles. Additionally, the 1970s witnessed creation in the IEEE of a new special section to promote reflection on the social implications of technology, and the editor of the flagship journal *IEEE Spectrum* began to refer to a "new professionalism" to be "based not only on traditional high standards of technical achievement but that embraces concern for the impact of technological developments on society as well" (Christiansen, 1972, p. 17). Again, such concern seemed to invite dialogue with philosophers, some of whom in the applied field had also become critics of technological transformations in society.

One outcome of the NEH project, in collaboration with the new EVIST program, was a number of publications oriented toward the more philosophical teaching of a more philosophical engineering ethics. The textbook that most integrated philosophy into engineering was by the team of Mike Martin (philosopher) and Roland Schinzinger (engineer) titled *Ethics in Engineering* (first edition, 1983). As the authors explained in their introduction, they took engineering ethics to be "the discipline or study of the moral issues arising in and surrounding engineering" and to involve "normative (evaluative) inquiries, conceptual (meaning) inquiries, and descriptive (factual) inquires," with normative inquires being central. The text itself then developed a challenging notion of engineering as social experimentation that required adaptation of the principles of free and informed consent (from biomedical ethics) and highlighted a primary professional concern for safety. It went on to examine the ways engineering is embedded in organizations and engaged with management and philosophically explicated both the responsibilities and rights of engineers. It concluded with philosophical reflections on career choices. This was the first full book in English that could accurately be described as bringing philosophical ethics to bear in engineering, and through multiple editions (1989, 1996, and 2005) it significantly influenced the field.

Over the course of the 1980s and 1990s engineering ethics courses became increasing features of engineering education. Indeed, the primary philosopher-engineer connection was between philosophy and engineering professors rather than with working engineers. In 2000 ABET, the organization that accredits U.S. engineering programs, began explicitly to list "an understanding of professional and ethical responsibility" as one of 11 required educational outcomes.

The primary way this understanding came to be taught was not so much through the kind of critical philosophical reflection promoted by Martin and Schinzinger as by the teaching of professional ethics codes and case studies. Indeed, professional and public discussion of the case of the space shuttle *Challenger* from 1986, and the way in which engineer Roger Boisjoly had opposed the disastrous launch, helped stimulate ABET accreditation policy. In respect to codes and cases, a second textbook, midwifed this time by NSF support, became exemplary: Charles Harris (philosopher), Michael Pritchard (philosopher), and Michael Rabins (engineer) titled *Engineering Ethics: Concepts and Cases* (1995, with subsequent editions in 2000, 2005, 2008, and 2013). Case studies were also important features of the Martin and Schinzinger text. It is worth noting that code and case study teaching was, as well, typical of the few pre-1970s engineering ethics courses, taught mostly by senior engineering faculty seeking to share their experiences with a younger generation. Post-1970s the case studies just became more fully developed and carefully analyzed. They became less like didactic "war story" lessons and more designed to provoke reflective discussion.

However, in a reflective review of the achievements of engineering ethics, Paul Durbin, a philosophy professor who had been involved in developing and teaching engineering ethics, questioned whether the philosophy–engineering interaction had realized its promise. It is certainly the case that engineering ethics never became as prominent a discourse as bioethics. Philosopher Stephen Toulmin (1982) once argued that "medicine actually saved the life of ethics" insofar as it push philosophers to become more involved with substantive issues. Professional philosophers, at least in the English-speaking world, had become involved almost exclusively with increasingly abstract questions related to such topics as the language of morals. The challenges of biomedicine asked them to deal again with substantive issues of good and bad, right and wrong, in real-life situations. It was unfortunate, Durbin argued, that engineering ethics had not been able to develop into as robust a pursuit. According to Durbin,

> The recent history of engineering ethics in the USA is not a happy one. Philosophical engineering ethics is an almost complete failure, largely because the efforts of engineers and their professional societies are too limited in both scope and impact. With Robert Baum and Albert Flores—in their original hopes for the National Project of Philosophy and Engineering Ethics—I believe that the way to go is through collaborative efforts involving philosophers and engineers. But I would qualify my optimism about the approach by saying that its success depends on significant behavioral changes. The engineers and their professional societies need to broaden their focus, moving beyond a focus on individual misconduct to broader social responsibilities, and to welcome a broader range of people into the dialogue. [Additionally,] philosophers, social critics, reporters and editors, environmental activists, and the like need to be less confrontational and more willing to dialogue. [This would create a better conception] of engineering ethics than a definition that focuses mainly on the potential misconduct of individual engineers and technical professionals.
>
> (Durbin, 1997a, p. 82)

Only by going beyond a focus on individual responsibility and such issues as whistle-blowing, can engineering ethics make an impact on society comparable to the impact made by engineering and technology themselves. Again, Durbin's plea can be framed as calling for a movement from ethics to politics.

GLOBALIZATION

The American approach to engineering ethics is much less deeply philosophical (in the sense of being engaged with major figures and themes of the Western philosophical tradition in a scholarly manner) than the German. But despite the fact that it arrived on the scene two decades after the German version, the American has arguably been the leading influence globally. Engineering ethics as it has been pursued and practiced in developed and developing countries alike has often echoed the American approach of adopting codes of conduct privileging the responsibility of individual engineers to protect public safety, health, and welfare.

To cite a few representative examples from the developed world:

- The ICE in England, after almost 150 years without a code, in 1963 adopted a set of "Rules for Professional Conduct"; the 2010 version of this set of

rules equates acting ethically with acting honorably and obligates "All members [to] discharge their professional duties with integrity [, competency, and with] full regard for the public interest, particularly in relation to matters of health and safety."
- The *Conseil National des Ingénieurs et des Scientifiques de France* (CNISF), which incorporates the *Société Centrale des Ingénieurs Civils* (founded 1848), in 2001 adopted a *Charte d'Ethique de l'Ingenieur* proclaiming "engineers are citizens . . . involved in civic actions aiming for the common good."
- The Engineering Society of Finland (founded 1880), in 1966 adopted a "Code of Honour" calling on members "to be of service of both [their] country and mankind as a whole." The Institution of Engineers, Australia (founded 1919), in 1981 adopted a "Code of Ethics" that states that "the responsibility of Engineers for the welfare, health and safety of the community shall at all times come before their responsibility to the Profession, to sectional or private interests, or to other Engineers."
- The Association of Professional Engineers of Ontario, Canada (founded 1922), in 1984 adopted a "Code of Ethics" that states, "A practitioner shall regard his duty to public welfare as paramount."

In modest contrast, while professional associations of engineers in developing countries have also created ethics codes, they have more commonly stressed obligations to enhance professional reputation. Two examples:

- The Institution of Engineers (India), established in 1920 (royal charter 1935), in 1944 created a code of ethics that stresses how the member "should scrupulously guard his professional reputation and avoid association with any enterprise of questionable character."
- The *Colegio de Ingenieros de Chile* (founded 1958) in 1981 adopted a code of professional ethics that aims to promote at once the professional reputation and national subservience of engineering; it is, for instance, contrary to the code "to permit actions or omissions that favor or permit the unnecessary use of foreign engineering for objectives and work for which Chilean engineering is sufficient and adequate."

Finally, transnational or globalizing engineering associations have likewise been influenced by the American model:

- The *Unión Panamericana de Asociaciones de Ingenieros* (UPADI or Pan American Federation of Engineering Societies, founded 1949) in the 1980s adopted a code of professional ethics that stressed professionalism but then in 2003 created a new code stressing activity that benefited "clients, society

and the environment, optimizing the use of resources and with reduced generation of wastes or any types of pollution."
- The European Federation of National Engineering Associations (FEANI, founded 1951) in 1988 adopted a "Code of Conduct" that obligates all members "to be conscious of the importance of science and technology for mankind and of their own social responsibilities when engaged in their professional activities."
- The World Federation of Engineering Organizations (WFEO, founded 1968) in 2001 adopted a "Model Code of Ethics" in which "professional engineers shall hold paramount the safety, health and welfare of the public and the protection of both the natural and the built environment in accordance with the Principles of Sustainable Development."

All three types of engineering ethics globalization—engineers in advanced countries other than the United States adopting ethics codes, engineers in developing countries formulating ethics codes, and transnational engineering associations creating ethics codes—reflect influences from the United States, although not exclusively. They highlight in general terms obligations to some version of the common good and especially to public safety, health, and welfare or professional loyalty to clients or employers. There are few if any codes on the German model in which engineers are obligated to contribute to technology assessment, nor are there similar efforts to ground engineering ethics in epistemological or general philosophical reflection.

There are perhaps four reasons for the prominence of American over German influence. First, the fact that English has become a more global language than German makes American discussions more readily communicated. Second, the stigma of German engineering involvement with World War II may continue to exercise some negative influence. Third, ABET is in the process of becoming a de facto global accreditation agency; a number of engineering programs in other countries are now seeking ABET accreditation and thus having to address the ABET criterion for engineering ethics learning. Finally, the fact that the German approach requires engagement with a philosophical tradition of depth and complexity that runs from Gottfried Leibniz through Kant and Hegel to Karl Marx, Friedrich Nietzsche, Martin Heidegger, and Jürgen Habermas makes it inherently more difficult to imitate. Although engineering ethics in the United States has involved philosophers, the kind of philosophy applied (as, e.g., pragmatism) exhibits less historical depth and is simply less demanding than that present in the German tradition. Indeed, in most other countries the engagement of engineers with philosophers has also been minimal.

There are, however, a few important exceptions to this generalization: three examples worth noting are Denmark, the Netherlands, and China. In

Denmark, initiatives to reform engineering education in the late 1990s led to the introduction of a requirement in 2000 that all technical and engineering curricula at the bachelor's level include a course in the philosophy of science for engineers to be implemented no later than 2004. In response, a number of philosophers of technology from the United States were invited for consultations, and Danish philosophers themselves undertook to work with engineering educators to develop appropriate courses. One remarkable effort in this general area was spearheaded by Steen Hyldgaard Christensen and led to publication of a textbook on *Philosophy in Engineering* (2007). Another created the *Companion to the Philosophy of Technology* edited by Jan Kyrre Berg Olsen, Stig Andur Pedersen, and Vincent F. Hendricks (2009).

The Netherlands is home to what is arguably the most intensive pursuit of the philosophy of engineering and technology in any country. In a nation that is a geographical artifact designed and maintained by hydrological engineers and which appropriately enough has one of the most well-developed communities of science, technology, and society (STS) scholars, it was natural that philosophers at technological institutions of higher education would engage with engineers. In 2007 they established the interinstitutional 3TU Centre for Ethics and Technology to bring together expertise of the philosophy departments from three universities—TU Delft, TU Eindhoven, and Twente University—"to advance understanding of ethical issues in engineering and technology development" through interdisciplinary applied research, fundamental research, teaching, and public outreach (https://ethicsandtechnology.eu). The 3TU Centre—which in 2018 added the University of Wageningen to become a 4TU Centre—quickly became the most integrated, interdisciplinary engagement of philosophers and engineers in the world. Among its many publications are Ibo van de Poel and Lambèr Royakkers, *Ethics, Technology, and Engineering: An Introduction* (2011).

In China, as in the Netherlands, initiatives to link engineering and philosophy emerged at universities dedicated to engineering and often in conjunction with STS programs. From the late Qing Dynasty (1644–1911) the modern engineering professional emerged in China as part of an effort to defend against Western imperialism. The first technical school was the Fujian Shipping School, founded in 1866 in response to Chinese defeats in the two Opium Wars (1839–1842 and 1856–1860) in order "to learn the skills of the barbarians in order to fight the barbarians." Because the explicit goal of early Chinese engineering development was to acquire Western science and technology to defend Chinese culture, technical education necessarily included what might be called a philosophical component. From the founding of the People's Republic of China in 1949, this took the form of technical education that included a significant component of Marxist ideology in order to create

"Red engineers." In a study of the premier technological university of China, one American social scientist describes Tsinghua as

> China's consummate trainer of Red engineers.... [T]he university's party organization is renowned for grooming political cadres [so that] Tsinghua graduates occupy key positions in the upper echelons of the party and state bureaucracies, and one-third of the members of the Political Bureau's Standing Committee ... are alumni.
> (Andreas, 2009, p. 6)

Independent of this historical version of engineering ethics that promoted engineering loyalty to the Communist Party, a number of technological universities, including Tsinghua, also created STS and engineering studies programs that have promoted further engagements between philosophers and engineers. Such engagements are encouraged by increasing recognition of the historically unprecedented character of technological transformation and the need for general reflection on the contributions engineers are making to the remaking of China. There has even been an effort to collaborate with and use the Dutch 4TU Centre to establish a similar Chinese interinstitutional center for ethics and technology based at Dalian University of Technology.

FROM ETHICS TO POLITICS

Once ethical reflection ceases to focus primarily on guidelines for the behavior of individual engineers or what is best for the engineering profession, reflection readily follows the path mapped classically by Aristotle and Confucius, whose discussions of ethics blend into discussions of politics. Such a movement is manifest in multiple ways in North America, in Europe, and in Asia—with the political being given different conceptualizations and expressions in different contexts and traditions.

One possible general conceptualization might configure the movement from ethics to politics as another instance of globalization, in a secondary meaning of the term. Most commonly, globalization refers to external processes that lead to new local challenges and/or ever greater economic, political, and cultural interactions across national borders—something clearly represented by the emergence of transnational engineering ethics codes. However, globalization can also involve expanding some previously narrow perspective into a more holistic one. Taking a global perspective on investing in a new technological innovation, for instance, would involve going beyond the economic interests of shareholders to include multiple benefits and risks related to all stakeholders, including environmentalists. Globalization in this sense would naturally reimagine engineering in terms larger than the technical professional occupation.

In North America, for instance, engineering has been mostly understood in the narrow sense as a historically self-defining group constituted by occupation, discipline, and profession. From such a perspective, engineering ethics is equivalent to the professional ethics of engineers. This view has dominated engineering ethics not only in the United States but in many other countries. But such a view places a heavy burden of responsibility on individual engineers, often calling on them to exercise moral heroism as whistle-blowers in the face of economic, managerial, or political pressures to compromise technical standards that undermine functionality or safety. In response, a number of scholars have sought to consider some aspect of the broader political context in which engineers work.

Engineer-philosopher Joseph Herkert (2001) summarized such efforts using the distinction, original proposed by John Ladd (1980) between micro-ethics (dealing with relationships individual engineers have with each other, their employers, and clients) and macro-ethics (addressing issues of collective social responsibility of the engineering profession as a whole). Herkert compared the efforts of two philosophers (Ladd and Richard De George), two engineers (G. F. McClean and Willem Vanderberg), and an STS scholar (Richard Devon) to conceptualize the macro-ethical context and argued that none successfully integrated the micro- and macro-ethical perspectives. Herkert's own proposal was simply for more research on the responsibilities of professional societies as a whole; specific suggestions were the need for professional engineering societies to establish institutional supports for individual whistle-blowers and to develop statements on public policy issues, such as product liability. The idea that professional societies should support individual whistle-blowers had actually been argued for some time by Stephen Unger (1982 and 1994).

While important, Herkert has a thin view of the political. He makes only the most limited reference to the ways in which engineering transforms the political, as argued by Langdon Winner (1980) or seeks to establish some form of technocracy (see Olson 2015). Related efforts to conceptualize engineering in its broader social and political context can be found in work on STS and engineering ethics (Johnson and Wetmore, 2008), humanitarian engineering (Mitcham and Muñoz, 2010), and engineering and social justice (Lucena et al., 2010; Lucena, 2013).

Beyond these isolated efforts, a more systematic approach to the political is found in Philippe Goujon and Bertrand Hériard Dubreuil, eds., *Technology and Ethics: A European Quest for Responsible Engineering* (2001), which expands the engineering ethics perspective in both form and content. In regard to form, its collaborative character—with contributions from engineers, philosophers, sociologists, and historians from across Europe—is more extensive than anything previously attempted in North America. In regard to content, this was perhaps the first engineering ethics textbook to attempt a broad contextualization of engineering. In three major sections, the volume moves from considerations of (a) problems related to engineers in technical institutions, through (b) technical systems and technical decision-making, to (c) technical development as a social issue. The tripartite structure of each major section—historical and sociological description, case studies, and philosophical reflection—stimulates thinking about engineering in more than narrow professionalism terms. In fact, the very title of the volume suggests a need to link engineering ethics with the ethics of technology, something that, as one summary of the North American field noted, has been largely ignored.

The distinctive achievement of the European quest is indirectly highlighted in the introduction to a reprint collection of 57 articles from philosophy and social science journals that would "provide in a single volume the most important essays on engineering ethics in a form that should be useful to a scholar unfamiliar with the field" (Davis, 2005, p. xx). The editor notes the absence of an intersection between the philosophy and ethics of technology and engineering ethics. In his words, although "the two fields might seem to have much in common" the fact is that in North America "the philosophy of technology tends to focus on technology itself rather than on those who make it and, even when attending to those who make it, tends to lump engineers with other 'knowledge workers' in the omnibus of 'technologists,' ignoring the special standards of engineering as a distinct profession" (p. xvi). The European textbook not only seeks explicitly to connect philosophy of technology and engineering ethics but also works to bridge multiple disciplines and language communities while placing engineering in broad social and political contexts.

Another expansive effort to move from ethics to politics is that undertaken in China by Li Bocong at the Center for Engineering and Society of the University of Chinese Academy of Sciences. Li Bocong argues that engineering ethics needs to be complemented by the sociology of engineering and that the professional engineer needs to be understood as but one member of a more expansive engineering community. The engineering community is, in turn, established by an engineering project. Li's (2008) idea of an engineering "ethics of coordination" could also be read a form of engineering politics.

Finally, in two volumes on *Engineering Education and Practice in Context* (Christensen et al., 2015), the most common approaches to engineering emphasize engineering as a profession and/or as design. But engineering is broader than any one profession or occupation. There are "engineers" who are neither professionals nor designers (e.g., persons with engineering degrees who work as managers or investors) and that there are nonengineers who are involved in engineering projects (e.g., persons who have learned from apprenticeship and practice or have degrees in physics, chemistry, or even the social sciences). In recognition of these facts, Li Bocong (2010) argues for anchoring an understanding not in engineering as an isolated profession but an aspect of engineering projects. To do this, of course, suggests the need for some concept of a project so that an engineering project can be distinguished from, say, a political, economic, or artistic project.

In English the word "project" as a noun commonly references a large undertaking, often involving significant amounts of money, many personnel, and major equipment; it is planned out in advance. As a transitive verb, the term can mean "to propose," "to throw," "to set forth or calculate" (something in the future). As an intransitive verb, it can mean "to extend or protrude," "to use one's voice so as to be heard at a distance," "to produce a clear impression of one's thinking or personality," and in psychology, "to ascribe one's own feelings, thoughts, or attitudes to others."

Simple etymology deepens the appreciation of these straightforward linguistic uses. Its roots are in the Middle English *project(e)*, meaning "plan," from the medieval Latin *projectum*, *projectus*, past participle of *proicere*, to throw forward, extend, from *pro* (preposition, in favor of, for) + *jacere* (verb, to throw). The English word thus connotes a somewhat forceful imposition not just on the future but also into the present.

A political project could be exemplified by the action of establishing a colony on newly explored land or the creation of a political party to seek control of the government. Economic projects are associated with the founding of corporations or investments in money-making activities. Artistic projects create not just a single painting or sculpture but a collection of paintings or/and sculptures, buildings with a certain flair, museums. What is remarkable about engineering projects is the degree to which they partake of politics, economics, and material construction, even with aesthetic dimensions. To rethink engineering ethics from the perspectives of these various engagements is an effort that is only now emerging—and will need to include politics.

CONCLUSION

From its prephilosophical political beginnings in association with the emergence of engineering as a profession, philosophical engineering ethics has exhibited a trajectory that runs from discussions in two distinctive countries to discussions in many countries. Some attempts to thematize prospects for this historical process foreground the phenomenon of globalization (for one example, see Murphy et al., 2015; Luegenbiehl and Clancy, 2017). Our own proposal seeks to highlight how ethics points toward politics and the need for political philosophy. The movement could thus be summarized as one from politics to ethics to political philosophy. Our overview from this historico-philosophical perspective has been built around the following five theses:

1. Originally engineering ethics did not involve philosophy; instead, it was pursued by engineers alone for political purposes.
2. Two initial collaborations between engineers and philosophers took place in Germany (1950s–present) and in the United States (1970s–present). In Germany the engagement among engineers and philosophers included more than ethics; in the United States engagement tended to focus more narrowly on ethics alone (and reflecting a pragmatist philosophical heritage).
3. The influence of the German approach to engineering ethics has been less influential in other countries than the American approach.
4. As engineering ethics has become a point of discussion in many other countries, it remains the case that in most contexts engineering ethics has not been significantly involved with philosophers. Three exceptions are Denmark, the Netherlands, and China.
5. Finally, echoing the philosophical connection between ethics and political philosophy as found in Aristotle, Confucius, and others, engineering ethics would benefit from expanding critical appreciation of the political aspects of engineering.

CODA: TOWARD A POLITICAL PHILOSOPHY OF ENGINEERING

As we have presented to different audiences and colleagues our argument that the philosophical engagement with engineering needs to expand from ethics to politics, we have regularly been asked for more specifics about how this might work. The following reflections are a provisional response.

Engineering ethics commonly draws on different ethical theories or traditions to help individuals think about nontechnical problems they encounter in engineering practice. The result is to produce ethical analyses that involve, for example, consequentialism, deontology, and virtue ethics in traditions associated with Aristotle, Confucius, Jeremy Bentham and John Stuart Mill, Immanuel Kant, and others. As many have noted, however, the results often create new problems. One concerns how individual engineers can live up to various ethical ideals. Another is the challenges that arise when different ethical perspectives lead in different directions. Such problems are of a political character, the kind often dealt with in political theory. Yet most attempts to address ethical problems that call for a shift from thinking about individual behaviors (the primary focus of ethics) to thinking about the behavior of groups and the structures of social institutions (which is the focus of political science) have not yet made much effort to draw on the political philosophical work of the same philosophers and philosophical traditions referenced in ethical analyses.

One example could note Herkert's argument for the importance of a macro-engineering ethics that would stress the responsibilities of professional societies not just individuals. This proposal could be advanced by political philosophical reflection on how professional societies are themselves structured. Plato, for instance, argues that the social institution known as the state should be governed by philosophers or those who most embody reason. Aristotle argues that the best structure for a state is a mixture of aristocracy and democracy. Confucius argues that the state should be ruled by virtuous individuals who govern through example more than through law. Bentham and Mill propose a representative democracy that creates laws on the basis of a utilitarian calculus. Might it not be useful to critically examine the structure of professional engineering societies from these various philosophical perspectives? What might diverse theories of the state or democracy (the ideological common ground of the West) have to say about the structure of professional engineering societies?

Another example might examine what Li Bocong calls the engineering community—consisting not only of engineers but also workers, investors, and others—from the perspective of political philosophy. How might such a community function differently in Plato's ideal state, Aristotle's *polis*, a Confucian regime, or liberal democracy? Moreover, although the sociology of the engineering community will enhance our understanding of engineering and engineering projects, sociology is not a normative science. But the engineering community and its numerous projects should be subjected to normative criticism. The philosophical challenges from engineering concern not only how to conduct engineering projects in the right way but what are the right engineering projects to undertake. The political ideal of justice and the political philosophical traditions of reflection on the nature of justice have implications for engineering. Engineering ethics and the sociology of engineering must be complemented by political philosophical reflections, especially when dealing not just with the construction of dams and cities but especially when confronting the problems of climate change and proposals for geoengineering of the planet.

A final case would address directly the question of individual versus group or collective responsibility. The problem of the diffusion of responsibility in large-scale organizations and complex technological projects is one that repeatedly arises in engineering. Psychologists and sociologists have noted how people are less likely to assume responsibility for a problem when others are present; by themselves individuals who see problems are more likely to take action than when they are members of a group confronted by the same problem. Ethical arguments for collective responsibility of all members of a group for the bad behavior of some of the members, insofar as all ignore or tolerate the bad behavior even when they do not actively collaborate in it, need to be complemented by political arguments for institutional structures that promote such responsibility. Theories of corporate social responsibility have struggled with how to get firms to do more than simply act in the narrow, profit-making interests of the firm or ways required by law. In all such cases, it is reasonable to propose that critical reflection could be advanced by drawing on political theory.

What we suggest here, however, is only a beginning. Our basic thesis remains simply that engineering ethics could benefit from recognizing the ways that ethics is a prologue to political philosophy. Critical reflection on the challenges associated with engineering ethics deserves to be advanced by broadening the scope of discussion to include politics and political philosophy.

Part Three

Philosophy

(after Lydia Davis)

Philosophy is often not about real things, and then, when it is about real things, it is often at the same time taking the place of some real things.

Essay 17

Engineering Policy

Exploratory Reflections

Although it is common to talk about science policy, the term "engineering policy" is something of an anomaly. *Wikipedia* has a substantial article on science policy but none on engineering policy. A Google search for the term "science policy" (March 2017) turns up almost 3 million hits whereas "engineering policy" yields only 150,000 (~5 percent as many). Two other related terms, "technology policy" and "innovation policy," come in respectively at 2.2 million and 600,000 hits, making "engineering policy" something of a stepchild. Although engineering associations, such as the Royal Academy of Engineering in London and the National Academy of Engineering in Washington, DC, have web pages devoted to engineering policy, they provide little in the way of general reflection on the topic. They nevertheless invite consideration of the extent to which the engagements between science and public affairs to which science policy refers might involve engineering as well, perhaps even more than science. What follows is no more than a preliminary effort to explore this hypothesis.

CONCEPTUAL ISSUE: WHAT IS POLICY?

"Policy" is a term of relatively recent, primarily English language provenance. It is closely related to but not the same as politics. So close is the relationship that in French, the English "politics" and "policy" are equally translated as *politique*. In Spanish, "policy" is translated, depending on the context, as *política*, *norma* (rule), and *póliza* (as *póliza de seguros* or "insurance policy"). In German, "policy" is translated with even more differentiation: for example, as *Politik*, *Police*, *Regel*, and *Taktik*. According to political scientist Gerhard Vowe (2008), "policy" has entered the German technical vocabulary as a foreign word, to be distinguished from "polity" and "politics": the first is a social structure or an institution; the second is a "power struggle between players" inside this institution; and the third is to "the planned formation of social domains such as economy, environment, or education through collectively binding decisions" (Vowe, 2008, p. 620). Similar instances of "policy" as a loan word can be found in other languages. (Note also the close connection between the English "policy" and "police.")

According to the *Oxford English Dictionary*, "policy" first occurred in the 1400s; although closely associated with politics, it seldom (if ever) appears in English translations of classic texts such as Plato's *Republic* or Aristotle's *Politics*. Its linguistic presence has nevertheless climbed steadily from 1800 (with a Google n-gram of 0.0060 percent) to 1980 (when its n-gram was 0.0220 percent—roughly a fourfold increase).

Simplifying for present purposes, politics concerns power, whereas policy concerns reason. (Google n-grams again provide the modest empirical support for this analytic distinction: "power politics" is three times more likely than "power policy"; "rational policy" is five times more likely than "rational politics.") Political actions are characteristic of the state, and the state, according to Max Weber, is defined not by any distinctive ends but by its distinctive means: physical force. "In the past, the most varied institutions . . . have known the use of physical force as quite normal." Violence was an accepted part of family life and in religious institutions. "Today, however, [the] state is a human community that (successfully) claims the monopoly of the legitimate use of physical force within a given territory" (Weber, 1946, p. 78). The successful monopolization of force can be justified on the basis of tradition and/or raw power or the threat of its use. In all cases, however, political decisions are ultimately based on force: either by a group that has a monopoly on weapons or by a democratic majority (aggregate force). In politics, reason remains in the background.

In policy, however, reason comes to the fore: an effort is made to replace violence with reason. Policy decisions claim to be rational decisions; they speak truth to power. They can result in the use of force, but are not made on the basis of force, except the nonphysical force of reason. Insurance companies write their policies on the basis of statistical information about the likelihood of certain events. Democratic legislatures make laws, primarily on the basis of interest group pressures. In these laws, power is often delegated to some community of experts to determine the precise policies that need to follow. A law may be passed to make transport safer or protect the environment, but the formulation of precise policies for transport regulation or environmental protection is delegated to government agencies with specialized knowledge and expertise. A useful overview of the workings of this process and its problems in the U.S. Environmental Protection Agency and Food and Drug Administration is provided by Sheila Jasanoff (1990).

The European Enlightenment witnessed a historic effort to replace politics with policy—that is, to replace tradition (especially religious tradition) and physical violence with reason, especially the reason of science, as a key determinate of social and governmental decision-making and action. To this end, the state established agencies to provide it with scientific information about such things as the size and health of the population (biopolitics), economic activity, and more. State support of religion was replaced with state support of science (see, e.g., Ezrahi, 1990), as exemplified especially in the American and French republics. One argument for this replacement was simply effectiveness. Science provides the kind of knowledge that can make political decisions more effective. Political decisions made by people simply because they have power (or the will to power) often fail in the pursuit of their goals, as in the failures of the Crusades, the Spanish Armada, and Napoleon's invasion of Russia. A strong will is not enough to guarantee success, even when the strength of will is based in democratic agreement and determination.

The policy promise can take three overlapping forms. First, scientific research should provide information on how politically determined goals might be operationalized. Second, a scientific assessment of political (i.e., nonscientifically determined) goals should be able to veto or modify any attempt to pursue goals that are not feasible. More positively, science can provide background knowledge for the political formulation of goals that are feasible. An example of the first type would be a political decision to provide safe drinking water to a city, with scientists to provide standards, since politicians do not have the necessary knowledge to determine what constitutes safe drinking water (e.g., what microorganisms and chemicals might be present in the water and what might be harmful in what ways to the users). An example of the second type might involve scientific criticism of a political decision to send humans to Mars. Given the technical means available at present and what we know about human physiology, a human mission to Mars would not work; the astronauts would die. Third, the same knowledge base would be consulted in advance to establish the realistic feasibility of different space exploration projects.

Laws are rules created by the state to regulate its members and are ultimately enforced by power. But laws created simply by power alone are often ineffective. In contrast, policies may be considered to be knowledge based and therefore may serve as rational guidelines for behavior for the more effective realization of well-chosen goals and clearly designated outcomes. Insofar as science makes a dominant claim to knowledge and reason, "science (scientific) policy" is a pleonasm. Of course, reason can also be mistaken; scientifically based policies are based on fallible reason.

In all its promises, science nevertheless remains ultimately subservient to politics. Science concerns means, not ends. Politics or power decides what goals to pursue, then science advices whether, to what extent, or how such goals are able to be pursued effectively.

Beyond such promises concerning the choice of means in public affairs, some have made a more expansive claim. For many scientists, the scientific way of life is a good that they propose as a goal deserving state support. John Dewey (1927) and Michael Polanyi (1962), for instance, have argued that the methods of science should serve as a model for politics. Additionally, some scientists argue that the scientific study of human beings or human evolution provides knowledge about the true goals of human behavior. In political regimes where many citizens question the influence of science (as in the United States), this third claim is extremely contentious.

As a first general observation, note that the science referred to in some of these examples is as much or more engineering. Providing advice on safe drinking water standards or human flight to Mars involves engineering knowledge, even though this is not commonly acknowledged in the science policy terminology.

BACKGROUND: CLASSICS IN SCIENCE POLICY

Although the movement from politics to policy is rooted in Enlightenment political theory, the development of policy theory had to await the twentieth century. Taking note of this background can be done by referencing three classic formulators of science policy discourse: Harold Lasswell (1902–1978) and Harvey Brooks (1915–2004) in the United States and Jean-Jacques Salomon (1929–2008) in France.

Political scientist Harold Lasswell coined the term "policy sciences" to refer to all sciences insofar as they can be made relevant to public policy formulation. The concept grew out of his positivist view of political science as the study of power dynamics in society. Lasswell's *Politics: Who Gets What, When, How* (1936), which contributed to the behavioral research program in political science, saw elites as the primary power holders. In numerous other works (1927, 1930, 1948, and 1969), he examined how elites could beneficially influence publics, and argued especially for the social sciences to apply themselves to serving a democratic commitment to social justice.

Most important for the present context is Lasswell's 1951 article on "The Policy Orientation." There, he contrasted policy and politics by describing "'policy' [as] free of many of the undesirable connotations clustered around the word political, which is often believed to imply 'partisanship' or 'corruption'" (p. 5). The policy sciences are further described as the content of all sciences appropriately marshaled for increasing the intelligence of decision making, along with the scientific study of the decision-making process itself. The need for both arose from the increasing complexities of contemporary political life under Cold War tensions and engineered technological change.

Lasswell names multiple social science disciplines—economics, psychology, sociology, anthropology, social work, social geography, history—as contributors to the development of the policy sciences. Relevant as well is "the knowledge of atomic and other forms of energy which is in the possession of the physicists and other natural scientists" (Lasswell, 1951, p. 14). He further describes "the problem attitude"—that is, a focus on solving problems—as central to the policy orientation. Despite the fact that nuclear engineers know more about atomic energy than physicists, and that engineers self-describe their profession as dedicated to problem solving, engineering is conspicuous by its absence in the Lasswellian constellation of the policy sciences. (This is also largely true in the broader field of policy studies.)

Harvey Brooks—physicist, dean of Engineering and Applied Sciences at Harvard (1957–1975), and member of the Science Advisory Committee in the administrations of Presidents Eisenhower, Kennedy, and Johnson—introduced a variant of Lasswell's distinction between content and method as one between "science in policy" and "policy for science." Science for policy is concerned with bringing science to bear in public policy decision-making, whereas policy for science deals with examining and optimizing "the mechanisms, institutions, and operating principles through which federal resources are channeled into scientific and technological activities" (Brooks, 1968, p. 254). In contrast to Lasswell, Brooks is more concerned with the latter than the former. From this perspective he considers, for instance, the strengths and weaknesses of the decentralization of governmental science funding versus the establishment of a federal ministry of science, or the funding of basic versus applied research.

Criticizing the positive views of science for policy (as in Lasswell) and policy for science (as in Brooks), the French philosopher and OECD administrator Jean-Jacques Salomon articulated the first extended criticism of the notion. In *Science and Politics* (1973), Salomon sought "to denounce [the] misuse of science and technology [and] the complicity of most scientists [in politics] " (p. 255). This denunciation proceeds through a broad brush history of the relationships between modern science and society to a critical examination of "politics in science" and "science in politics." The latter involves scientists attempting to exercise political power and naively failing to appreciate their incumbent responsibilities at both national and international levels. For Salomon, it is an illusion that scientific policy can ever replace politics, because science itself is infused with politics.

SCIENCE, TECHNOLOGY, AND ENGINEERING

In their positive stances, neither Lasswell nor Brooks makes strong distinctions among science, technology, and engineering; instead, they tend to subsume technology and engineering within science. Folk philosophy distinctions, though contested, are still useful here: science produces knowledge, engineering produces technologies (from large-scale structures and infrastructures to consumer goods). Communities of scientists and engineers clearly differ, and technologists seem always to be ranked a little lower in the society in which everyone nevertheless desires technologies. The n-grams for these three terms reveal a relatively steady English linguistic presence of "science" from 1800 to the present, a meteoric rise in "technology" from the 1940s (from 0 percent to the point where it is on par with "science"), and a slow rise of "engineering" (from very low in the mid-1800s to about one-third as prominent as "science"). Insofar as "technology" refers more to artifacts than to knowledge, as our lifeworld is transformed into a techno-lifeworld suffused with technological objects, "technology" just naturally increases in usage.

These distinctions are mirrored in common policy parlance. Using *Wikipedia* again (accessed March 2017, as a source that reflects common intellectual opinions), science policy is defined as concerned with "understanding the processes and organizational context of generating novel and innovative science and engineering ideas"; technology policy as the "public means for nurturing [technology or the 'capabilities, facilities, skills, knowledge, and organization required to successfully create a useful service or product'] in the service of national goals and the public interest" (quoted from American science policy advisor Lewis Branscomb, 1995, p. 186); and industrial policy (a term whose n-gram numbers eclipse both "technology policy" and "science policy") as an "official strategic effort to encourage the development and growth of part or all of the manufacturing sector." As already mentioned, there is no article on, and hence no *Wikipedia* definition for, engineering policy. But insofar as all these common beliefs focus on some forms of policy for promoting science, technology, and industrial development, they implicitly include engineering. Engineering policy would then be constituted by efforts to promote engineering education, research, development, and construction for the "use and convenience" of national goals and national commercial enterprises.

The extent to which science policy discourse has become focused on policy for science, engineering, and technology at the expense of science for policy reflects the degree to which science, engineering, and technology have become integral to the techno-lifeworld. In the science for policy arena, even when unthematized, engineering for policy is pervasive. Yet, applying

Brooks' analysis, "engineering for policy" and "policy for engineering" point in different directions. Engineering for policy includes both engineering advice to policy makers (where best to site a dam or airline safety certifications) and the use of actual engineering to achieve policy goals (designing, constructing, and operating an electric grid, water system, or public transport infrastructure). By contrast, policy for engineering focuses on how much to promote engineering education or to encourage high standards of engineering construction or ethics.

Despite their differences, the common marginalizing of "engineering" in Lasswell, Brooks, and Salomon all contribute to an on-going failure to appreciate the importance of engineering. There is a failure to appreciate the extent to which science policy is what it is often only insofar as science takes practical form in engineering. For one example, take the handbook on *Science of Science Policy* (Fealing et al., 2011), which aims to examine how science policy really works. Throughout chapters on theory, empirical research, and practice—with repeated references to the need to promote innovation and solve problems—engineering remains at best a marginalized stepchild; the term "engineering" does not even occur in the index. The only chapter that begins to acknowledge the central role of engineering to enhancing public policy decision-making is on "Technically Focused Policy Analysis" written by Granger Morgan, the founding director (1976–2014) of an academic engineering and public policy program at Carnegie Mellon University. It includes a short paragraph reference to "engineering analysis" and notes that "simple ignorance or misunderstanding of the natural world or of engineered systems will . . . lead to silly and ultimately unrealistic policy outcomes" (Fealing et al., 2011, p. 127).

So far then, the conceptual argument has proceeded by criticizing the adequacy of the understanding of what science really is in science policy. The science in science policy almost always includes engineering, without acknowledging it as such. For example, the so-called science policy decision made by President Kennedy in 1961 to send humans to the Moon should more accurately be called an engineering policy decision. It is time to take off the mask of science policy advice and recognize it for what it often is: engineering policy advice. Against this conceptually clarifying background, the argument can now shift to more normative reflections on how engineering policy might best be practiced.

NORMATIVE ARGUMENTS: HENRY PETROSKI

In *The Essential Engineer: Why Science Alone Will Not Solve Our Global Problems* (2010), civil engineer Henry Petroski extends an argument threaded through more than a dozen books published since 1985. His thesis is that, although fraught with costs as well as benefits, engineering is a uniquely human activity with special abilities to make the world more humanly habitable. In the present instance, the argument is deployed with reference to policy questions involving climate change, energy, and related challenges. Petroski argues that the utility of engineering to policy requires an appreciation of its distinction from science and the range of engagements that engineering has with politics.

Petroski begins with a review of the ubiquity of risks in human affairs and maintains that risks are what science and engineering attempt to overcome. Using the risk of asteroid impacts, he argues that "scientists warn, engineers fix." Engineering is nevertheless more complex than it commonly appears. As much or possibly more than medicine, engineering deserves credit for two centuries of increases in human health. Although science can sometimes precede engineering, the opposite is often the case.

For example, scientists use engineering techniques when they construct hypotheses and the instruments for testing them. Additionally, even when engineering fixes, the fixes can have unintended consequences that need their own fixing. Petroski calls these "speed bumps," noting that speed bumps themselves illustrate the problem: while slowing traffic, they increase fuel consumption, pollution, and noise as cars slow and resume speed, and impede emergency vehicles. Good speed bump design requires systems engineering that, *plus respicere*, takes more into account than the speed bump itself.

Petroski then takes up specific public policy challenges and considers how science and engineering might address them. Leading off is a discussion about energy, with a broad overview focused on the public policy context in the United States over the past half century that has effected energy development related to nuclear, wind, solar, geothermal, batteries, oceans, pedestrian power, biofuels, conservation, fuel cells and hydrogen, and natural gas. In considering such a plethora of energies, Petroski reiterates his brief for a systems engineering approach by quoting "the legendary engineer-educator Hardy Cross" to the effect that engineering practice is involved with three trilogies: "The first is pure science, applied science, engineering; the second is economic theory, finance, and engineering; and the third is social relations, industrial relations, engineering" (p. 145). However, engineering is more related to social problems than to pure science, because "engineering is all about designing devices and systems that satisfy the constraints imposed by managers and regulators" (p. 155).

A further discussion of complex systems draws initially on the history of dam construction and its discontents to point up how science–engineering–society relations are becoming increasingly complex: first science supports dams as sources of power, then it criticizes them as causes of environmental damage. The "windshield wiper" input from science is equally well illustrated by a back-and-forth movement in healthcare debates: one study points up benefits from something that another study indicates is harmful. In truth, "the solution to problems involving complex systems can be expected to require the involvement of complex systems of people and approaches" (p. 172). Using the examples of earthquakes and hurricanes, Petroski distinguishes between uncertainty in science and engineering. "Generally speaking, the responsibility of the scientist qua scientist ends with the warning, which is where the responsibility of the engineer begins" (p. 185). Scientists can predict earthquakes and hurricanes with some level of probability, to which engineers can respond with designs utilizing safety factors, which are in effect efforts to mitigate uncertainties. But then policy makers and politicians must decide how to allocate resources among competing predictions, designs, and financial pressures.

Petroski concludes by reviewing engineering achievements of the twentieth century, challenges for the twenty-first century, and the newly emerging science and technology policy of seeking to meet challenges by offering large financial prizes for doing so. The review of achievements, as cataloged by the U.S. National Academy of Engineering (NAE), highlights the extent to which electrification, the automobile, airplanes, and more all depend on interdisciplinarity, can never be perfect or finished, and do not come without costs. "These are important lessons to remember when engineers look to tackling and are looked to for tackling the global problems that threaten planet Earth" (p. 211) as itemized in another NAE list of 14 Grand Challenges judged critical to future human flourishing. However, "as much as the inadvertent harmful by-products of technological achievement might be blamed for everything from local smog to global warming, it is also solid engineering and enlightened public policy that will be necessary to reverse the negative effects and bring forth new achievements for a new time" (p. 212).

Enlightened public policy becomes such, in Petroski's view, through the acceptance of engineering policy advice. An example regarding energy policy is the vision of a "2000-watt society" advanced by an engineering faculty at ETH Zurich. The aim is to reduce energy consumption to the world average of 2,000 watts per person in Europe (where usage is 6,000 watts per person) and the United States (a 12,000-watt society), while allowing countries such as China (a 1,500-watt society), India (a 1,000-watt society), and Bangladesh (a 300-watt society) to increase consumption. No serious technological breakthroughs are needed to achieve the goal, since Switzerland actually had a 2,000-watt society as recently as the 1960s. The only problem is the enlightened political will to realize the vision. But what evidence is there that such a political will exist or is likely to exist? While engineering is essential, even more so, Petroski suggests, is an enlightened public. Yet there is nothing in his argument that gives much reassurance that the many will become enlightened and either listen to the warnings of scientists or enact the design fixes of engineers.

NORMATIVE ARGUMENTS: ROGER PIELKE, JR.

Further normative issues are raised by considering the work of science policy analyst Roger Pielke, Jr. In *The Honest Broker: Making Sense of Science in Policy and Politics* (2007), Pielke argues for recognizing four different ideal type approaches to science for policy, that is, for scientists offering advice to politicians and policy makers: the pure knowledge exponent, the advocate, the arbiter, and the honest broker. To what extent does Pielke's analysis apply to engineers who might make design recommendations for addressing a public problem?

To illustrate his distinctions, Pielke imagines people asking for help in finalizing dinner plans. The scientist who acts as a pure knowledge exponent responds like a detached bystander lost in his own world; he describes the physiology of digestion and chemistry of nutrition, which may be interesting but is not helpful. An issue advocate, by contrast, acts like a salesperson and immediately argues for Joe's Steak House right down the street, perhaps with a peculiarly scientific justification about the special nutritional qualities of bovine muscle tissue. In a third case, an arbiter functions more like a hotel concierge. She asks inquirers about their dinner preferences: healthy nutrition, good economic value, quiet ambiance? Then she provides reliable knowledge to inform choice of options.

The arbiter supplies knowledge only after engaging with a client. Elaborating on this model, Daniel Sarewitz and Pielke (2007) describe it in terms of "reconciling supply of and demand for science." In those cases where expert advisory committees have worked most successfully (e.g., at the Food and Drug Administration), they have been structured so as to draw on the supply of science to meet a specific demand for science. Pielke thinks such good science arbitration is common and exemplified more or less well but deserves to be more clearly promoted.

Finally, honest broker scientists distance themselves from the immediate needs or interests of any inquirers and, without asking for contextual details, offer a matrix of information about restaurants in the area covering, for instance, nutritional value, price range, ambience, distance, and more. The effect will often be to stimulate re-thinking on the part of inquirers—perhaps a reconsideration of the needs or interests with which they may have been operating, even if they were not consciously aware of it.

In considering the strengths and weaknesses of each ideal role, Pielke admits that ideal types are seldom pure and often mixed. Along with the science arbiter, he nevertheless defends the importance of the honest broker, and is especially critical of what he calls a "stealth advocate": that is, the policy adviser who claims to be an honest broker, but is actually arguing for a particular action or set of actions. To what extent might Pielke's argument apply to engineers rather than scientists advising policy makers? To what extent might thinking about the applicability of his analysis to engineers even challenge Pielke's distinctions?

When applied to engineering, insofar as engineering can be well distinguished from science, the honest broker role seems peculiarly inappropriate if not impossible. In the first place, it is not clear that Pielke's ideal is possible. Are there any scientists who are not influenced by their own values in the kinds of research they do and the conclusions they draw and/or think significant enough to communicate to others? Honest broker scientists may exercise some detachment from the immediate needs or interests of any particular inquirer but will find it difficult to be as detached from their own personal interests. Scientists are unlikely to be doing research on the physiology of nutrition without some personal interest, and certainly what to include in any informational matrix will reflect what they judge to be significant. For instance, even an honest broker will be unlikely to include the political or religious affiliation of restaurant owners, rejecting this information as scientifically irrelevant, even though there are numerous Americans who would be quite interested in using this information to help inform a dining decision.

Additionally, it is not clear that someone seeking advice from scientists or engineers really wants an honest broker to just lay out a matrix of alternatives. As more than one policy scholar has noted, science can actually upset politicians and policy makers. President Harry Truman, for instance, is alleged to have objected to economic advisers who would say "on the one hand" and "on the other hand." Instead, Truman said what he really wanted was "a one-handed economist" (Haas, 2005, p. 386). At the same time, are there not occasions when politicians and policy makers might be better served by having adversarial advocates than honest brokers?

With regard to engineers, first, seldom will any engineer be tempted to act as a pure knowledge exponent. Indeed, this is probably a reductio ad absurdum option even among scientists. Engineering as a profession has built into it a rejection of knowledge for its own sake, always wanting to use whatever knowledge about the world will work to help solve a particular problem.

Second, insofar as engineers work in and for particular companies or institutions, they are unlikely to be able to function except as issue advocates for their specific practical skills and expertise. How could an aeronautical engineer working for Boeing Airplane Company, if invited to advice on constructing a national transport system, recommend anything other than air-transport? Aeronautical engineers would not know enough about either automobiles or trains to advocate for them—not to mention the fact that, were a Boeing engineer to venture such advocacy, it would likely be at the sacrifice of current employment. Instead, aeronautical engineering advocacy would need to be complemented by engineering advocacy from competing transport sectors.

Third, even more than scientists, it is hard to imagine engineers functioning as honest brokers. The job of engineers is to design particular, real-world solutions to problems that have been specified for them in advance. In the real world of decision-making, two (or more) handed engineers are even less welcome than two handed economists.

Finally, as with science, perhaps the most likely model for successful engineering policy advice is that of technical arbiters reconciling supply and demand: supplying an engineering pathway to meet an externally specified need. This is also to some degree a more natural activity for an engineer than a scientist. Engineers, by virtue of being professional engineers, necessarily communicate with some public, whether inside the company or out. Engineers qua engineers are strongly guided by well-expressed needs or interests from clients or the public.

NORMATIVE ARGUMENTS: NATASHA McCARTHY

In 2014, the Royal Academy of Engineering published a study examining interactions between engineers and government in the procurement of large public projects, specifically in information technology and communications (Royal Academy of Engineering, 2014). Working with the project was philosopher Natasha McCarthy, subsequently Head of Policy at the Royal Society. In a brief reflection on that report, McCarthy identifies "the challenge of bringing engineering knowledge and engineering practice into policymaking" as "rooted in the relationship between the technically focused specification of systems and the socially, economically and politically shaped statement of requirements" (McCarthy, 2017, p. 140).

Drawing on the work of Thomas Kuhn, McCarthy considers way to bridge the incommensurability between these two descriptions. She does not think it is realistic to attempt to create a common language, precisely because of the incommensurable character of terms such as "risk" and "value" as used by engineers and politicians. She also considers the idea of creating a trading zone pidgin language, as when engineers and physicists collaborate in laboratories, according to Peter Galison (1997). The problem here is that "pidgins are by nature simplistic and limited in use, and might fail" if used to design a large complex system (p. 148). Instead, what McCarthy proposes is a "systems architect who can see where communication could break down, and where and why divergent expectations emerge among the different stakeholders" when undertaking the government procurement of a large public project. The systems architect would, in effect, constitute a "third man" to mediate between engineering and politics.

CONCLUSION

The modest exploration here, developed from both conceptual and normative perspectives, has been that science policy is often actually engineering policy. Normatively, engineering advice is critical to good public policy decision-making, often more so than scientific advice. Just as with science, however, there are problems with the public acceptance of engineering advice. In many instances, public policy needs may be restated in technical specifications that import a bias toward engineering, which, in effect, can constitute a weak form of technocracy. Appeal to the mediation of systems architects may not necessarily meliorate this difficulty. The argument here thus remains fundamentally incomplete.

Essay 18

Energy Constraints
(with Jessica Smith)

Technical and public discussions of energy are unreasonably narrow. Although virtually everyone today thinks that the engineering of energy production and its commercial utilization are important, conceptualizations and responses to energy issues differ wildly—and yet remain remarkably constrained in what they take into account. The framework for discourse about increased production (whether in oil, natural gas, or even renewables), profitability, environmental protection, and tax regimes has not fundamentally altered in the West since energy became a central theme in policy debates. Yet the issue of energy—precisely because it has become so intertwined with the ways of life and self-understandings in highly engineered societies—deserves much broader reflection. Consider just two approaches that could help enhance public and academic intelligence about this issue: anthropology and philosophy.

ANTHROPOLOGIES OF ENERGY

It was anthropologist Leslie White in the 1950s who, developing ideas of Lewis H. Morgan, gave the first extended expression to what is arguably the most widely accepted view of the energy-society relationship: the idea that particular forms of society are dependent on amounts of energy input. According to White, "Everything in the universe may be described in terms of matter and energy, or, more precisely, in terms of energy" (1959, p. 33). What is true of physical systems is true for biological and sociocultural ones as well. He quotes with approval from Nobel chemist Frederick Soddy, who was among the first to imagine nuclear power and maintained that the laws of energy are not only important in physics. They are fundamental "in the whole record of human experience, and they control, in the last resort, the rise and fall of political systems, the freedom or bondage of nations, the movements of commerce and industry, the origin of wealth and poverty, and the general physical welfare of the race" (White, 1959, p. 39). White might also have referenced chemical engineer A. R. Ubbelohde (1955), who maintained that the ideal political system, which he called "Tektopia," would rest on the back of large numbers of "inanimate energy slaves" as replacements for traditional human and animal slaves. A more well-developed, qualified, and policy-sensitive contemporary engineering analysis can be found in the extensive research and many publications of multidisciplinary geographer Václv Smil (from 2006 to 2017).

White put his own historical anthropology in quasi-engineering terms, with the formula $E \times T = P$, where E is energy, T is the technology of its production, and P is product (or goods and services). "A culture is high or low depending upon the amount of energy harnessed per capita per year" (White, 1959, p. 42). He proposed five basic stages of cultural development insofar as energy is derived from humans themselves, from domesticated animals, from plants (in agriculture), from natural resources (coal, oil, gas), and from nuclear energy. He summarized his view with a "law of cultural development: *culture advances as the amount of energy harnessed per capita per year increases, or as the efficiency or economy of the means of controlling energy is increased, or both*" (p. 56, italics in original). It is easy to see this view reflected in an American ideological refusal to consider reductions in energy production or use as anything other than a threat to the American way of life. Such a refusal reflects an incredible cultural reductionism.

White's grand narrative was formulated in the face of an emerging, now more prominent approach in anthropology to treat all cultures in their own terms and to abandon unilineal theories of progress. In opposition to White, for instance, Margaret Mead (1953) argued that although less technological cultures have something to learn from more technological ones, more technological cultures can also learn things from less technological ones. This observation has been reinforced in many ways, one being the simple quantitative finding that different countries produce and consume enormously different amounts of energy, with some lower consumption countries scoring higher on quality of life indicators than those with higher per capita energy coefficients. Yet Mead and others paid little attention to energy per se—and White's vision has connected in a remarkably self-serving, reinforcing manner with the dominant trend in the West to justify and promote a commitment to the ever-increasing production and consumption of energy.

If the anthropologists who followed White did not completely buy his thesis linking energy and progress, they and historians such as David Nye (1990 and 1998) did uphold his more general observation that transformations in energy use engender transformations in society and culture. Adding a critical edge, social scientists and humanities scholars contributing to *The Culture of Energy* (Rudiger, 2008), the Rice University "Cultures of Energy Initiative" (2011–present; see http://www.culturesofenergy.com), and a recent anthropological collaboration on *Cultures of Energy* (Strauss et al., 2013) document the complexity of energy-culture relationships. Ethnographers in the *Cultures of Energy* volume especially point out how the production and consumption of energy simultaneously distributes social and cultural power unevenly among world populations.

A strength and limitation of this research is that it tends to focus on particular sources or sectors rather than energy in general. Concerning nuclear, for example, anthropologists study scientists and engineers along with Native American uranium miners and communities living in national sacrifice zones such as open pit mines, garbage landfills, and chemical or nuclear waste disposal areas. Other anthropologies learn about oil from indigenous communities dependent on the land that is irrevocably changed by wells and waste sites, state officials grappling with aspirations of modernity, and corporate personnel attempting to position their operations as socially responsible. Still others trace the impact of U.S. dependence on coal by examining the well-being of workers, activists, and the environment. Anthropologists have been on the frontlines of the fracking boom in gas and oil production as well as efforts to expand renewable energy. In an exception to the pattern of sector-specific research, anthropologists investigate the impacts of climate change for the livelihoods and structures of meaning of the world's most vulnerable people.

Starting from the insight that people's use of energy shapes and is shaped by their understanding of it, anthropologists also explore different energy worldviews. Americans, for example, switch between religious, magical, and technical registers when discussing and attempting to define energy (Rupp, 2013). A similar sense of magic surrounds oil and its intoxicating promises of wealth (Weszkalnys, 2013). Rural electrification projects in the developing world bring to light the articulation of energy with established religious beliefs and practices and social institutions of marriage and kinship. In rural Zanzibar, for example, people associated electricity with Islamic ideals of purity and safety, even as the ability to stay up watching television past sunset results in some people missing morning prayers. The safety of lights relaxed but did not erase restrictions against men and women sharing social space (Winther, 2013).

PHILOSOPHIES OF ENERGY

A quite different but complementary effort to appreciate relationships between the engineering of energy and society exists in philosophy. Philosophers have paid little systematic attention to the phenomenon of energy, but what they have paid is revealing. First, the concept of energy is not as simple as is often presumed. Although we can have direct experience of burning wood, coal, and oil, energy itself is elusive. Engineers define energy as the capacity to do work and distinguish kinetic (motion) from potential (position) energy. For physicists, however, energy is a fundamental aspect of matter defined by the formula $E = mc^2$. Yet as Richard Feynman says, "It is important to realize that in physics today, we have no knowledge of what energy is" (1963, sec. 4–1). There may be "formulas for calculating some numerical quantity," but this leaves energy itself as something of an ontological mystery.

For physicist and historian Jennifer Coopersmith (2010), energy is a "subtle concept." However, the subtlety is greater than even Coopersmith argues, since there are notions of energy functioning in biology, medicine, and psychology that deserve to be related to those found in the physical sciences and engineering. Going further afield is the term 气 (*qi*, traditional form 氣) in Chinese acupuncture, which is commonly understood to reference an exceptionally subtle "energy."

The historico-philosophical analysis of the concept of energy in the West from Aristotle to Einstein suggests the need for more careful analysis than is usually found in talk about energy policy and politics. Aristotle's ἐνέργεια (*energia*) or active reality is only remotely related to the energy of early modern natural philosophy and mechanics. At the same time, David Hume in the 1700s found the terms "power," "force," and "energy" quite "obscure and uncertain" (*Enquiry Concerning the Human Understanding*, §49). Philosophical discussions in the 1800s and early 1900s postulated a *vis viva* present in both nonliving and living entities (see, for instance, Henri Bergson's concept of *élan vital*).

Second, the philosophical analysis of various social commitments to energy production and use can invoke a variety of ethical frameworks. White and others generally argue in a consequentialist or utilitarian manner that energy production increases human power and thereby raises the quality of life, sometimes understood in a circular manner as measurable in terms of energy consumption. (Does it make no difference what the energy is used for?) There is also an occasional suggestion that increasing energy use by humans is natural. Such a view is fundamentally teleological, arguing that inherent to human nature is a drive toward or attraction for energy, that energy production and use realizes or perfects human nature. From this perspective, efficient energy production takes on the character of a virtue. From a deontological perspective, it could be argued that rationality legislates a categorical obligation to maximize energy production. Giving White's argument a slightly different twist, it is possible to develop a philosophy of history that sees expanding energy productivity and use as the core of historical change. Last but not least, someone could propose an aesthetics of energy as beautiful. Certainly it is the case that large-scale energy explosions and energy projects such as dams evoke as sense of the sublime (Nye, 1995).

Taking this last possibility to an extreme is a theory of Georges Bataille (1991), as explored in an article by Robert-Jan Geerts and colleagues that is in one of the few truly philosophical approaches to energy. For Bataille, energy is not a means for the satisfaction of needs but satisfying needs is a way to dispose of excess energy. "Rather than understanding energy as a resource to help society advance, Bataille perceives society as the result of an energy surplus, and a surplus that is explosive and impossible to fully control" (Geerts et al., 2014, p. 113). Could this explain what can often seem, when examined with detachment, like a pathologically wasteful production and use of energy?

Finally, third, there are any number of economic and political philosophical questions that bear on energy production and use. Is the production of energy more properly managed by private corporations or public agencies? To what extent should states create institutions that foster cheap energy for citizens and consumers? To promote energy conservation as well as utilization? How should the dangers of explosive energy releases (whether chemical or nuclear) be managed? To what extent and how should human harms and risks of energy production and use be adjudicated and regulated? How are trade-offs between the degradations of the natural environment by energy production (pollution) or use (waste and climate change) to be measured and assessed? What is really meant by such apparently idealistic terms as "sustainable energy," "green energy," or "alternative energy"?

One of the most general ethical issue of energy production and use falls under the rubric of energy equity and justice. It is important from the perspective of justice to consider to what extent those who most benefit from an energy regime pay their fair share in terms of harms and risks and whether those who are most subject to harms and risks fairly benefit. There are issues of free and informed consent in regard to energy production and use just as in medical knowledge production and healthcare.

TYPE I VERSUS TYPE II ENERGY ETHICS

In the exploration of such a cluster of questions, it is important not just to promote analytic precision with regard to specific cases but to reflect on alternative ways to frame issues. There are at least two quite different frameworks that bear directly on and can easily modify common productive, economic, environmental, and political attitudes toward energy. For want of better names, call these Type I and Type II frameworks. The belief that there is a linear relation between energy and culture constitutes Type I. It necessarily assumes that energy production and use is a fundamental good. Skepticism with regard to such a linear relationship is the foundation of a Type II framework.

During the energy crisis of the 1970s—and the term "energy crisis" deserves more questioning than it usually gets—the radical sociocultural critic Ivan Illich challenged prevailing beliefs with a little book on *Energy and Equity* (1974). Although he admits the value of energy production and use up to a point, in counterfoil to White, Illich attacks the ideology of never ending growth and criticizes energy policies in the advanced and much of the developing world. "For the primitive, the elimination of slavery and drudgery depends on the introduction of appropriate modern technology, and for the rich, the avoidance of an even more horrible degradation depends on the effective recognition of a threshold in energy consumption beyond which technical processes begin to dictate social relations" (Illich, 1974, p. 8). Beyond a threshold abstractly defined as that between enough and too much, energy production and consumption begins to undermine the abilities of people to lead their own lives. Illich tries to get specific with regard to energy used in transport and argues that beyond about 15 miles/hour persons increasingly become passive consumers of travel. The simple comparison of the authentic auto-mobility of walking with riding in a so-called "auto-mobile" or flying in an airplane calls attention to increasing degrees of passivity and dependence on engineering and its intrusions.

For Illich the issue of equity is not the same as equality, fairness, or justice. Equity implies some level of ownership or engagement, as when one holds equities or stocks in a corporation. The problem with advanced forms of energy production is that they progressively depend on expertise and the alienation of a majority—turning citizens into consumers.

> What is generally overlooked is that equity and energy can grow concurrently only to a point. Below a threshold of per capita wattage, motors improve the conditions for social progress. Above this threshold, energy grows at the expense of equity. Further energy affluence then means decreased distribution of control over that energy.
>
> (p. 5)

High-level energy production and consumption necessitates technocracy—a tendency most evident with nuclear energy but present as well in any high-tech energy production system. Only engineers can know what is really going on, and even their knowledge will be limited by the complexities of what they have created. The energy system becomes as opaque as the inside of a computer.

Whereas Type I pro-energy ethics rests on acceptance of the validity of one or more arguments for the energy-civilization coefficient, Type II energy ethics raises one or more questions in the same regard. Energy is argued to be at most a qualified rather than an unqualified good; as perhaps necessary, but only up to a point, beyond which it can in multiple ways become counterproductive. In the form of a consequentialist or utilitarian argument, after crossing a certain threshold, increasing energy production and use reduces the quality of life. In teleological terms, stabilized or balanced energy use by humans is more natural than unrestricted increases. From a deontological perspective, humans are rationally obligated to limit not only their utilizations of energy but also its production. Historically, there are clearly questions to be raised about whether the grand narrative of human change can be characterized as simply one of progressive energy development. And surely there are instances in which energy is ugly—even sublimely ugly, horrible.

According to Illich,

> The energy crisis cannot be overwhelmed by more energy inputs. It can only be dissolved, along with the illusion that well-being depends on the number of energy slaves a man has at his command.
>
> (1973, p. 10)

Slave holding, of inanimate as well as animate slaves, has an inescapable effect on the slave owner. Energy gluttony is just as vicious as, and not so different from, overeating obesity or sexual lust. Today the radical question of how and to what extent energy production and consumption influences opportunities for leading the examined life, the only one that (according to Socrates) is truly human, has been largely suppressed in favor of the pursuits of efficiency or renewable energy.

To claim that the highly engineered way of life has become deeply intertwined with energy production and use is not the same as the energy determination thesis of White nor of popular self-understandings and assumptions. It is to admit that this is indeed the way many people today think. Some have chosen the image of themselves as energy-dependent beings—while bemoaning the extent to which energy production and consumption are constrained by economic, ecological, or political factors. But could it not be that energy production and use, when examined from the limited perspectives of engineering, economics, or politics, is itself a constraint on leading the good life? Anthropology and philosophy suggest that life is more than energy production and use. Are there not other perspectives from history to art, poetry, psychology, and religion that could further deconstrain and enrich the way we think about energy?

Essay 19

Can Philosophy Be Engineering?

As counterpoint to the claim that engineering should be infused with philosophy is the argument that philosophy deserves to be transformed by engineering. This is a thesis argued by William Wimsatt's *Re-Engineering Philosophy for Limited Beings: Piecewise Approximations to Reality* (2007). The volume collects and supplements 30 years of work that, beyond serving as a capstone summary for Wimsatt's philosophy of science, opens up another perspective on philosophy of engineering.

LEARNING FROM TRYING

Reacting against what he considered his professor father's messy research in biology, Wimsatt began university studies searching for certainty in physics, frustratingly moved into philosophy, and finally circled back to biology from a philosophically informed perspective. Along the way he did some engineering work for the adding machine division of NCR and picked up a BA in general studies and philosophy (Cornell, 1965) and a PhD in philosophy (University of Pittsburgh, 1971), where he focused on the philosophy of evolutionary biology, before assuming a position in philosophy at the University of Chicago. Although most well known as a philosopher of biology, his central argument is that the key to what goes on in both scientific research and biological evolution is something that takes place most prominently in engineering: modular heuristic bootstrapping (not his term). After being compelled to do his own bootstrapping at NCR, he formulated this as a general epistemic methodology and then used it to discover a similar process going on in nature.

> Theorists and methodologists of the pure sciences have much to learn about their own disciplines from engineering and the study of practice, and from

evolutionary biology, the most fundamental of all (re-)engineering disciplines. Our cognitive capabilities and institutions are no less engineered and re-engineered than our biology and technology, both collections of layered kluges and exaptations.

(Wimsatt, 2007, p. 6)

Wimsatt's fundamental intuition, which he repeatedly defends, is that human cognitive abilities are quite limited while being situated in a reality that is extremely complex. When seeking knowledge of nature, there is a serious mismatch. The best response is the kind of trial-and-error bootstrapping (take something that works in one context and try it in another) that goes on in engineering design, where there is no perfect solution to a design problem but only trade-offs that are, with luck or persistence, good enough for who or what they're for. Humans are limited beings who can only grasp reality in piecemeal approximations, that is, limited but nonetheless workable cognitions. Instead of saying that models lie (Cartwright, 1983), he takes them as partially (but genuinely) true.

What is remarkable is that using this method, Wimsatt also found it operating in biological evolution (as if confirming Hegel's view: "To one who looks rationally at the world, the world looks rationally back"). Just as there are no perfect designs for engineered artifacts, there are no perfectly evolved biological species or ecosystems. Nature is "a reconditioned parts dealer and crafty backwoods mechanic, constantly fixing and redesigning old machines and fashioning new ones out of whatever comes easily to hand." All engineering is really reengineering, which "has profound implications for the character of evolutionary products" (p. 10). Species are no more fixed (or natural) than artifacts. They can and are always open to being repurposed.

The general name for the bootstrapping method common to engineering, science, and biological evolution is heuristics. This is perhaps the single most deployed technical term in Wimsatt's book (which includes two appendices, one summarizing key properties of heuristics, another describing common reductionistic heuristics).

Heuristic principles are most fundamentally neither axioms nor algorithms, though they are often treated as such. As a group, they have distinct and interesting properties. Most importantly, they are re-tuned, re-modulated, re-contextualized, and often newly reconnected piecemeal rearrangements of existing adaptations or exaptations, and they encourage us to do likewise with whatever we construct.

(p. 10, Wimsatt's italics)

Not only are heuristics everywhere in human knowing and acting, they are the way biological evolution works: "It's heuristics all the way down" (p. 328).

Given this argument, it is remarkable and revealing that he makes not a single reference to Billy Vaughn Koen's (2003) rationalist argument for the primacy of heuristics. Wimsatt's approach is empirical not rationalist; his argument is a running criticism of rationalist methodologies of any sort. In this he makes common cause with what he calls "the softening of the hard sciences" in STS studies (to which he makes only passing allusions), while strongly objecting to any "generalized rejection of objectivity and espousal of a [deep] valuational (and usually socialized) conception of relativity" (p. 319). Instead, his view is simply that the "experimentalist in any science (pure or applied) has to be part engineer—to know what sorts of instrumentation are possible, and how to design, build, calibrate, and maintain it, how to tell when it is not working, or biased, how to improve it, and how plausibly to redesign, extend, or adapt equipment and experimental designs to other purposes" (p. 336).

Re-Engineering Philosophy concludes with some amusing autobiographical anecdotes illustrating the multiple entanglements of pure science, applied science, and engineering—selected stories from an atypical philosophical career. These experiential narratives of chapter 13 by a reflective practitioner provide a livelier impression of what goes on at the bench level of engineering work than observational (but highly theorized) descriptions by anthropological observers. In this, he complements with a more gritty voice STS scholars such as Harry Collins and Trevor Pinch (2002) and engineering narrators such as Henry Petroski (e.g., 1996). He summarizes his path as "from simplistic reductionism as a hopeful engineering physicist to a fascination with heuristic methods for dealing with complexity." Later, "experiences in theoretical applied physics and engineering design work stimulated my interests in methodologies for complex systems" (p. 313).

Rather than the often criticized idea that engineering is applied science, Wimsatt almost suggests that science is applied engineering—or perhaps more accurately, that science and engineering are both, under different circumstances, applications of each other. As he summarizes the "engineering perspective" in a glossary:

> A cluster of theses derived from the assumption that theory has much to learn from practice and application. Teleological: Design is design for an end. View scientific activities as functional, and evaluate their designs for that supposed end. . . . Relation to practice: Focus not only on theory and *in principle* arguments, but on the practical implications of a view of science, how to apply it, and how it must be adjusted or qualified to do so. The central role of heuristics as fallible inferential tools, rather than sources of certainty. Applied not only to our theories and methods as instruments, but also to our mental capabilities and inferences. Most engineering is re-engineering, recognizing that we rarely start from scratch, but will use what comes readily to hand, as quicker, cheaper, more

convenient. . . . This view is profoundly instrumental, but denies any necessary tension between instrumental usefulness and truth or realism.

(p. 354)

Wimsatt's philosophy places engineering at the center of a proper understanding of both science and biology, promoting a distinctive approach to the philosophy of science and evolutionary biology. It also has implications for issues in philosophy of engineering, although he does not directly consider these. To no small degree, Wimsatt does philosophy of engineering without saying so.

TOWARD AN ENGINEERING EPISTEMOLOGY AND METAPHYSICS

Epistemology and metaphysics have been fundamentally transformed in the modern period by the rise of natural science. What science tells us about the nature of reality are now taken as a normative standard for certified and reliable knowledge. This subservience is furthered by the fact that epistemology and metaphysics are (along with logic) branches of philosophy that are inherently tipped toward rationalism (more so, at least, than ethics, political philosophy, and aesthetics). This is true not only with a philosopher such as René Descartes; even David Hume's empiricist theory of knowledge relies on a reflective introspection of consciousness and thought experiments—more than experiments. The philosophical analysis of modern scientific knowledge stresses conceptual analysis to clarify the nature of knowledge, truth conditions, and scientific methodology.

With regard to metaphysics, Immanuel Kant again proceeds by means of conceptual analysis to deny cognitive salience to abstract entities such as God. Pushing Kant further, an analytical pragmatist such as Willard van Orman Quine argues for the "regimentation" of theory in a way that yields what he calls a "desert ontology" in which sets are the only abstract entities. The epistemology of science is paired with minimalist ontological commitments. According to scientific naturalist Chet Raymo (2008), Quine once commented on Hamlet's complaint that "there are more things in heaven and earth" than are dreamt of in philosophy with, "Possibly, but my concern is that there not be more things in my philosophy than are in heaven and earth."

Insofar as natural science is taken as the paradigmatic form of knowledge, knowing of any other type, including engineering, is analyzed in terms of its relation to scientific knowledge. The philosophical examination of technological knowledge, as with Mario Bunge (1967) and many others (see Houkes, 2009, for a brief overview), brings science into engineer. Wimsatt proposes to work the other way around: to take engineering as a norm and use it to examine scientific knowing. (Joseph Pitt, 2013, offers a complementary but weaker argument along this line.)

There is a challenging richness to Wimsatt's critical reflections on, and extensions of engineering that could be described as a philosophy of engineering from within. One of the distinctive claims of philosophy is that it is self-reflective. For Wimsatt, engineering is at least as self-reflective.

> Engineering . . . practice is part of its subject matter. Engineering design bridges . . . making, studying, designing, maintaining, manufacturing, and repairing objects and processes that show adaptive complexity, and also studying and creating methods for accomplishing these tasks. These methods are also engineered objects, thus part of the subject matter. And our bodies and minds are naturally engineered objects—evolved cognitive capabilities and all. Reason and rationality, traditional domains of the philosopher, are no less engineered than the rest. As [Daniel] Dennett suggests: "Biology is not *like* engineering—biology *is* engineering."
>
> (pp. 313–14)

For the quoted reference, see Dennett (1995, chapter 8, "Biology Is Engineering"). Somewhat before Wimsatt, Dennett made his extended argument regarding the unity of biology and engineering, one that complements Wimsatt and can be read as another brief for engineering as philosophy. However, by interpreting Darwinian evolution as an inflationary theory of everything, Dennett's project has its own problems (see Hösle and Illies, 2005), some of which may also have relevance for Wimsatt. Two leading influences on Wimsatt's philosophy of engineering have been Herbert Simon and Donald Campbell: "sciences of the artificial" and evolutionary epistemology. Now, as he says, "the legacies of Simon, Campbell, and others from the 1950s" have in the early 2000s returned to some prominence (p. 314). A major contributor to their return has been Wimsatt's philosophical explication of engineering.

Regarding sciences of the artificial: discussion of design methods has been a significant theme in engineering education at least since the 1950s and has been a theme in the philosophy of technology since the 1990s. At the same time design discourse has expanded across a whole range of activities: from architectural, engineering, and industrial to artistic, interior, advertising, fashion, and more. Simon's pioneering effort to provide a general theory of design was, according to Wimsatt, "double edged: he advocated formal and foundational methods but sought to build a theory of human problem solving on satisficing and heuristics" (p. 364n1). Wimsatt disagrees with the first edge in order to push forward the second. For Wimsatt, engineering design is just satisficing and heuristics, which cannot be formalized by foundational methods. Engineering design simply takes given elements and utilizes them in new contexts to address previously unanticipated problems.

As for evolutionary epistemology, Campbell saw evolution as a process present not just in biological systems but also in human cognitive abilities and cognition itself. "Campbell saw selection processes as the only methodologically acceptable explanations for fit between a system and its environment—embracing all levels of our cognitive, social, and cultural experience." Although not emphasized, "technological selection and evolution were implicit in his views"—views that have strongly influenced historians of technology and cultural evolution theorists.

> Engineering is full of heuristic methods applied to solve and rationalize the most complex construction problems we know. Engineering shows—writ large—the robust pragmatic realism . . . and other heuristic elements permeating methodology as practiced in all sciences . . . but often obscured in their more formal statements.
>
> (pp. 314–15)

For Wimsatt, a broad "'engineering' epistemology" will recognize and utilize "the characteristics of evolved adapted systems, and how they gain information from their imbedding in their co-evolved environment" (p. 134).

As for metaphysics, Wimsatt's "ontology of complex systems" (pp. 193–240) replaces Quine's desert ontology with a "rainforest ontology" with different levels of organization, perspectives, and what he calls "causal thickets." Wimsatt proposes an ontology based on an engineering-like criterion of robustness: "the use of multiple independent means to detect, derive, measure, manipulate, or otherwise to access entities, phenomena, theorems, properties, and other things we wish to study" (p. 37) to identify the locally real.

> Rather than opting for a global or metaphysical realism (an aim that bedevils most of the analyses of "scientific realists"), I want criteria for what is real that are decidedly local—which are the kinds of criteria used by working scientists in deciding whether results are real or artifactual, trustworthy or untrustworthy, objective or subjective (in contexts where the latter is legitimately criticized—which is not everywhere). When this criterion is used, eliminative reductionism is seen as generally unsound, and entities at a variety of levels—as well as the levels themselves—can be recognized for the real objects they are.
>
> (p. 195)

For Wimsatt, "the kinds of criteria used by working scientists" are those more clearly exposed in engineering. Engineering for robustness or safety, for instance, relies heavily on "multi-trait-multi-method matrix" (p. 196) of safety factors that for Wimsatt provides a model for how scientists want confirmation of a phenomenon with multiple measures.

As an aside, one question that does not seem adequately addressed in his rainforest ontology, perhaps because he considers it specious, is a distinction between natural objects and artifacts. Wimsatt uses the word "artifact" to refer not to human-made objects but to illusions or distortions in knowing that need to be corrected through better cognitive engineering. But is it not useful to distinguish between a humanly engineered bed and a naturally engineered tree? In one parenthetical comment, he simply notes the exceptional complexity of biologically evolved constructions for which "we lack any design plans" and have to practice "reverse engineering" to figure out how they work "and why they are designed" the way they are. But this commonly yields only a

> simplified and partial account. Because these "designs" are layered kluges, we can't expect to find a "unitary" design plan "from the ground up."
> (p. 315)

But since humanly designed artifacts are often, as in biological evolution, layered kluges, this seems an inadequate way to distinguish an engineered bed versus from a natural tree.

Earlier in the book, Wimsatt also argued that neither genetics in nature nor genetic engineering design from scratch. Genetically engineering molecules "are not examples of *ab initio* constructions, but rather examples of the conversion of naturally occurring organic factories to the production of other products." "There is some assembly to be sure, but it is assembly of the jigs on the production line and sometimes rearrangement and redirection of the line—not construction of the factory" (p. 202). What engineering does is to assemble "complex systems out of simpler parts, a process that can be iterated" (p. 206). The result is a vision of technology that imitates nature that imitates technology—in ways that would seem to undermine any privileging (Aristotelian or other) of nature over technology.

THE QUESTION OF ENGINEERING

Despite provocative and initially credible arguments, there are some gaps in Wimsatt's philosophical reengineering program. One is the just mentioned slighting of the ontology of artifacts, a topic of long-term philosophical interest and recent attention in the philosophy of technology (see, e.g., Hilpinen, 1992; Kroes, 2012; Newberry, 2013; and Franssen et al., 2014). However, perhaps the most serious is a failure to take into account the contested concept of engineering itself. Relying mostly on personal experience and contacts, Wimsatt adopts what might almost be called a folk concept that fails to engage with substantial philosophical efforts to think through the whatness of engineering. Since its emergence as an explicit profession (beginning c. late 1700s), what engineering is about when it constructs, manages, and studies artifactual products, processes, and systems has been (self)understood in various not mutually exclusive terms but with differential emphases.

The classic, often quoted or adapted definition is that engineering is "the art of directing the great sources of power in nature for the use and convenience of [humans]" (Charter, Institution of Civil Engineers, 1828). This is a definition with three key components—art, power in nature (versus, e.g., political or economic power), and use and convenience—each is able to serve as the fulcrum for further conceptual refinement. Economist Thorstein Veblen (1921) sees engineering as the pursuit of efficiency; Byron Newberry (2015) subjects this idea to critical examination. Philosopher scientist Mario Bunge (1967) conceives of engineering as applied science; historian of technology Edwin Layton (1971 and elsewhere) has strongly criticized the adequacy of this conceptualization. In dissent from cultural historian Lewis Mumford's (1967) criticism of engineering as the pursuit of power, engineer-philosopher Taft Broome (1997) thinks of the engineer as on a heroic quest. Philosophers Michael Davis (2010) and Heinz Luegenbiehl (2010) argue for sociological and functional definitions, respectively. Even within the community of engineers and philosophers who essentialize engineering as design, as with Simon (and Wimsatt), there are considerable debates about the whatness of design. (For a good overview of the state of the discussion, see Franssen et al., 2018, sections 2.3 and 2.4.) Engineering as essentially design cannot simply be taken as an unproblematic given.

One key issue here is related to the relationship between artisanal craft and engineering, which Wimsatt minimizes. When Wimsatt describes engineering design he emphasizes the elements of skilled contingent selection, jerry-rigging, and bricolage that are inevitably involved but sidelines engagements with and dependencies on science. In engineering, properly so-called these more intuitive practices are typically subordinated to or tightly linked with more systematic and scientific ones. For instance, in Bunge's (1967) account of engineering as applied science, there are two ways that science can be applied: by drawing on knowledge already established in science or by utilizing the methods of science to created new and more directly relevant knowledge and practices (the engineering sciences). In either case, the selection of particular types of scientific knowledge on which to draw or specific methods to be adapted from mathematical modeling and empirical confirmation will involve some creative and often layered kluges and exaptations. Yet because of the involvement of engineering with modern science, a new kind of making has come on the historical scene that is replete with philosophical challenging questions. Walter Vincenti's influential *What Engineers Know and How They*

Know It (1990) identifies six types of knowledge involved with engineering design work, none of which seem especially characteristic of craft making. Newberry's (2013) unpacking of the key features of engineering artifacts sees them as dependent on clearly specified functional requirements articulated in ways that do not take place in craft design.

QUESTIONS OF ENGINEERING ETHICS AND POLITICS

Finally, there is the question of engineering ethics, a theme crucial to any philosophy of engineering but conspicuous by its absence in Wimsatt. Engineering ethics can mean two quite different things. One would be a philosophical account of ethics in general as an engineered construction. This would simply apply Wimsatt's understanding of evolution as engineering to evolutionary ethics. The term "ethics" does not occur in Wimsatt's index, but in one of the two places where it does occurs in the text (p. 35), there is a hint that ethics could be understood as another heuristic. (The other place is a passing, autobiographical mention of having read Mill before he had "warmed to ethics.")

Another meaning would be a normative philosophical account of how to do engineering. This is certainly what Wimsatt wants to provide for a re-engineered philosophy of science. As he puts it, through "a species of realism" he aspires to

> provide both better descriptions of our activities and normative guidance based on realistic measures of our strengths and limitations. . . . A philosophy of science for real people means real scientists, real engineers, historians or sociologists of real science and engineering, and real philosophers interested in how any of the preceding people work, think about their practice, think about the natural worlds we all inhabit, and think about what follows reflectively and reflexively from these facts.
>
> (p. 5)

The normative guidance comes from descriptively identifying what truly functions to produce reliable knowledge.

It is a little strange that Wimsatt aspires to do this for science but not for engineering. Since he provides some heuristic selection guidance for science, why not for engineering? However, since he sees science as a kind of engineering, perhaps he simply thinks the normative guidance is the same for both.

But is this normativity enough? The idea of norms for science or engineering that are only internalist ignores some of the most pressing issues regarding both science and engineering. Even for science, internalist norms for reliable knowledge production are not sufficient. There is no discussion of the conditions under which science (or engineering) should be practiced, funded, or politically supported and for what economic, social, or political ends. Wimsatt's is a remarkably thin normativity that effectively assumes science is a good in itself, failing to engage with the new political philosophy of science as found, for example, in the work of Philip Kitcher (2003 and 2011).

The dangers here apply to science and engineering, to philosophy, and to the world. First, with regard to science: It is hard to imagine physicist Chet Raymo inspired by Wimsatt's engineering epistemology to find in "empirical knowledge of the sensate world" a deep revelation of its beauty. To focus so strongly on local, good enough production of piecewise knowledge runs a serious danger of losing contact with wisdom—of living in a rainforest ontology without being able to appreciate the flora and fauna. Something similar goes for engineering: becoming so caught up in constructive tinkering that one forgets to ask for what purpose.

Second, a philosophy that doubles down too hard on piecemeal engineering-like solutions to narrow analytic questions at the expense of larger reflections runs the danger of abandoning its own heritage. Plato's *Theaetetus* concludes (210b-d) with Socrates declaring the result of the inquiry into the nature of knowledge as having so far given birth only to "wind-eggs" and departing for the law court to respond to the suit brought by Meletus.

Finally, as engineering and its collaborator modern science enlarge human power via nuclear, biological, genetic, nano-, and geo-engineering, it is incumbent on engineering to recognize and to respond appropriately to the threats that scientifically enabled heuristics have placed at the threshold of the Anthropocene. The true philosophy of engineering should not be more of the same.

CONCLUSION

Reengineering philosophy, as advanced by Wimsatt, is primarily a reengineering of the philosophy of science. Arguing that humans exist in a complex world with cognitive abilities that are only partially successful in grasping reality, Wimsatt's adapts heuristic (do-whatever-you-can until it works at least good-enough-for-whomever-it's-for) methods from engineering. Engineering never builds perfect structures, but the ones it successfully constructs are quite real. In like manner, science produces knowledge that, while never certain, can be quite reliable. It truly does get hold of reality, although only piecemeal. Then using this method, science discovers that biological evolution works in much the same way, heuristically, not to produce perfect organisms, but ones that do fit into their worlds, so that we can have reasonable confidence that human cognitive abilities are good enough, if not perfect, for knowledge production. There is a virtuous circle at work in Wimsatt's philosophy.

As presented by Wimsatt, however, his circle has at most only reengineered a small portion of philosophy—and even in this portion, it is not clear that the reengineering has depended on engineering. Wimsatt's heuristics may well have truly got hold of something, piecewise, but, like most heuristics, has not succeeded perfectly. A philosophy focused almost solely on the theory of knowledge in science lacks the robustness with regard to safety that engineering requires of its constructions. If philosophy is to be reengineered, even more should engineering be philosophized.

Essay 20

In Conclusions

There is no simple conclusion. There are only comments—salted with partial conclusions.

As noted from the beginning, there is an unfortunately stepping-stone, non-systematic, reinventing-the-wheel character to these essays. A more accurate title might have been *Stumbling toward a Philosophy of Engineering*. This reflects the limitations of a very ordinary brain but perhaps also the extent to which engineering itself is a stumbling block. We do not yet know where we are engineering ourselves to—or how to think about it in a big way. Nevertheless, that synthesis has not been possible does not mean it should be completely abandoned. Comments and cross references may serve as pointers.

WHERE TO BEGIN?

"Science, Technology, Engineering, and the Military" is an opening plea for paying more attention to engineering as the source of technology and its deep involvement with warfare, actual and potential. As the addendum further notes, contemporary developments in weapons engineering make this relationship ever more fraught. The engineered world is a world becoming one of continuous, hidden warfare in which we are all—engineers and nonengineers alike—implicated if not complicit. Civilianized engineering is warfare by other means.

The collection of conceptual and ethical-political questions with which the essay ends provides a first cut template for any comprehensive philosophy of engineering. Missing from the template, of course, are others related to the epistemology, ontology, and aesthetics of engineering and the engineered. What kind of knowledge is provided by the engineering sciences? On what distinctive knowledge does engineering design rest? What is it about technological knowledge that enables it to be so readily captured by military (and capitalist) interests? What ontological and aesthetic distinctiveness might adhere to the engineered? A number of these left-behind questions turn up later.

Stepping beyond the opening gambit, a pair of essays on "Ethics into Design" and "From *Dasein* to Design" seek to illuminate a central element of modern techno-agency: engineering design. The first develops a distinction between two types of design (aesthetic and engineering), considers a spectrum of possible relationships between engineering design and ethics, and then suggests thinking of engineering design, because of the way it brackets itself from the larger world, as a kind of technically learned playfulness. When playing games, the first principle is to follow the rules. A second is not to let the game get out of hand, to become wholly self-contained, an end in itself, and to remember there is a more encompassing historical lifeworld. Insofar as engineering design is not able to take full cognizance of the materials and contexts within which it works, a fundamental ethical counsel would be: "Remember the materials."

This counsel might seem to anticipate and endorse Peter-Paul Verbeek's arguments (2005 and 2011) for a philosophy of technology that pays positive attention to artifacts via material hermeneutics and a theory of mediation (even though it fails fully to credit closely related work from predecessors such as Karl Marx, Lewis Mumford, and others). To some extent this is the case. There is much to be praised in Verbeek's attention to the detailed interactions of human-artifact-world relationships and the synthesis he constructs from the work of Don Ihde, Bruno Latour, and Albert Borgmann. In fact, however, the argument here and elsewhere implicitly questions a too ready philosophical enthusiasm for design (engineering and nonengineering alike) as manifested in much postphenomenology. The intimate relationships between the design industry and its manipulations of mass commodity fetishism, the power of the marketing industry, and critical consumer culture studies all give reason for prudent skepticism. As one example of relevant investigative reporting scholarship, see the critical historical-sociological exposures of Stuart Ewen (1976, 1988, and 1996).

This counsel is a restricted version of a principle proposed in the second essay and defended later with regard to English-speaking engineering as a whole: the practice of a duty *plus respicere*, to take more into account than is commonly done. Rhetorically (mis)appropriating a term from Heidegger, essay four sketches a phenomenological interpretation of designing as a special form of being-in-the-world, and then considers prospects for distinguishing authenticity from inauthenticity in the new engineering way of world making. Given the problematics of design and the enormous powers and dangers inherent in engineering, its retention and proper exercise call for a change in norms, demanding a proportional expansion of responsibility. Protection of public safety, health, and welfare (the standard paramount principle for conduct of English-speaking engineering) is not enough. The counter proposal for a duty *plus respicere* is returned to and further defended in essay nine.

There are a number of weaknesses with these two design-focus arguments. One concerns the *poeisis/praxis* distinction that crudely fails to reflect the nuances of Plato and Aristotle, not to mention our experiences today. The catalog of possible standard ethical theory engagements with design ignores virtue ethics, which has made a significant revival in relationship to technology; see, for example, the concept of technomoral virtues as developed by Shannon Vallor (2016).

The relationship between aesthetics and engineering design work should also have considered the industrial design program of the Bauhaus (1919–1933) under the philosophical leadership of Walter Gropius to integrate the arts, crafts, architecture, and engineering. Post–World War II, the heritage was reconfigured in the Ulm School (1953–1968) with a vision to reconstruct Germany that became inflated, especially under the philosophical inspiration of Tomás Maldonado, to a point where design was postulated as the master discipline for world creation. According to Maldonado,

> The designer will be the coordinator. His responsibility will be to coordinate, in close collaboration with a large number of specialists, the most varied requirements of product fabrication and usage; his will be the final responsibility for maximum productivity in fabrication, and for maximum material and cultural consumer satisfaction.

Speaking for designers themselves, he described "the task entrusted to us by society [as] the reconstruction of human environment in the new eras of scientific humanism" (both quotations from a memorial by Jorge Frascara, 2019, p. 95). In this vision, there are many elements to endorse—especially the insistence on knowledge as a basis for action. At the same time, this inflationary vision incubates a *hubris* that is a major contributor to the very wicked problems that engineering design now seems fated to engage. The vision seems all too clearly mirrored in what must be judged as the grand but unfortunately illusory design thinking of such public intellectuals as Bruno Latour and Peter Sloterdijk: the noble lie replaced with a noble fantasy. More modestly, enthusiasm for the potential of design thinking to alter the world for the good has spilled over into piecemeal design enthusiasms associated with technoscience.

The issue of design in nature, which was alluded to in the two design essays, is more complex than recognized but comes up again in essay 19. Design by simulation calls for deeper reflection. Finally, there is the issue of moving from designing artifacts to designing actions. The isolated reference to Edward Bernays suggests a need for critical engagement with his work; there is also relevant work by, among others, Michael Bratman (1987).

Prescinding from such qualifications, Robert Frodeman and Adam Briggle (2016) have applied the notion of a duty *plus respicere* in their critical account of the modern institutionalization of science. They argue for replacing a common ideological appeal to curiosity as a justifying value for much academic research in both science and humanities with "curiosity *plus respicere*."

> What we have in mind is a model that still allows humanities scholars to follow their curiosity, but to do so in a way that takes more into account. They would be asked to consider the audiences they want to influence, the kinds of change they want to inspire, and the mechanisms or pathways that might bring those changes about.
>
> (p. 148)

To extend Frodeman and Briggle's own extension, it could be proposed that *plus respicere* applies not just to engineering because engineering has become infused into the contexts in which science, the humanities, and even philosophy now exist. In a world transformed by engineering, there is no free lunch curiosity with free-standing serendipitous benefits.

Sandwiched between the two design essays, "The Importance of Philosophy to Engineering," is a simplified (if not simplistic) brief for philosophy as useful for engineers. More than any other, except for essay 15, it attempts to think with rather than against engineering. It is also the point of strongest contact with *Thinking through Technology* (1994) and its earlier effort to think engineering.

The fifth essay, "Professional Idealism among Scientists and Engineers," bypasses the methodological sophistications that have enveloped STS in order to call attention to instances of technosocial idealism in the technoscientific community. To some degree, this complements essay one by highlighting the fact that despite major military funding, there have been push-backs from scientists and engineers themselves, although with minimal (if still worthy) results.

Because "Can Engineering Be Philosophical?" was originally a plenary provocation at a conference on philosophy and engineering, it is a slight departure in tone. More clearly than other essays, it represents something central to a truly philosophical philosophy of engineering: thinking both with and against engineering. The philosophy of engineering is both for and not for engineers and, as such, is for all of us. The argument is a recurrent theme, that in some sense we are now all engineers and have an obligation as thinking humans to practice philosophy—as a distinctively human way of life—in the new techno-human condition in which we find ourselves.

"Convivial Software" attempts a constructive proposal by adapting the concept of conviviality (from Ivan Illich) to an assessment of computer software design. This constitutes the introduction of another norm into engineering, one that nevertheless is faced with almost insuperable barriers to implementation. Not mentioned in the essay is how the ideal also echoes the Aristotelian and Confucians ideals of friendship (e.g., *Nicomachean Ethics* VIII and IX; *Analects* 1, 1) and the incredible difficulties of negotiating the attractions of the faux friendships of facebook.

The final essay in part one, "Comparing Approaches to the Philosophy of Engineering," complements "The Importance of Philosophy to Engineering." There are more possibilities for philosophical engagement with engineering than essay three mentioned. The importance of allowing philosophical reflection to be stimulated by engineering language resonates especially in the repeated references to "use and convenience."

CONTINUING

Essay 9, "A Spectrum of Ideals in Engineering, Simplified," opens part II and its circling of issues related to engineering ethics and engineering education. It is the longest essay in the volume. It aims to map out a historico-ethical heuristic for interpreting engineering ethics in the United States and the contemporary principle of holding paramount the protection of public safety, health, and welfare. Its substantive argument, continued from essay four, again endorses a duty *plus respicere*, to take more things into account.

A serious weakness here and elsewhere with regard to assertions about a close linkage between engineering and capitalism is the absence of any deep discussion of capitalism. Capitalism as a distinctive economic-political formation originated with a disembedding of the economy (as analyzed by Karl Polanyi, 1944). Traditional societies are characterized by markets in which social relationships trump monetary ones. The disembedding of economic from social relationships has taken form variously in merchant capitalism, industrial capitalism, laissez-faire capitalism, managed or regulated capitalism, state capitalism, finance capitalism, and more. Each capitalism has manifested a close relationship with engineering that, on the surface at least, seems to exhibit a greater codependency than with, for example, craft production, the arts, religion, politics, or agriculture (until the development of agricultural engineering). The twenty-first-century emergence of supply-chain capitalism, in its dependency on engineered transport and communication speed and efficiency, is the most recent case in point. One hypothesis to account for the evident synergies between capitalism and engineering is what might be termed a conjunctive disembedding. Just as the market is a form of disembedded economics, so engineering is a disembedded making, which is precisely why each takes off with a kind of independence that shreds traditional cultures. Albert Borgmann's philosophy of technology and Peter Sloterdijk's *In the World Interior of Capital* (2013) offer some support for such a hypothesis.

"The Concept of Sustainability" examines the problematic popularity of this notion. The concept has become increasingly important to engineering and has strong normative implications. Whether or not sustainability can live up to the hopes placed on it deserves epistemological, ontological, ethical, and political philosophical attention—although essay 10 just barely scratches the epistemological and ontological surface.

The absence of the sustainability question in the political ecology of Bruno Latour is quite remarkable. Why in all his work on political ecology (2004, 2017a, 2017b) does he not even deign to criticize it? Adapting a description from his work, however, one might say that sustainability has undergone an "ecological mutation" and shifted philosophical focus from the plane of ethics and political philosophy to epistemology and ontology while creating a new fourfold hybrid. Whereas the traditional ethics-politics-epistemology-ontology fourfold practiced receptivity to what is real, an engineering mutation has made it projective of the real. No longer focused on aspiring to understand and live in harmony with what is, the projective or engineering fourfold asks what kind of reality are humans constructing and how does this come about. As a contributor to the philosophical prosecution of this projective (or design) ontology, Latour has been joined by his German colleague agent provocateur Sloterdijk as well as a number of others. Even if sustainability finds no place in discourse on the question of cosmopolitics, cosmopolitics deserves a place in sustainability discourse—a place it has yet to be fully accorded.

Essays 11 and 12 focus again on engineering ethics. "Engineering Ethics Education in the American Context" offers a selective overview of developments in engineering ethics education in the United States in order to make a case for extending ethics into the contexts of policy and political philosophy. A weakness is the lack of any examination of engineering education in its own larger sociopolitical context. It would have been good to relate the argument to the kind of general history found in Lawrence Grayson's "A Brief History of Engineering Education in the United States" (1980) or his more expansive book version (1993) combined with some of the new, less celebratory and more detailed, critical histories that are emerging in engineering studies scholarship (e.g., Jesiek and Jamieson, 2017).

"Notes on Engineering Ethics in Global Perspective" contains complementary observations and comparisons of the function of professional ethics in diverse sociocultural contexts. (Discussion of the situation in Germany is revisited in essay 16.) This analysis could have been strengthened by taking into account critical reflections on the globalization not only of engineering ethics but also of engineering. Byron Newberry, for instance, has called attention to how the inherently localizing character of engineering—that engineering projects are *"objectively local*, each one comprising a unique set of requirements, constraints, and influencing factors, independent of the methods used to address the problems" (Newberry, 2005, p. 9)—can, without proper qualification, make "engineering globalization" an oxymoron. Engineering may be a major factor in the globalization of transport and communications networks, but this does not make for globalized engineering, which continues to be, more than science, the handmaid of nationalist economics, power, and politics—though engineering tends to drag science along with it into the political and capitalist vortex.

Humanitarian engineering, the theme of essay 13, was invented in the late 1990s during parallel crises in humanitarianism and in engineering. At the end of the Cold War a humanitarian crisis arose on the international scene, with efforts to promote human rights as development independent of former political rivalries. The "humanitarian intervention" (war) in the Balkans in order to stop ethnic killing was in conflict with humanitarian opposition to war. At the same time, engineering was struggling to recruit students (especially women and minorities) into the discipline. The idea of humanitarian engineering was more appealing to some of these students than corporate or nationalist projects, and engineering action seemed inherently more compatible with humanitarian than military action.

One remarkable thing about humanitarian engineering is the late arrival in engineering quarters of discussions about human rights. Post World War II, commitments to human rights became a controlling political and moral ideology of the West. Following the 1948 adoption of the Universal Declaration of Human Rights, in the mid-1970s the American Association for the Advancement of Science (AAAS) took an initiative (see Edsall, 1975) that eventually led to a program for defending the human rights of scientists and using science to defend the human rights of everyone (essay five already referenced this issue). In 2005, the AAAS sponsored a conference that led four years later to the formation of a broader Science and Human Rights Coalition, which as of 2019 had 26 member or affiliated organizations.

Remarkably, the Science and Human Rights Coalition includes no engineering societies. Despite more explicit commitments in professional engineering to public welfare, engineers have lagged behind scientists in expressing concerns for human rights. It is mostly outside professional engineering that humanitarian engineering programs have promoted human rights. This is a failure that deserves more attention than it has so far received—and a shortcoming of the "Humanitarian Engineering" essay. The complex relationships between engineering and medicine (which is increasingly conceivable as a regionalized form of engineering) also calls for critical reflection.

Essay 14, "A Philosophical Inadequacy of Engineering," uses a problematic distinction between strong and weak professions to question the character of engineering itself. Strong professions (such as medicine and law) incorporate ideals into professional education and practice. Weak professions (the military and business) either lack such ideals or only weakly incorporate them in education and practice. Engineering aspires to be a strong profession but is functionally weak. The debate with David Goldberg recorded in the addendum could be opened much further by including discussions of ethics in the professional communities engaged with computers and information, media and communication technologies, and more. Imbrications between ethics in engineering agency and its delegation (intentionally or not) to artifacts agencies are equally ignored here.

Essay 15, "The True Grand Challenge for Engineering," is another effort to think more with than against engineering. The U.S. National Academy of Engineering Grand Challenges is a praiseworthy effort by the professional community to assume a measure of control over its own historical trajectory. This assumption nevertheless deserves to be challenged for its failure to think more deeply about what is happening within the engineering community itself and to the world at large as the world comes to be salted with engineering ways of living (one of which, of course, is thinking in terms of engineering grand challenges). At the same time, the idea of a New Axial Age has been proposed by a number of interpreters of the contemporary historical situation and needs more careful comparison between alternative descriptions if it is not to become a cliché.

Confucian philosopher Tu Weiming, for instance, asserted at the turn of the millennium that "there should be a broader and more effective communication among the different countries, cultural traditions, and civilization systems, so as to form a global dialogue of civilizations and even to bring about another axial age in human history" (Tu Weiming et al., 2000, p. 387). For Tu, the New Axial Age emerges from cultural globalization. By contrast, religious philosopher Karen Armstrong, after a restatement and elaboration of Karl Jaspers' original thesis (see interview with Roemischer, n.d.), proposes Western Enlightenment and its global spread as a Second Axial Age. A collection of studies by Shmuel Eisenstadt (2012) offers critical background for a more fully developed idea of engineering as the core of a New Axial Age.

Essay 16, "From Engineering Ethics to Politics," compares the emergence of explicit philosophical discourse about engineering in Germany during the 1950s and in the United States during the 1970s and then calls attention to global interactions. But given the broad political context within which engineering agency exists and its complex socio-environmental impacts, the adequacy of limiting philosophical discourse to ethics is questioned. Political theory has become as relevant as ethical theory to normative discourse concerning engineering. Just as different ethical frameworks such as consequentialism, deontology, and virtue ethics can differentially enlighten individual responsibilities in engineering practice, so can different theories of the state such as liberalism, conservatism, and socialism provide illuminating perspectives on engineering projects.

AND MORE

The theme of connections between engineering and politics is picked up and reexamined in the first two essays of part three: one on relationships between science policy and engineering, another on energy politics and policy. Essay 17, "Engineering Policy," as a modest exploratory effort, begins with a conceptual analysis of policy, interpreting policy as an effort to replace power politics with rational politics, and then does a truncated introduction to three classic contributors to the science policy literature, before reflecting on the ideas of a contemporary science policy theorist, an engineer's brief for engineering policy, and a philosopher's proposal for addressing engineering policy incommensurabilities.

It does not require much deconstruction to notice the hidden place of engineering in science policy discourse. Another text that should have been referenced is Vannevar Bush's influential post–World War II report to the president, *Science: The Endless Frontier* (1945). In proposing what has become the default framework for science policy, Bush maintained that basic research was necessary to "insure our health, prosperity, and security as a nation in the modern world" and that science must work "as a member of a team" (p. 5). Although "engineering" is mentioned only four times, it is an unstated assumption that engineering is the key element in this team (Bush was himself an engineer), the activity that enacts the industrial exploitation of scientific knowledge and procedures. Correcting Bush's title while advancing his argument, Sunny Auyang (2004) concludes with a quotation from Bush's subsequent book, *Endless Horizon* (1946):

> The impact of science is making a new world, and the engineer is in the forefront of the remaking. . . . He builds great cities, and builds also the means whereby

they may be destroyed. Certainly there was never a profession that more truly needed the professional spirit, if the welfare of man is to be preserved.

(Auyang, 2004, p. 299; Bush, 1946, p. 141)

The post–World War II attention to science policy and the engineering salience of science contributed to a broader shift in social prestige from science to engineering and technology that has been documented by historian of technology Paul Forman's widely cited historiographic analysis. "In the epochal global transformation from modernity to postmodernity [technology acquired], beginning about 1980, the cultural primacy that science had been enjoying for two centuries worldwide, and in the West for two millennia" (2007, p. 2). There are nevertheless questionable assumptions in the optimism in this transformation, as essay 17 concludes by trying to suggest; engineering policy aspirations may be deficient in appreciating limitations in the political conditions into which it inserts itself.

Essay 18, "Energy Constraints," suggests this even more strongly by challenging the adequacy of common engineering perspectives on energy production and use. Anthropological arguments that privilege energy as the foundational metric of civilization are ideological in the classic Marxist sense of rationalizing power interests. Reflections on energy regularly fail to appreciate the fundamentally problematic character of the concept. An alternative energy policy analysis would begin with a distinction between Type I and Type II energy ethics: the former of which accepts a fundamental obligation to increase energy production and use but only questions the means (e.g., in terms of sustainability), the latter of which questions the idea of any and every rising need for energy. It is as important to break the constraints on energy thinking as on energy engineering.

Finally, essay 19, "Can Philosophy Be Engineering?," takes up the engineering-philosophy relationship from a distinctly different angle, one that challenges the argument of essay six and to some extent a hypothesis underlying this whole collection. Unfortunately this essay does little more than call attention to philosopher William Wimsatt's provocative thesis that common processes are present in human cognition and engineering construction. For Wimsatt, the philosophy of science and epistemology must be reconceived in engineering terms. Because, remarkably, it is fundamentally the same engineering process that is at work in natural evolution, ontology too needs to be rethought from an engineering perspective.

Peter Galison's argument for the emergence of a new "engineering way of being within the sciences" along with Alfred Nordmann's thesis that the Enlightenment scientific enterprise is now confronted by the rise of regime of technoscience can complement Wimsatt. Galison focuses on physics, where he finds the opposition between realism versus instrumentalism (ontology versus anti-ontology) giving way to ontological indifference. This new engineering ethos in science is illustrated with an imaginary exchange between a critic and a nano-physicist. The critic complains that the nano-scientist is not doing real physics but only engineering. The nano-physicist replies,

> I want to make things, to construct devices at the smallest possible scale. . . . I am concerned about robustness and scalability—not whether something is "real." Ontology is simply irrelevant to me.
>
> (Galison, 2017, p. 19)

Galison thinks this attitude has even found purchase in string theory in "an engineering way of being that values the making and linking of structures with little regard for the older fascination with existence for its own sake" (p. 25).

Nordmann's conception of technoscience exhibits similarities (in fact, he quotes Galison). Technoscience is not interested in an Enlightenment progress toward truth but with innovation.

> For technoscientific innovation, the future is merely a repository of technical possibilities that await to be realized. In the meantime its task is framed entirely by the present as one of matching up supply and demand—technoscientific capabilities and the societal or environmental problems that call for technical solutions. . . . Technoscientific research proceeds in a design or engineering mode, and the hallmark of good technoscience is the acquisition and demonstration of basic capabilities and, beyond that, the creation of robust technical systems.
>
> (Nordmann, 2011, p. 28)

In a subsequent analysis Nordmann describes technoscience as an "inverted Platonism." "The Platonism of scientific theorizing [searches] for structures behind the appearances [whereas] the inverted Platonism of technoscience theorizing . . . refers to objects just beyond the grasp of engineering" (Nordman, 2017, p. 63).

The ontological indifference that Galison finds in high theoretical physics and Nordmann in technoscience meshes with an aspect of Wimsatt's attitude toward the ontological issue in biology regarding the concept of species. For Wimsatt, however, the indifference toward the problem of clearly conceptualizing species being is not so much dismissal as a willingness to accept as species whatever seems to work; species is just whatever maintains itself in some environmental niche over some reasonable period of time. Engineers, too, seldom worry about such issues as whether computers can think or whether robots are human. "Let's just build them and see."

Essay 20

CONCLUDING UNSYSTEMATIC POSTSCRIPT: TOWARD A TECHNO-HUMAN CONDITION OR CLASH OF ANTHROPOLOGIES?

Three philosophical themes have occurred repeatedly in this stumbling collection. The first is that engineering is the core of technology. The philosophy of technology tends to overlook the dynamicism that is the technological present.

Second, engineering is not just what engineers do but is a way of life for which engineers are the leading edge. Engineering (at least in its English-speaking determination) is a distinctive way of being in the world, an anthropology, that began to flourish at a particular time in history. It is not coeval with being human. And like any anthropology, it implicates an ecology as well, a way that is world.

Because of this, third, the philosophy of engineering is not just for engineers or for philosophers but for all of us caught in the engineered and engineering maelstrom. As such, the philosophy of engineering is a thinking both with and against engineering.

As engineering comes to be in the midst of preengineering ways of life we are witnessing a clash of anthropologies (and ecologies). Debates about whether to engineer or not human life (genetic engineering) and the planet (geoengineering) and about the ways of spreading or not engineering thinking and acting into previous forms of thinking and acting (speaking a language, preparing food, taking care of one's body, honoring family and friends, educating one's self and future generations) all come down to questions of what it means to be human.

One historico-philosophical framing for such themes is the idea a new (techno-) human condition. An early effort to think this historical mutation is Hannah Arendt's *The Human Condition* (1958). Arendt framed her analysis with a distinction among three basic types of human activity: labor, work, and action.

Labor denotes those repetitive bodily behaviors of the human as biological animals that bind us to nature: getting up and going to bed; finding or growing, preparing, and eating food; washing and tending to body and clothing. For the *animal laborans* it is the species life that predominates, and individuality barely exists. The species requires and only continues through labor, which takes form in cyclical patterns that echo those of nature: day and night; eating and eliminating; spring, summer, fall, and winter.

Work fabricates with the hands material things not found as such in nature and which exhibit a measure of nonnatural individuality and durability that reacts back on and into their makers. Members of the species begin to manifest individuality in the mirror of the particular things they make: *homo faber* becomes a jeweler or painter, a potter, an iron worker, a carpenter, a maker of household furnishings and weapons of war. The durables of tools and buildings construct a world within which individuals are born and die and in the process pass from one generation to the next in a more-than-biological culture: language, customs, traditions. Yet culture remains influenced or conditioned by the materials with which *homo faber* comes to be through differential material engagements. Desert peoples differ from mountain peoples, farmers from urban dwellers. As Arendt puts it, humans

> are conditioned beings because everything they come in contact with turns immediately into a condition of their existence. The world in which [humans live] consists of things produced by human activities; but the things that owe their existence exclusively to [humans] nevertheless constantly condition their human makers. In addition to the conditions under which life is given ... and partly out of them, [humans] constantly create their own, self-made conditions, which, their human origin and their variability notwithstanding, possess the same conditioning power as natural things.... The impact of the world's reality upon human existence is felt and received as a conditioning force. The objectivity of the world—its object- or thing-character—and the human condition supplement each other; because human existence is conditioned existence, it would be impossible without things, and things would be a heap of unrelated articles, a non-world, if they were not the conditioners of human existence.
>
> (Arendt, 1958, p. 9)

Latour may speak of artifacts as "society made durable" and of the impossibility of artifacts that do not incorporate social relations, but four decades earlier Arendt had already and with much less rhetorical excess described the emergence of durability and the mutual conditioning between humans and their world. The Churchill principle of Borgmann is a clear and sober reprise and deepening of Arendt's phenomenology of *homo faber* 50 years on, in a way that has the potential to nourish and inform a broader political discourse.

For within the world of artifice there emerges a third human activity, action, constituted by discourse engagement among humans that is not subordinate to laborious cooperation in the family or apprenticeships of making in the home, field, or workplace. Among those who live in common beyond the ties of kinship there emerges *homo politicus* who, through a new kind of making, the making of laws, establishes an impersonal web of human relationships with its own distinctive but distinctly fragile durability.

> Labor assures ... the life of the species. Work as its product, the human artifact, bestow a measure of permanence and durability upon the futility of mortal life and the fleeting character of human time. Action, in so far as it engages in founding and preserving political bodies, creates the condition for remembrance, that is, for history.
>
> (Arendt, 1958, p. 8)

Action can become remembered not just through poetry or laws in ways that established a historical web of human relationships but also through contesting with—that is, acting into—a rival web. Indeed, rivalry in word and deed is coeval with the emergence of political action. And in a world in which action has, through engineering, become "action into nature" (Arendt, 1958, p. 324; see also Cooper, 1991) there is an ever-rising moral imperative for political discourse that can incorporate this new dimension of making action that Arendt points up but does not yet analyze in any detail other than to note its novelty.

There is, however, a deep and troubling irony of the engineered world in the early twenty-first century. A neoliberal idolatry of the individuality found in the engineering mirror constitutes a troubling resistance to constructing and enacting the collective imperatives required by its corresponding ecology. Individualists refuse to listen to or even acknowledge the existence of any warning from Gaia. "Don't tell *us* what to do."

Two decades later, without any reference to Arendt, Bernard Gendron (1977) argued a superficial Marxist comparison of technological impacts on the human condition in what he termed utopian, dystopia, socialist forms. Marxist thematizations of the techno-human condition have struggled for public efficacy, even in the much more sophisticated philosophy of Andrew Feenberg.

Three more decades on, again minus reference to Arendt, *The Techno-Human Condition* by Branden Allenby and Daniel Sarewitz (2011) offers a substantive analysis of three types of technological-society cause-effect relationships. Level I causal relationships are straightforward technological fixes. Yet such relationships often depend on networks of Level II relationships. The airplane is a Level I fix to the problem of fast long-distance transport; but air transport requires Level II airport construction, air traffic control systems, financial investments, commercial management, and government safety regulations. Moreover, Level II air transport systems have Level III effects on the global environment: they threaten stratospheric ozone and become vectors for the rapid spread of disease. Level I relationships are rather easy to understand or manage, Level II relationships much less so, while Level III can become almost incomprehensibly complex and engender wicked problems.

The basic challenge in an engineering conditioned techno-human world is that "we inhabit Level III, but we act as if we live on Level II, and we work with Level I tools" (Allenby and Sarewitz, 2011, p. 161). In this confederacy of incompatibles, the authors advance a case for humility and a suite of related normative responses that run from giving up the quest for definitive "solutions," relying on pluralism over expertise, playing with scenarios, increasing the frequency of decision-making, questioning predictions, promoting continual learning, and more (pp. 162–69).

Another take on this is provided by the French philosophical anthropologist Louis Dumont. For Dumont (1986), individualism as ideology is the fate and threat of the West. It is, as well, entangled with a modern commitment to what he terms "artificialism" or an ontology that ranks artifice above nature:

> This artificialism is surely failing to some extent, in that it sees itself confronted by limits which it had not predicted and which result from its very power, limits which must force it at least to change course and complicate itself.... At the level of opinion, the ecology movement illustrates this point most clearly, but the issue is much broader.... One may perhaps doubt whether specialists always know well enough what they are doing; one may ask whether power is not tending to emancipate itself from knowledge, which used hitherto to be considered as essential.
>
> (Dumont and Delacampagne, 1981, p. 6)

Or as he elaborates with regard to the issue of complexity: given the contrast between our problems and "the simplicity of the common fund of ideas and values which we have at our disposition" any "model" for action requires complication. "In one sense, that means becoming an adult, or if you like, a philosopher; that's the heavy burden that falls on everyone today" (p. 6).

In marked contrast to such sober uneasiness there has emerged since the early 2000s a celebratory enthusiasm for the artificial. A simplified tracking of American public attitudes toward science and engineering can distinguish, first, the authority of and trust in science and engineering that held sway post–World War II from, second, a rising skepticism during the 1960s under the influence of nuclear threats, environmental pollution, and a neoliberal attack on social institutions and praise of commodification. Of course, the latter did not completely replace the former, but it modified the context in which the former was manifested. (The increased presence of flashy commercial advertising in the pages of *Science* and *Nature* is one example.)

As this second attitude took hold policy legitimations of science and engineering shifted and there emerged various theories of "post-academic science," "post-normal science," a "triple helix of entrepreneurial science," and "mode-2 knowledge production." Although "science" was the primary term of reference, scratching the rhetorical surface reveals engineering as deeply implicated insofar as the new social epistemologies readily applied as well or better to engineering. English-speaking engineering has from its origins been post-academic, post-normal, triple helix entrepreneurial, and operated more or less explicitly in a mode-2 knowledge production framework. (Compare, for instance, the historical profile of commercial advertising in *IEEE Spectrum* versus in *Science* and *Nature*.) The engineering way, it may be argued, was coming to be seen as applicable to science just as much as to engineering. Within this polynominal ambiance a seriousness and sobriety nevertheless remained.

In the early 2000s, this mutated as a complementary third stance began to take hold with regard to science and engineering, especially in engineering. Roots can be traced back to the emergent cultural capital of Silicon Valley and the digital revolution more generally, in the personal computer and computer games, in the progressive media-ization of communications and entertainment along with the miniaturization of electronic devices of all sorts. Expression of this attitude was anticipated by civil engineer Samuel Florman's now all but classic *The Existential Pleasures of Engineering* (1976), which has become a touchstone in engineering apologetics. To put it in the vernacular, being a geek became cool and engineering became fun. In the words of one set of insightful observers, "we [moved] from a period of disenchantment, rationalization, and intellectualization to an age where the technoscienfitic world of our own creation [became] an enchanted, magical place" (Nordmann et al., 2011, p. 8).

This alternative anthropology-ecology (or culture-nature) is well exemplified in Don Ihde's celebratory philosophy of technology. Although Ihde does not focus as much on engineering productive agency as does Florman, his careful phenomenological analyses of the ways engineered devices multiplicitly embody and expand the experiences of users both results from and supports a fundamental affirmation of the human as *homo faber*, as reinterpreted and clarified so as to challenge Arendt's more restricted usage. Together Ihde and anthropologist Lambros Malafouris build a philosophical bridge between postphenomenology (phenomenology that thematizes the ever present mediation of technologies in contemporary human experience) and material engagement theory (which emphasizes the fundamental role played by tool making and using in human evolution). In Ihde's words, the human mode of being today is constituted by "a continuum of human-prostheses inter-relations." For Malafouris, human becoming took place through "a long-term commitment to the discovery of new varieties of material forms, . . . through a saturated, situated engagement of thinking and feeling with things and form-generating materials" (Ihde and Malafouris, 2017, p. 196; both quotations come from previous works, with Malafouris's text italicized in the original). The "longview" of Material Engagement Theory and "the contemporary view of postphenomenology" complement each other to yield an enriched conceptualization of *homo faber*.

> We are *Homo faber not just because we make things but also because we are made by them*. People are both changing and changed by technology. [Our argument] is not one favoring technological determinism or utopianism but one that emphasizes the active role of material engagement in the enactment and

constitution of human life. Materiality and the forms of technical mediation that humans make and use are not passive or neutral but actively shape what we are in a given historical moment. The challenge for us is understanding in which ways and to what degree human beings are shaped and constituted by the stuff they make.

(p. 209)

The tool maker is equally the tool made. It is not tool using that is distinctively human (many animals use tools) or even tool making (some animals do this as well) but the being made and used by tools. Although Ihde here and elsewhere (see 1986 and 2008) strongly rejects both utopian and dystopian interpretations of technology, since there are few technologies not to like in one or another of their multistable manifestations, his evident high comfort level effectively constitutes a kind of soft utopianism.

Peter-Paul Verbeek carries this soft utopianism into the world of design. With somewhat more circumspection than Ihde and a more explicit commitment to "moralizing technology" (2011), Verbeek nevertheless appears to share a questionable confidence in the ability of designers and engineers to co-construct the good life. Verbeek (2005) explicitly rejects the technological alienation of such classical European philosophers of technology as Karl Jaspers and Martin Heidegger, who, he maintains, were looking backward rather than forward and thus failed to engage with technological materiality in its complexity and rich diversity. In so doing he seems to overlook new forms of alienation present in the hegemonic power manipulations shot through contemporary material culture, although in response to critics he candidly agrees that more attention is needed to the social and political dimensions (Verbeek, 2009). Engagement with critical media philosophy might also be relevant (see, e.g., Floyd and Katz, eds., 2015). In all these frames, disenchantment with engineering and technology tends to give way to the appreciation of new possibilities, which, in the apotheosis of innovation as the solution to all problems, feels like a recreation of the heroic industrial age in digital technology form.

In the philosophy of science, there is a recurring fundamental debate between realism and anti-realism. In the philosophy of engineering one can find an analogy between artificialism and anti-artificialism. But in a world so saturated with artifice as the present, anti-artificialism is much more difficult to think than anti-realism. Still, in both cases, philosophy must encourage measured consideration of the strengths and weaknesses, bright and dark sides, of difference stances. With regard to engineering, the judgments of history and politics may also eventually play decisive roles than is the situation with science.

HEMINGWAY AND MARINETTI

There is another story about Gertrude Stein and Ernest Hemingway. Sometime after *The Sun Also Rises* Hemingway was visiting 27 Rue de Fleurus and, looking at a small Marinetti painting, remarked that he wasn't sure he understood it. Standing to one side, but as if speaking for Stein, Alice B. Toklas replied that he didn't and wouldn't.

Appendix

(after Lydia Davis)
Just as you were going out the door, what did you say?

On Engineering Use and Convenience

Revisiting earlier periods in a historical trajectory can occasionally contribute to a deeper understanding of the present. This is especially the case when the earlier period was a time of founding, as in the founding of a political regime or religion or a cultural tradition and there exist founding documents (such as laws or sacred texts). Beginnings can never provide the sole basis for understanding the present, but sometimes things were revealed in the past with a clarity that has been forgotten. With this in mind, consider the founding of the English-speaking tradition of engineering and one of its key documents, the Charter of the Institution of Civil Engineers (ICE).

THE CHARTER OF THE INSTITUTION OF CIVIL ENGINEERS

The classic definition of civil (nonmilitary) engineering is given in the Royal Charter granted by King George IV in 1828 to the Institution of Civil Engineers. According to the charter, engineering is "the art of directing the great sources of power in nature for the use and convenience of man." The insight and influence of this definition, decreed by a monarch who represented a rising, self-confident bourgeois class proud of its recently acquired and expanding economic and political clout, deserves more attention than it commonly receives. Terms in this definition that warrant special note include "art (of directing)," "great sources of power in nature," and "use and convenience (of man)." Among these "use and convenience" stands out today as more enigmatic or puzzling than the other two, although in various permutations it has been at the root of subsequent engineering self-understandings. The history of engineering ethics, for instance, can be read as a string of alternative interpretations of what more particularly constitutes use and convenience. Indeed, implicit in the definition as a whole, and especially in the conceptualization of the end of this uniquely modern form of human agency, are elements of anthropology, ontology, and epistemology along with ideas about art, labor, and aesthetics.

The continuing importance of the definition is testified to by its retention across Supplemental Charters (1887, 1896, 1922, and 1962), amendments (1968 and 1972), and the approval of a new Charter by Queen Elizabeth II in 1975. The value of revisiting it is further witnessed by an ICE General Assembly commemorating the 150th anniversary of the Royal Charter, which hosted talks by leading engineers and other dignitaries on three themes: great sources of power in nature, use and convenience, and man. In a cultural context (mid-1970s), where engineering and technology were being challenged by social criticisms, speakers acknowledged problems but resolutely reaffirmed the centrality of use and convenience as a fundamental value of engineering. As one speaker put it, although he did not recall when he first heard the phrase, it "seemed to me then—as it still does seem—to describe admirably the ambitions" of engineering, even while "opinions on 'use and convenience' have changed since [the] definition was formulated" (ICE, 1978, pp. 64 and 82).

What follows then is an effort to reflect on the 1828 definition, particularly on that compound term which the ICE placed at the center of its own celebratory reflections, as an exercise in recovering an idea deeply relevant to both practice and theory across a 200-year history of English-speaking engineering and, in the process, to cast new light on the present. A framing for this reflection will be a simplified narrative of the career of Thomas Telford (1757–1834), the first ICE president and person who presented to the king the definition incorporated into the Royal Charter.

IMMEDIATE ORIGINS OF THE CHARTER

The definition of engineering is actually derivative of another document commissioned by the ICE and written by Thomas Tredgold (1788–1829). This document exists as a handwritten entry in the *Minutes of the Proceedings of the Institution of Civil Engineers, for Facilitating the Acquirement of Knowledge Necessary in the Profession and from Promoting Mechanical Philosophy* available in the ICE archive. (I am grateful to the ICE library staff for facilitating my 2008 research visit, responding to subsequent inquires, and allowing quotations from the records.)

As recorded in the *Minutes*, vol. 1 (1818–1823), the first formal meeting was held on January 2, 1818, at the Kendal Coffee-House, Fleet Street, and began with a presentation by Henry Robinson Palmer, today considered the founder of the institution (Watson, 1988, p. 171).

> It is a remarkable fact that notwithstanding the extensive advancement of Science and the general increase of means for an acquaintance with it, while the principles of systematic education for most of the learned and scientific professions have been and still are actively encouraged, not even an attempt seems to have been made towards the formation of any special source of information or instruction for persons following or intending to follow the important profession of a Civil Engineer.

Palmer continued by explaining how such an organization was needed to support "a profession in the practice of which the utmost skill of man is called forth," requiring "a general acquaintance with some of the most abstruse branches of science [with a familiarity that] will enable the practitioner to apply them to immediate use." This profession further demands "not only a knowledge of one leading branch of science but many, not only of one leading art, but of an indefinite number." From this perspective,

> An Engineer is a mediator between the Philosopher and the working Mechanics, and like the interpreter between two foreigners must understand the language of both.
> The Philosopher searches into nature and discovers her laws, and promulgates the principles on which she acts. The Engineer receives those principles and adapts them to our circumstances.

Palmer was but 23 years old and the average age of the six founding members was only 25. The ICE originated as a vision of young and ambitious men. After a year of regular meetings, in which they exchanged contributions to an emerging body of technical knowledge, it was proposed (February 3, 1820) that Telford, an engineer much their senior and in whose firm Palmer was employed, be solicited as president, with others "respected for their learning" invited to become honorary members or associates. The institution was to be given more social capital by enrolling others who already possessed it.

The next month (March 16, 1820) it was announced that the 63-year-old Telford had not only agreed to serve but also donated a set of books from his library to establish a research collection. In a formal acceptance (March 21, 1820) Telford appealed to what would now be called a libertarian streak by pointing out how the ICE was different from related institutions in other countries. "In foreign countries similar establishments are instituted by Government, and their members and proceedings are under its control, but here [because a different course is adopted] it becomes incumbent on each individual member to feel that the very existence and prosperity of the Institution depends in no small degree on his personal conduct and exertion" in order to avoid those "incurable inconveniences" that often perplex such societies.

Following seven years of stable growth through individual initiative, the society determined to seek formal legal status, and after debate, decided to apply for a Royal Charter. Taking the Charter of the Royal Society of Literature as "a very good model" (December 14, 1827), the well-known author and honorary member Tredgold—who had recently published a major study of *The Steam Engine* (1827)—was invited (December 29, 1827) "to define the object of the Institution of Civil Engineers." Tredgold's white paper, which can be assumed to represent a strong consensus of the thinking that had emerged among the leading members of the profession about its nature, was submitted on January 4, 1828. A charter was then (over a series of meetings on January 8, 11, and 15, 1828) drafted, edited, and approved—and submitted to the king. On June 3, 1828, George IV formally granted the Royal Charter.

Before taking up the Telford narrative, which will form a spine of interpretation, the text of Tredgold's short 600-word white paper is given here in full.

DESCRIPTION OF A CIVIL ENGINEER:
BY THOMAS TREDGOLD

Civil Engineering is the art of directing the great Sources of Power in Nature for the use and convenience of man; being that practical application of the most important principles of natural Philosophy which has in a considerable degree realized the anticipations of Bacon, and changed the aspect and state of affairs in the whole world. The most important object of Civil Engineering is to improve the means of production and of traffic in States, both for external and internal Trade. It is applied in the construction and management of Roads—Bridges—Rail Roads—Aqueducts—Canals—river navigation—Docks, and Storehouses for the convenience of internal intercourse and exchange;—and in the construction of Ports—Harbours—Moles breakwaters—and Light Houses, and in the navigation by artificial Power for the purposes of Commerce. Besides these great objects of individual & national interest it is applied to the protection of property where natural powers are sources of injury, as by embankments for the defence of tracts of country from the encroachment of the sea or the overflowing of Rivers; it also directs the means of applying Streams and rivers to use, either as powers to work machines, or as supplies for the use of Cities and Towns, or for irrigation; as well as the means of removing noxious accumulations, as by the drainage of Towns and Districts to prevent the formation of malaria, and secure the public health. This is however, only a brief Sketch of the objects of Civil Engineering, the real extent to which it may be applied is limited only by the progress of Science; its scope and utility will be increased with every discovery in Philosophy, and its Resources with every invention in Mechanical or Chemical Art, since its bounds are unlimited, and equally so must be the reaches of its professors.

The enterprizing Hollanders towards the close the 16th Century first separated Civil Engineering from Architecture under the Title of Hydrolic Architecture. This example was followed in France towards the end of the 17th and soon afterwards was systemized in the great work of Belidor on Hydraulic Architecture.

One of the great bases on which the practice of Civil Engineering is founded, is the Science of Hydraulics, every Kingdom, every province, every town has its wants which call for more or less acquaintance with this Science. Water which is at once the most useful of the necessaries of life and the most dangerous element in excess—when limited by the laws of this Science is rendered the best of Servants; the rolling cataract which spends its powers in Idleness may be directed to drain the mine—to break the ore—or is employed in other works of Labour for the use of man; the streams are collected and confined in canals for inland Traffic, Harbours are formed to still the raging of the waves of the ocean, and offer a safe retreat to the Storm driven mariner, and ports are provided with Docks to receive the riches of the world in Security: hence arose the term Hydraulic Architecture, but it was too limited;—the various applications of water had rendered the natural supplies inadequate to the wants of man, till he discovered that combined with heat, it formed a gaseous element induced with energies not less powerful than the falling cataract its stream confined and directed by Science became a new Source of Power which in a few Years altered and improved the condition of Britain, and we are every day witnessing new applications as well as the extension of the older ones to every part of the globe.

There are at least seven points in this text that can help illuminate the special character of English-speaking engineering, although the hypothesis is that "use and convenience" is at the overlooked center.

THOMAS TELFORD AND THE INSTITUTION IN CULTURAL CONTEXT

Telford was born in 1757, to a sheep-herding household in rural Scotland midway between the great city of Edinburgh and the mountainous border with England. His father died at age 33, less than a year after Thomas's birth, exemplifying the harshness of life in the rough countryside.

At that time in Edinburgh the Scottish Enlightenment was flourishing, led most famously by the philosopher and gentleman David Hume (1711–1776), who would pass away the year that witnessed publication of Adam Smith's great rationalization of bourgeois capitalist enterprise, *The Wealth of Nations*.

So Telford's birth came at a propitious period in British, in European, and indeed in world history.

In reaction to the kind of hard existence into which Telford was born, Hume developed a radical criticism of Christianity. In a paradox of its modern transformation, traditional Christianity had once rationalized subjection to the kind of fate experienced by Telford's family and then, through the Reformation and reactions against it, introduced into public affairs a violence that engulfed much of Europe.

Barely had the Peace of Westphalia of 1648 been concluded, in which European states renounced warfare between themselves in the name of religious purification, than civil war broke out in England. In 1649, in the name of religious purification, King Charles I was beheaded in London and Oliver Cromwell proceeded to prosecute a campaign against his fellow countrymen in order to stamp out a false Christianity that polluted the realm. It was Hume's aim to stamp out such stamping out, with the pen rather than the sword.

When the crown was restored to Charles II in 1660, among his first acts was to formally charter the Royal Society to "encourage philosophical studies, especially those which by actual experiments attempt . . . to shape out a new philosophy." The goal was "to extend not only the boundaries of the [British] Empire, but also the very arts and sciences" by promoting "the sciences of natural things and of useful arts" so that they "may shine conspicuously amongst our people." "At length," proclaimed the king, "the whole world [should] recognize us . . . as the universal lover and patron of every kind of truth." Truth in biblical religion was in the process of being supplemented with (eventually to be superseded by) truth in natural philosophy and its utilities for power and improvement.

According to Thomas Sprat's early *History of the Royal Society*, its inspiration was "one great Man, who had the true Imagination of the whole extent of this Enterprize, as it is now set on foot; and that is, the Lord Bacon." Sprat would have preferred that, he wrote, "there should have been no other Preface to the History of the Royal Society" than Bacon's own works (Sprat, 1667, p. 35). When the ICE was founded, the Royal Society was an obvious precursor. In the second sentence of his text, Tredgold references "that practical application of the most important principles of natural Philosophy [i.e., modern science] which has in a considerable degree realized the anticipations of Bacon, and changed the aspect and state of affairs in the whole world."

TELFORD'S PATH:
FROM STONE MASON TO ENGINEER

It was into this transformational milieu that Telford came of age, yet with roots in his "native district of Eskdale," as he put it in a posthumously published autobiography (Telford, 1838, p. 2). Taking local employment as a mason, he developed "the natural senses of seeing and feeling requisite in the examination of materials [along with] the practiced eye, and the hand which has experience of the kind and qualities of stone, of lime, of iron, of timber, and even of earth, and of the effects of human ingenuity in applying and combining all these substances," and acquired "the habits and feelings of workmen" (p. 3). In the process, he also learned to assess the "convenience and usefulness" of "improvements" in buildings (pp. 2 and 4).

Then "at the age of 23 [considering himself] master of [his] art . . . and having . . . an opportunity of visiting Edinburgh, [he dismissed his] border" home and moved to the city (p. 15). There he put his journeyman trade to work helping construct the New Town—now the city center and Princes Street—which retains a neo-classical and Georgian architectural heritage. Picking up "a general knowledge of drawing, and particularly of its application to architecture, and having studied all that was to be seen in Edinburgh" (pp. 16–17), he effectively became the first in his family to acquire a higher education. Yet little more than a year later, in early 1782, when invited to ride a horse to London to deliver it to a friend, Telford judged it advisable himself "to proceed southward, where industry might find more employment, and be better rewarded" (p. 19).

Such was the beginning of Telford's single-minded (he never married) professional life, one ever dedicated to construction, that would take him on multiple journeys throughout Great Britain. In a personal supplement to the posthumously published autobiography, its editor described Telford as having lived "like a soldier, always on active service" (p. 283). The spare narrative of the autobiography, which is mostly dedicated to cataloging works, is complemented (and drawn on here) by L. T. C. Rolt's (1958) eponymous volume as well as a 250th anniversary exhibition at the Ironbridge Museums (2007).

Separate from the details of his geographical peregrinations, the most significant journey took place within his professional self-conception: from architect to engineer. The professional migration in Telford's career was from stone artisanship to architecture, then from architecture to engineering, to become one of the most renowned civil engineers of his day: responsible for the design and construction of many roads and bridges and especially the canal system of the United Kingdom. In the first biography three decades after Telford's death, Samuel Smiles, in the second volume of his *Lives of the Engineers* (1867), introduced his subject with a history of roads and travelling in Great Britain. From the perspective of history, however, Telford's achievements can be seen as marking "the end of the golden age of architecture in England" (Rolt, 1958, p. 14).

One reason Telford is not as well known today as he might have been is that just as he was completing the canal system it was superseded by the railroads engineered by George Stephenson (1781–1848) and Isambard Kingdom Brunel (1806–1859). By one of those historical contingencies that often wash over individuals professionally committed to creative destruction, Telford's great achievement was undermined by what followed. Like James Watt (1736–1819), a fellow countryman with whom he occasionally worked, Telford thought steam power was destined to be limited on land to stationary emplacements.

Having arrived in London on horseback, Telford took a natural next step in professional advancement and began to put himself forward, not simply as a master stone mason, but as an architect. Over the course of relocations from London to Portsmouth in the south and then to Shrewsbury in the west, he worked at and became progressively known as a designer and builder of residences and civic structures. In the self-education recorded in his "Architectural Commonplace Book" Telford copied and commented on texts from Colen Campbell's *Vitruvius Britannicus* (3 volumes, 1715–1725, which, of course, echoes the Roman Vitruvius's classical *De architectura*, from the first century BCE), Bernard de Montfaucon's *Antiquity Explained and Represented in Diagrams* (15 volumes, 1721–1725), James Bentham's writings on Gothic architecture in England (mid-1700s), James Stuart and Nicholas Revett's *Antiquities of Athens* (1762), and William Gilpin's *Observations, Relative Chiefly to Picturesque Beauty* (1786) (Rolt, p. 35 et pass.). In letters written to his closest friend back in the Scottish countryside, he described how he was always "surrounded by Books, Drawings, Compasses, Pencils and Pens." It was as an architect that Telford designed houses and the new churches of St Mary Magdalen, Bridgenorth, which the architectural historian Sir Nikolaus Pevsner has described as of "remarkable design, of great, and apparently done in full awareness of recent developments in France" (Pevsner, *Architectural Guides: Shropshire*, 1958, as cited in Paxton, 2008).

Then in the early 1790s—in his own early thirties—during a sojourn at Shrewsbury in the West Midlands, a further change took place. Telford ceased to describe himself as an architect and began to call himself an engineer. This was historically significant, not just in his life but in the history of technology and construction of what has become our engineered and engineering way of life. Telford's decision echoed James Watt's desire from the previous generation to alter his occupational profile from mechanic to mechanical engineer.

At the time Telford moved to Shrewsbury, "engineers" were still military men, members of the Corps of Royal Engineers that traced their lineage back to those William the Conqueror brought with him to England in 1066, although the Corps of Engineers as such was not established by the Royal Regiment of Artillery until 1716. In Shakespeare's *Troilus and Cressida* (1602) the Greek warrior Achilles is described as a "rare engineer" (Act II, scene 3, line 7); and in *Hamlet* (1603) there is "the engineer hoist with his own petar" (Act III, scene 4, lines 206–207).

John Smeaton (1724–1792) was apparently the first Englishman to describe himself as a "civil engineer." In 1771 at the King's Head tavern in London's West End, Smeaton had established an informal Society of Civil Engineers for occasional social gatherings. Yet architecture continued to be the profession of choice among those dedicated to building. The epitaph for the great pre-Telford canal builder James Brindley, who died in 1772, describes him simply as one who

> made Canals, Bridges, and Locks,
> To convey Water; he made Tunnels
> for Barges, Boats, and Air-Vessels;
> He erected several Banks,
> Mills, Pumps, Machines, with Wheels and Cranks;
> He was famous t'invent Engines.

Only *ex post facto* was Brindley denominated an engineer. It was not a term he applied to himself—but it was one that Telford adopted.

In his autobiography Telford records that between 1790 and 1796 he built some 40 road bridges in Shropshire (of which Shrewsbury is the county seat). Major examples were bridges across the Severn River at Buildwas, Bridgenorth, and Bewdley. At the end of his life, Telford had built six of the 13 roadway bridges across the Severn between Shrewsbury and the sea, only one of which has since been replaced. The significance of this bridge building goes beyond straightforward physical achievement.

In *De architectura*, the first century BCE compendium of classical architectural learning, bridges are conspicuous by their absence. For Vitruvius, the architect is concerned first and foremost with the building of civil public spaces such as the forum, the basilica, the theater, the gymnasium, porticos, and baths. *De architectura* deals at length with what is called the fitting orders in temples, doorways, and the acoustics of the theater. From considerations of how buildings should be adjusted to harmonize with different climates, it provides guidelines for private houses, the functions of different rooms, wall decorations, provisioning for water, and more. In the whole ten books, there are only eight spare references to road building and not one to bridge construction. The architect is one who humanizes space rather than provides for transport via roads and bridges, which are more necessary to military and commercial affairs than to inhabiting a land. Although urban habitation depends on the transport of water into the city, it does not require commercial transport via waterway canals, which are repeatedly referenced in Tredgold's essay and in the charter definition to which it contributed.

Architecture is more concerned with movement within an essentially sedentary life than with movement across great distances for commercial gain. Famously, *De architectura* proposes three principles of good building: *firmatis*, *utilitas*, and *venustatis*. Buildings should be durable, function well for those living with them, and manifest beauty in ways that lift their spirits. In order to realize these principles "the architect should be equipped with knowledge of many branches of study and varied kinds of learning" (I, 1). They must be able to work with both hand and mind, having studied history, geometry, medicine, law, astronomy, music, and philosophy. Up until the mid-1790s, having initially acquired the skills of the hand in his native place, Telford made efforts in many of these fields. He studied not only science and mechanics; he read and wrote poetry. He was an admirer, for instance, of his fellow countryman Robert Burns and well recognized the ways that the commercial spirit could undermine the aesthetic. In one of his best poems he lamented the result of Burns leaving the countryside and moving to the city:

> The Muses shall that fatal hour
> To Lethe's streams consign,
> Which gave the little slaves of pow'r,
> To scoff at worth like thine.
>
> But thy fair name shall rise and spread,
> Thy name be dear to all,
> When down to their oblivious bed,
> Official insects fall.

At the same time, without fully appreciating it, Telford became himself an element in that insect-creating bourgeois maelstrom that was the Industrial Revolution.

For Adam Smith, the Industrial Revolution depended on specialization and division of labor. *The Wealth of Nations* opens with Smith's famous analysis of the pin factory. Only when pin fabrication is cut free from the craft tradition by means of analytical decomposition into, for example, spooling wire and cutting it to length, sharpening one end and soldering a head on the other, and packaging it up could one person be assigned to the monotonous repetition of one isolated step in this process, another person to another (cut, cut, cut; sharpen, sharpen, sharpen, etc.) and could productivity be increased to yield greater use and convenience. The fact that the process of specialization makes the work itself less fulfilling for workers, who perforce become what Karl Marx termed "wage slaves," must be set aside. Let the workers become laborers, who may then use whatever wages they are paid, to pursue the romance of modern consumerism (Campbell, 1987).

Telford and the profession of engineering applied this same principle of specialization to building. Whereas architects were broadly educated and sought to construct a durable, useful, and beautiful world for sedentary habitation, the English-speaking engineer began to set aside the pursuit, especially of beauty, in order to specialize in utility—a capitalist utility understood in the most basic terms as "use and convenience." In a panegyric to engineering that opens volume three of *The Lives of Engineers*, Smiles, quoting Bacon, claims that along with fertile soil and industrial production, the engineering of "easy conveyance for mean and commodities from one place to another" is the foundation of civilization. In further exemplification of this flattening of civilization in the professional life, Rickman testified to how in his youth Telford had been "tinctured with the then fashionable doctrines of Democracy," but admonished by Sir William Pulteney and as an engineer "silently abandoned politics . . . and during the remainder of his life . . . never conversed on political topics" (Telford, 1838, pp. 280 and 282).

Having thus eschewed participating in the contentious reconstruction of the British political order—Jeremy Bentham (1748–1832) was his contemporary—Telford favored instead the engineering of a new physical infrastructure convenient to capitalist expansion. In the end, having traveled widely in Britain and to Scandinavia, while undertaking consultancies as far away as Russia and Canada, Telford died quietly at his London home in early September 1834.

USE AND CONVENIENCE: BEFORE AND AFTER TREDGOLD AND TELFORD

As employed in the ICE white paper, "use and convenience" is more than a common or off-hand phrase but has technical significance, as can be seen by earlier occurrences in the work of Hume, the greatest British philosopher of the period. Hume's empiricist, this-worldly philosophy of possessive individualism, in both *A Treatise of Human Nature* (1738–1740) and *An Enquiry Concerning the Principles of Morals* (1751), founds moral and aesthetic judgment in "utility," especially when conjoined with "convenience." The *Treatise* argues that beauty in animals and other objects "is derived from the idea of convenience and utility" (II, pt. I, sec. 8, para. 2); the *Enquiry* proposes that all goodness in human artifacts rests on their "use and conveniency" (sec. II, pt. I, para. 5) and "ascribe[s] to their utility the praise, which we bestow on the social virtues" (sec. V, pt. I, para. 1). A popular essay, "The Stoic, or the Man of Action and Virtue" (1741), opens with a paragraph that all but anticipates Tredgold's argument: Nature has endowed man with a spirit and placed him in a situation that

> urges him, by necessity, to employ, on every emergence, his utmost art and industry.... Every thing is sold to skill and labour; and where nature furnishes the materials, they are still rude and unfinished, till industry, ever active and intelligent, refines them from their brute state, and fits them for human use and convenience.

Hume's usage, which is further developed by classical British political economists, additionally recalls John Locke's labor theory of value and property, in which convenience plays a key role. According to the theory of property in the *Second Treatise of Government* (1689), "God, who has given the world to men in common, has also given them reason to make use of it to the best advantage of life and convenience" (ch. 5, para. 26). For Locke, labor creates value and property in nature by rending nature useful and convenient. An earlier unpublished Latin text from Locke's journal of 1676 likewise speaks of *commodum* and *utile*, which can be translated as "convenience" and "useful" (Locke, 1954, pp. 180–81).

Pushing even further back, Thomas Hobbes described the state of nature as "incommodious" and presented the social contract as a pathway to "commodious living" (*Leviathan* I, 13). Locke too described "civil government [as] the proper remember for the inconveniences of the state of nature" (*Second Treatise*, ch. 13, para. 123). Historians of the Industrial Revolution in Britain such as Joel Mokyr (2002 and 2009) and Deirdre McCloskey (2010), by arguing the critical influence of the Enlightenment and bourgeois worldview, further implicate the Hobbes-Locke-Hume tradition on the theory and practice of engineering.

The etymology of "convenience" contributes further to appreciation of the uniqueness of the English Enlightenment usage in these texts. As Thomas Tierney has argued in a genealogy of the value of convenience, "convenience" is

> an adaptation of [the Latin] *convenientia*, which means "meeting together, agreement, accord, harmony, conformity, suitableness, fitness." The adjective *convenient* is based on the present participle of the verb *convenire*, which means "to come together."
>
> (Tierney, 1993, p. 39)

Before the 1600s English meanings remained close the Latin roots. "Something could be described as convenient or as a convenience if it was in accord or agreement with something such as nature or 'the facts', or if it was suitable or appropriate to a given situation or circumstance, or if it was morally appropriate," meanings that were largely superseded by the time Locke introduced the term into political philosophy. As Tierney (1993, p. 39) continues,

> The modern meaning of *convenience* is "the quality of being personally convenient; ease or absence of trouble in use or action; material advantage or absence of disadvantage; commodity, personal comfort; saving of trouble." And the current sense of convenient is "personally suitable or well-adapted to one's easy action or performance of functions; favorable to one's comfort, easy condition, or the saving of trouble; commodious."

Additionally,

> The difference between the obsolete and current meanings of these words lies not only in the modern addition of the sense of ease and comfort, but also in the fact that what remains of the older meaning's sense of suitability has shifted and narrowed. Convenience is no longer a matter of the suitability of something to the facts, nature, or a moral code; suitability in the modern meaning of convenience refers back to the person, the self. Something is a convenience or convenient in the modern sense of these words if it is suitable to personal comfort or ease.
> (Tierney, 1993, p. 39)

Reinforcing the significance of the shift in the meaning of convenience is another in the concept of comfort. From its Latin origins, "comfort" referred to "strengthening" (from *com-*, expressing intensive force + *fortis* "strong"). The meaning of "something producing physical ease" did not arise until the mid-1600s, at which point it also began to be associated with "convenience" (Giedion, 1948, pp. 260ff.).

Tiernrey's genealogy argues that the term becomes increasingly prominent in technological culture because of a transformation of values that seeks to escape rather than live with the body.

> The modern attitude toward the body, as reflected in the consumptive activity of the household, is quite different from the ancient Greek attitude toward the body. While the Greeks thought that the satisfactions of bodily demands required careful attention and planning throughout the household, modernity treats the body instead as the source of limits and barriers imposed upon persons. What these limits require is not . . . attention, but the consumption of various technological devices that allow people to avoid or overcome such limits.
> (Tierney, 1993, p. 38)

We now inhabit a world of labor saving and household conveniences, convenience stores filled with products in convenience packaging to support convenience shopping, all driven by a desire to escape the inconveniences of embodiment. Tierney's argument is pursued with an approach adapted from Max Scheler's phenomenology in critical reflections on American history, Weber, Nietzsche, and classic early modern political philosophy. He never mentions engineering, but the confluence of this cultural value with a new way of making oriented toward convenience in the products made is a striking symbiosis.

THE DISTRACTIONS OF CONVENIENCE

A complementary contribution to appreciating convenience as a formative value for engineering is Albert Borgmann's philosophical inquiry into the influence of technology on the character of contemporary life. In his phenomenologically inspired analysis, Borgmann identifies a distinctive pattern to the elements of material culture dominant in the techno-lifeworld, what he terms the (engineered) device. The paradigmatic feature of a device is disclosed by a comparison between two basic variations of the way a commodity such as heat can be made available for users.

Traditional availability comes directly from fire, perhaps in a wood burning stove, which may need to be kindled anew on winter mornings for a warmth that likely shares with a dwelling aromas from the particular material it burns. Using a wood stove to satisfy the demands of the body depends as well on the technics of wood cutting, perhaps even woodlot management, its curing and storing, and then attending to the stove to keep the fire burning properly and to avoid being burned by it.

In typically modern technological form, availability is delivered by a central heating plant with an automatic thermostat that produces a heat unremindful of its source (gas or electricity). The central heating plant will have been designed by engineers, installed in the house before purchase, to be serviced when necessary by technicians, so that users need not practice much in the way of skilled usage; attention is minimized to setting an indicator to the desired temperature (and paying a monthly utility bill).

In the former case, the process by which warmth is produced remains transparent, the functioning of the stove calls forth household engagement, and it is likely to serve as a physical locus for gatherings of family and friends. In the latter, the heat-producing mechanism is placed out of sight, from where it reliably and safely delivers heat evenly throughout a structure. Generalizing the comparison, Borgmann describes the central heating system as a *device* that hides its machinery for producing a purified commodity and the wood stove as a *focal thing* in which machinery and commodity remain transparently intertwined: the one leaving users free and unencumbered, the other calling forth user engagement. In Borgmann's words,

> A thing, in the sense in which I want to use the word here, is inseparable from its context, namely, its world, and from our commerce with the thing and its world, namely, engagement. The experience of a thing is always and also a bodily and social engagement with the thing's world.
>
> (Borgmann, 1984, p. 41)

By contrast, in the device paradigm, there is progressive decontextualization from our experiential world and use of a purified commodity is made conveniently available without demands and on demand. Utility bill paying can also be automated. In the "smart house" wired with the Internet of Things, convenience is further amplified by reducing control to voice commands or even given over to an artificial intelligence that anticipates desires.

On Borgmann's account, engineering can be described as a prosecution of the severance between mechanism and commodity in order to replace contextual engaging *thinghood* with dis-embedded *deviceiveness*. Once the severance has been made by functional decomposition, "engineering ingenuity" proceeds to draw on "scientific insight" to make a commodity ever "more available" (p. 43)—that is, ever more convenient. In a subsequent analysis Borgmann (1999) describes this same process at work in computers and information technology, where engineering digitalization (a platform enabling ever more radical decomposition) renders texts, speech, images, and music ever more conveniently available by means of devices such as printers, radios, recordings, movies, television, and electronic files.

Viewed through this prism, engineering agency can be seen to aim for use and convenience across a spectrum of structures, products, and processes. Roads and bridges and canals along with trains, automobiles, and airplanes enhance convenience in travel: mechanical engineering designs, not just machines for mass production but machine tools, machines that make machines, thus conveniently mass-producing mass production. The engineered mass manufacturing of products and modular construction yields further dimensions of convenience in commercial and built environments. Process engineering conveniently refines oil and other chemicals non-stop into fuels, foods, and pharmaceuticals. Communications and computer engineering extends the convenience trajectory by designing an Internet of (smart) Things (but not "things" in Borgmann's sense) that ever more deeply occludes mechanisms while creating ever more convenient material and information environments.

Alternative efforts to thematize a defining end for the techno-engineered lifeworld have proposed control or efficiency ("the one right way") as the animating values at the root of celebrations of improvement and progress. Safety and dependability have been others. Critics rightly point out the political character of control and how, because efficiency (as well as safety and dependability) depends on a contextual setting of parameters, such concepts are socially defined and other-value infected. To some degree this is true of convenience as well. What is convenient for one person and in one context may be inconvenient for and in another.

Still, independent of particulars cases, and in the name of convenience (insofar as it functions as a revealing dimension of efficiency or safety), the manifold of mechanisms on which contemporary material culture rests appears subject to a general and persistent agency of removal into a hidden or encapsulated (black boxed) background. What Don Ihde (1990) terms a "background relation" exercises, as it were, a kind of priority over mediated and alterity relations, insofar as mediating technologies of any complexity are black boxed and the otherness of alterity technologies is enhanced (if one tries to think about them) by the hiddenness of their mechanisms. Only specialists can repair cars or computers. We no longer really know where foods and clothes come from, except that they can be found on the shelves of stores or ordered from the Internet; users of airplanes and pharmaceuticals have little idea of how they function, although they are readily prized and praised for their convenience and criticized when they fall short and become threatening or burdensome. The quality of engineered advancement in the techno-lifeworld "can be likened to the step from a log cabin to a skyscraper, a move from a limited and laborious environment to a world of tremendous capacity, convenience, and speed" but in which users are also denied options to "take the stairs [or] even open a window" in their offices (Borgmann, 1999, pp. 164–65).

A CONVENIENCE PARADOX

What Borgmann recognizes as the promise of technology and its motivating power is a life freed from the experiences of disease and labor traditionally taken as inherent components of the human condition. It is a promise born of the Enlightenment vision of a new knowing of what is and a correspondingly new construction of what should be (meaning what we want). Crudely simplifying, the former would replace religious superstition with positive scientific knowledge; the latter aspires to exchange cut-and-try craft for scientific medicine (to reconstruct the human body) and scientifically allied engineering (to redesign the environment); together they are envisioned as transforming the human condition from one of accepting inconveniences to a techno-human condition of conveniently affluent liberty. Although such a modernity may never have been fully achieved, it is an aspiration that fully animates the modern.

But the world of conveniences gives rise not just to social injustice issues (from the 1800s forward) regarding the distribution of its realizations but to a broadly experienced paradox. Excessive convenience counter-produces its own inconveniences.

As conveniences and freedoms of choice multiply, their mass production presents user consumers with an affluence of commodities of all types: food stuffs, clothes, durable goods, housing, medical treatments, drugs, entertainments. We are not just offered the opportunity to choose our lives, we are effectively required to do so. Although celebrated as freedom this can also be experienced as its own distinctive burden—a burden of which still another form of engineering, the engineering of consent through marketing, aspires to relieve us. In all cases engineering tends to occlude the mechanisms behind and in the commodities, mechanisms that in the past provided some guidance for usage.

Through convenience we lose contact with the materialities and histories that once oriented our lives, and we become subject to an "experiential gap" (Briggle and Mitcham, 2009): that is, a gap between experience and reality. We experience ourselves as freed from dependencies while in reality becoming increasingly networked into hidden dependencies (as always on call through our excessively convenient smart phones). On grander scales, the convenience with which we now exercise world mutating power (nuclear weapons to climate change) is linked to outcomes we can scarcely imagine. Under such circumstances, philosophical counsel might well recommend a disciplined (ascetic) delimiting of convenience in order to retain some measure of focal practice. Yet philosophical resolution is a weak mast against which to bind ourselves in the midst of the Sirens' songs of use and convenience. The ideal of a new discipline of "focal engineering" that would eschew convenience (Moriarty, 2008) seems little more than a dream, however noble.

AN ENGINEERING PHILOSOPHY FOR ENGINEERING

Return again to Tredgold's conceptualization of engineering as enacted by Telford's career and illuminated by Borgmann's phenomenology. There are at least seven key points that invite reiteration relative to this newly founded agency. (1) Its name is engineering, definitionally packaged as (2) an art (ingenuity) of directing the (3) great sources of (scientifically grasped) powers in nature (4) for (a promised free availability of) human use and convenience. Projecting further, the text operationalizes use and convenience in civil engineering terms appropriate to 1800s England with a twofold linkage to capitalism: the promotion of commerce and protection of property. (5) The "most important" way for civil engineering to realize use and convenience at that point in time and place "is to improve the means of production and of traffic in States, both for external and internal trade," after which follows a list of more specific engineered contributions useful to transport: road bridges, rail roads, and more. Additionally, (6) there is need for "the protection of property where natural powers are sources of injury." Engineers not only construct a new material culture; they protect from the intrusive disruptions of an old one. Reiterating the importance and connection of points (2) and (3), the text further states that this "brief sketch of the objects of civil engineering" will be enlarged by the "progress of science; its scope and utility will be increased with every discovery in philosophy, and its resources with every invention in mechanical or chemical art, since its bounds are unlimited" (see "Description of a Civil Engineer"). The progressive production of positive knowledge will form the basis of engineering and material progress.

Finally, in a historical point (7) which is elaborated to half the essay and that helps disclose the distinctive feature of engineering agency, the text connects English-speaking engineering with companion disciplines in the Dutch- and French-speaking versions of the discipline. It was in the sixteenth century in the Netherlands, and then again in seventeenth century in France, that civil engineering was first separated from architecture "under the title of hydraulic architecture." Civil engineering is not only military engineering cut loose from warfare; it is the art of directing the great source of water power for use and convenience. The term "hydraulic architecture . . . was too limited," connoting as it did primarily the delivery of water to urban habitats and the decoration of fountains. The convenient harnessing of water power as a prime mover in mills and steam engines and the corresponding formulation of hydraulics as an engineering science, eventually repositions water as a fluid (H_2O) with dynamic powers to the controlled and exploited.

OBJECTIONS AND QUALIFICATIONS

There are numerous possible objections to privileging the ICE definition as a window on engineering and the engineered world. One would note the contemporary lack of prominence of this definition on the ICE website. In the decades immediately before and after the 1978 ICE celebration of its charter, this definition was ignored by engineers themselves when pursuing philosophical reflections on their profession. In a 1966 article, "The Philosophy of Engineering," electrical engineer Henry Greber argued for defining engineering as "the skill of utilizing knowledge, acquired from science and from practical experience, to produce devices for the use of mankind" (p. 113). Yet while the word "convenience" does not occur, the phenomenon seems clearly present.

In 1977 and again in 1982 another electrical engineer, George Sinclair, sought to engage both historians (Sinclair, 1977) and philosophers (Sinclair and Tilston, 1982) in a conceptual clarification of the relationship between engineering and technology, without any explicit reference to the ICE definition. Sinclair argued for two otherwise largely ignored concepts as central to engineering as a learned profession: courageous creativity and reliability. Sinclair was primarily concerned to ally engineering as a profession with the professions of medicine and law, where knowledge alone is never sufficient and practitioners have to rely on mature judgment. As Sinclair pointed out, "child prodigies [are] conspicuous by their absence in the practice of engineering, medicine, legal practice, etc., when they are able to make significant contributions to the natural sciences" (Sinclair, 1977, p. 688).

In still another philosophical reflection from 1983 by mechanical engineer G. F. C. Rogers, the nature of engineering was packaged as "the practice of organizing the design and construction of any artifice which transforms the physical world around us to meet some recognized need" (p. 51). This definition was later adopted by aeronautical engineer Walter Vincenti (1990, p. 6). Here the notion of engagement with science was simply dropped, and the end of use and convenience replaced by the fuzzy notion of human need. Insofar as Tierney is right, in the present historical context in the West, convenience functions as a primary value that makes the fuzzy concept of need modestly more concrete.

In no instance does the alternative description offer much insight into the ways that engineering agency and the dynamics of engineering influence and are influenced by the contemporary lived world. The Rogers definition provides no basis for distinguishing agricultural cultivation or the craft of throwing pots from the engineering of airfoils. A philosophy of engineering needs to identify a distinctive pattern in its productive agency. By virtue of its expansiveness, the Rogers definition serves mostly as a justification for engineering against its social critics. It is more ideological than philosophical.

Another objection might be that engineers are not in fact guided by this definition. It has no conscious application. Yet for a definition to be philosophical it need not exercise practical guidance; it need only enlighten our fraught experience. Moreover, whenever English-speaking engineers are pressed to reflect on what they do and provide a basic description, they repeatedly offer some variation on the structure of the Charter definition. ABET (which accredits engineering education programs in the United States), for instance, stipulates, "Engineering is the profession in which a knowledge of the mathematical and natural sciences gained by study, experience, and practice is applied with judgment to develop ways to utilize economically the materials and forces of nature for the benefit of mankind." Other permutations are found in most English dictionaries, including technical dictionaries.

The hypothesis has been that dropping "use and convenience" inconveniently deprives philosophical reflection on engineering of a significant early perspective for appreciating its distinctive agency. Although beginnings never reveal everything about their future development, on occasion elements from the past can throw new light on what has come to be. This need imply no claim that other lights might be equally appropriate.

Finally, there is the objection that this definition applies only to English-speaking engineering. This may well be true, but insofar as English-speaking engineering is related to other traditions of engineering, it is one that can be a basis for the interrogation of any tradition.

References

— According to Thomas Aquinas, the argument from authority is the weakest all arguments (*Summa theologiae* I.1.8.2us).
— Depends on your social ontology, doesn't it?

ABET. (1998) *Engineering Criteria 2000*. The most recent version of this document, re-dated for the current academic year, is available as "Criteria for Accrediting Engineering Programs" at www.abet.org. The original document is included as an appendix in Lattuca, Terenzini, and Volkwein, *Engineering Change: A Study of the Impact of EC2000* (Baltimore, MD: ABET, 2006).

ABET. (2007) "Criteria for Accrediting Engineering Programs: Effective for Evaluation during the 2007–2008 Accreditation Cycle". Available at www.abet.org.

Ableson, Philip H. (1991) "Sustainable Future for Planet Earth," *Science*, vol. 253, no. 5016 (12 July), p. 117.

Achterhuis, Hans J. (1988) *Het Rijk va de Schaarts* [The reign of scarcity]. Baarn, Netherlands: Ambo.

Achterhuis, Hans J. (1993) "Scarcity and Sustainability". In Sachs, ed., *Global Ecology: A New Arena of Political Conflict* (London: Zed Books), pp. 104–16.

Achterhuis, Hans J. (1994) "The Lie of Sustainability". In Zweers and Boersema, eds., *Ecology, Technology and Culture* (Cambridge: White Horse Press), pp. 198–203.

Achterhuis, Hans J., ed. (2001) *American Philosophy of Technology: The Empirical Turn*. Bloomington: Indiana University Press.

Adams, Henry. (1918) *The Education of Henry Adams*. Boston, MA: Houghton Mifflin.

Adler, Paul S., and Terry A. Winograd, eds. (1992) *Usability: Turning Technologies into Tools*. New York: Oxford University Press.

Akin, William E. (1977) *Technocracy and the American Dream: The Technocrat Movement, 1900–1941*. Berkeley: University of California Press.

Albertson, Maurice L., Andrew E. Rice, and Pauline E. Birky. (1961) *New Frontiers for American Youth: Perspectives on the Peace Corps*. Washington, DC: Public Affairs Press.

Alder, Ken. (1997) *Engineering the Revolution: Arms and Enlightenment in France, 1763–1815*. Princeton, NJ: Princeton University Press.

Alexander, Jennifer Karns. (2008) *The Mantra of Efficiency: From Waterwheel to Social Control*. Baltimore, MD: Johns Hopkins University Press.

Alger, Philip L. (1958) "The Growing Importance of Ethics to Engineers," *Electrical Engineering*, vol. 77 (January), pp. 42–45.
Alger, Philip L., N. A. Christensen, and Sterling P. Olmsted. (1965) *Ethical Problems in Engineering*. New York: John Wiley.
Allenby, Branden, and Daniel Sarewitz. (2011) *The Techno-Human Condition*. Cambridge: MIT Press.
American Society of Civil Engineers (ASCE). (2007) *The Vision for Civil Engineering in 2025*. Reston, VA: ASCE.
Anderson, Robert M., Robert Perrucci, Dan E. Schendel, and Leon E. Trachtman. (1980) *Divided Loyalties: Whistle-Blowing at BART*. West Lafayette, IN: Purdue University Press.
Anderson, Scott. (1999) *The Man Who Tried to Save the World: The Dangerous Life and Mysterious Disappearance of Fred Cuny*. New York: Doubleday.
Andreas, Joel. (2009) *Rise of the Red Engineers*. Stanford, CA: Stanford University Press.
Anonymous. (1942) "The Engineering College Research Association," *Science*, vol. 96, no. 2501 (December 4), pp. 510–11.
Anonymous. (1977) "Engineers, Philosophers to Explore Ethics," *Chemical and Engineering News*, vol. 55, no. 32 (August 8), p. 8.
Arendt, Hannah. (1958) *The Human Condition*. Chicago, IL: University of Chicago Press.
Arenson, Karen W. (1995) "Missing Relief Expert Gets MacArthur Grant," *New York Times*, June 13, p. A12.
Armstrong, Karen (2006). *The Great Transformation: The Beginning of our Religious Traditions*. New York: Knopf.
Arnett, Jeffrey. (2014) *Emerging Adulthood: The Winding Road from the Late Teens through the Twenties*. 2nd edition. New York: Oxford University Press.
Arthur, W. Brian. (2009) *The Nature of Technology: What It Is and How It Evolves*. New York: Free Press.
Association for Computing Machinery (ACM). (1992) "Association for Computing Machinery Code of Ethics and Professional Conduct". Available at www.acm.org.
Association for Computing Machinery (ACM). (1999) "Software Code of Engineering and Professional Practice". Available at www.acm.org.
Auyang, Sunny Y. (2004) *Engineering—An Endless Frontier*. Cambridge, MA: Harvard University Press.
Averill, Marilyn. (2005) "Unintended Consequences". In Mitcham, ed., *Encyclopedia of Science, Technology, and Ethics* (Detroit: Macmillan Reference), vol. 4, pp. 1995–99.
Babbage, Thomas. (1830) *Reflections on the Decline of Science and some of Its Causes*. London: B. Fellowes.
Baker, Charles Whiting, Charles T. Main, E. D. Meier, Spencer Miller, and C. R. Richards, Committee on Code of Ethics. (1913) "A Proposed Code of Ethics for Engineers," *Engineering News: A Journal of Civil, Mechanical, Mining and Electrical Engineering*, vol. 69 (January 2), pp. 29–30.
Barnett, Michael, and Thomas G. Weiss, eds. (2008) *Humanitarianism in Question: Politics, Power, and Ethics*. Ithaca, NY: Cornell University Press.

Baron, Marcia. (1984) *The Moral Status of Loyalty*. Dubuque, IA: Kendall Hunt.
Barsoux, Jean-Louis. (1989) "Leaders for Every Occasion (French Engineering Education)," *IEE Review* (January), pp. 26–27.
Barthélémy, Jean-Hugues. (2015) *Life and Technology: An Inquiry into and Beyond Simondon*. Trans. Barnaby Norman. Centre for Digital Cultures, Leuphana University, Lüneburg, Germany: Meson Press.
Bataille, Georges. (1991) *The Accursed Share*, vol. 1: *Consumption*. Trans. Robert Hurley. New York: Zone Books. Original French, *La Part maudite* (Paris: Éditions de Minuit, 1949).
Baum, Robert J. (1977) "Engineering and Ethics," *Hastings Center Report*, vol. 7, no. 6 (December), pp. 14–16.
Baum, Robert J., and Albert Flores. (1978) *Ethical Problems in Engineering*. Troy, NY: Rensselaer Polytechnic Institute. Second edition in two volumes: Vol. 1, ed. Flores, Readings; vol. 2, ed. Baum, Cases, 1980.
Bayet, Albert. (1907) *La Morale scientifique: Essai sur les applications morales des sciences sociologique*. Paris: Felix Alcan.
Bayles, Michael D. (1989) *Professional Ethics*. 2nd edition. Belmont, CA: Wadsworth.
Beder, Sharon. (1998) *The New Engineer: Management and Professional Responsibility in a Changing World*. South Yarra: Macmillan.
Bell, Sarah. (2011) *Engineers, Society, and Sustainability*. San Rafael, CA: Morgan & Claypool.
Bengoetxea, Juan Bautista, and Carl Mitcham. (2010) *Ética e ingeniaría*. Valladolid: Editorial Universidad de Valladolid.
Bensaude-Vincent, Bernadette, Sacha Loeve, Alfred Nordmann, and Astrid Schwarz, eds. (2017) *Research Objects in Their Technological Setting*. London: Routledge.
Bernays, Edward L. (1947) "The Engineering of Consent," *Annals of American Academy of Political and Social Science*, vol. 250, no. 1 (March), pp. 113–20.
Berrett, Dan. (2014) "A Curriculum for the Selfie Generation," *Chronicle of Higher Education* (June 2).
Bertalanffy, Ludwig von. (1968) *General System Theory: Foundations, Development, Applications*. New York: Brazillier, 1968.
Bijker, Wiebe E. (1995) *Of Bicycles, Bakelite, and Bulbs: Towards a Theory of Sociotechnical Change*. Cambridge: MIT Press.
Bijker, Wiebe E, Thomas P. Hughes, and Trevor Pinch, eds. (1987) *The Social Construction of Technological Systems: New Directions in the Sociology and History of Technology*. Cambridge: MIT Press.
Birsch, Douglas, and John H. Fielder, eds. (1994) *The Ford Pinto Case: A Study in Applied Ethics, Business, and Technology*. Albany: State University of New York Press.
Blockley, David. (2012) *Engineering: A Very Short Introduction*. Oxford: Oxford University Press.
Blockley, David. (2014) *Structural Engineering: A Very Short Introduction*. Oxford: Oxford University Press.
Blue, Ethan, Michael Levine, and Dean Nieusma. (2014) *Engineering and War: Militarism, Ethics, Institutions, Alternatives*. San Rafael, CA: Morgan and Claypool.

Boisjoly, Roger. (1991) "The Challenger Disaster: Moral Responsibility and the Working Engineer". In Johnson, ed., *Ethical Issues in Engineering* (Englewood Cliffs, NJ: Prentice Hall), pp. 6–14.

Bontems, Vincent. (2018) "On the Current Uses of Simondon's Philosophy of Technology". In S. Loeve, X. Guchet, and B. Bensaude-Vincent, eds., *French Philosophy of Technology: Classical Readings and Contemporary Approaches* (Dordrecht: Springer), pp. 37–49.

Booker, Peter Jeffrey. (1963) *A History of Engineering Drawing*. London: Chatto and Windus.

Boot, Max. (2006) *War Made New: Weapons, Warriors, and the Making of the Modern World*. New York: Gotham.

Borgmann, Albert. (1984) *Technology and the Character of Contemporary Life: A Philosophical Inquiry*. Chicago, IL: University of Chicago Press.

Borgmann, Albert. (1999) *Holding On to Reality: The Nature of Information at the Turn of the Millennium*. Chicago, IL: University of Chicago Press.

Bowen, Richard W. (2009) *Engineering Ethics: Outline of an Aspirational Approach*. London: Springer.

Boyle, Charles. (1984) "Military Technology". In Charles Boyle, Peter Wheale, and Brian T. Sturgess, eds., *People, Science, and Technology: A Guide to Advanced Industrial Society* (Totowa, NJ: Barnes & Noble Books), pp. 151–68.

Brand, Stewart. (1987) *The Media Lab: Inventing the Future at MIT*. New York: Viking.

Bratman, Michael E. (1987) *Intention, Plans, and Practical Reason*. Cambridge, MA: Harvard University Press.

Briggle, Adam, Robert Frodeman, and J. Britt Holbrook. (2006) "Introducing a Policy Turn in Environmental Philosophy," *Environmental Philosophy*, vol. 3, no. 1, pp. 70–77.

Briggle, Adam, and Carl Mitcham. (2009) "Embedding and Networking: Conceptualizing Experience in a Technosociety," *Technology in Society*, vol. 31, pp. 374–83.

Brooks, Harvey. (1968) *The Government of Science*. Cambridge: MIT Press.

Broome, Taft H., Jr. (1978) "Issues in Ethics for Engineering School Publications," *Professional Engineer*, vol. 48, no. 1, pp. 28–30.

Broome, Taft H., Jr. (1985a) "Engineering the Philosophy of Science," *Metaphilosophy*, vol. 16, no. 1 (January), pp. 47–56.

Broome, Taft H., Jr. (1985b) "Engineers Debate Hazardous Technology Issues," *Civil Engineering*, vol. 55, no. 8, p. 84.

Broome, Taft H., Jr. (1986) "The Slippery Ethics of Engineering," *Washington Post* (Sunday, December 28), p. D3.

Broome, Taft H., Jr. (1987) "Engineering Responsibility for Hazardous Technologies," *Journal of Professional Issues in Engineering*, vol. 113, no. 2 (April), pp. 139–149.

Broome, Taft H., Jr. (1989) "Can Engineers Hold Public Interests Paramount?," *Research in Philosophy and Technology*, vol. 9, pp. 3–11.

Broome, Taft H., Jr. (1990) "Imagination for Engineering Ethicists". In Durbin, ed., *Broad and Narrow Interpretations of Philosophy and Technology*, Philosophy and Technology, vol. 7 (Boston, MA: Kluwer), pp. 45–51.

Broome, Taft H., Jr. (1991) "Bridging Gaps in Philosophy and Engineering". In Durbin, ed., *Critical Perspectives on Nonacademic Science and Engineering* (Bethlehem, PA: Lehigh University Press), pp. 265–77.
Broome, Taft H., Jr. (1996) "The Heroic Mentorship," *Science Communication*, vol. 17, no. 4, pp. 398–429
Broome, Taft H., Jr., and Jeff Peirce. (1997) "The Heroic Engineer," *Journal of Engineering Education*, vol. 86, no. 1 (January), pp. 51–55.
Broome, Taft H., Jr. (1999) "Case Study: The Concrete Sumo," *Science and Engineering Ethics*, vol. 5, no. 4, pp. 542–47. Followed with commentaries and response.
Broome, Taft H., Jr. (2010) "Metaphysics of Engineering". In Van de Poel et al., eds., *Philosophy and Engineering: An Emerging Agenda* (Dordrecht: Springer), pp. 295–304.
Brose, Eric Dorn. (1992) *The Politics of Technological Change in Prussia: Out of the Shadow of Antiquity, 1809–1848*. Princeton, NJ: Princeton University Press.
Brown, Kathryn S. (2000) "Ecologists on a Mission to Save the World," *Science*, vol. 287 (18 February), pp. 1188–95.
Brown, Lester. (1981) *Building a Sustainable Society*. New York: W. W. Norton.
Bucciarelli, Louis L. (1988) "An Ethnographic Perspective on Engineering Design," *Design Studies*, vol. 9, no. 3 (July), pp. 159–68.
Bucciarelli, Louis L. (1994) *Designing Engineers*. Cambridge: MIT Press.
Bucciarelli, Louis L. (2003) *Engineering Philosophy*. Delft: Delft University Press.
Buchanan, Richard. (1989) "Declaration by Design: Rhetoric, Argument, and Demonstration in Design Practice". In Margolin, ed., *Design Discourse: History, Theory, Criticism* (Chicago, IL: University of Chicago Press), pp. 91–110.
Bugliarello, George. (1991) "The Social Function of Engineering: A Current Assessment." In Hedy E. Sladovich, ed., *Engineering as a Social Enterprise* (Washington, DC: National Academy Press), pp. 73–88.
Bulleit, William, Jon Schmidt, Irfan Alvi, Erik Nelson, and Tonatiuh Rodriguez-Niki. (2015) "Philosophy of Engineering: What It Is and Why It Matters," *Journal of Professional Issues in Engineering Education and Practice*, vol. 141, no. 3, pp. 02514003-1 to 02514003-9.
Bunge, Mario. (1967) *Scientific Research II: The Search for Truth*. New York: Springer.
Burck, Gordon M., and Charles C. Flowerree. (1991) *International Handbook on Chemical Weapons Proliferation*. New York: Greedwood Press (for Federation of American Scientists).
Burke, John G. (1966) "Bursting Boilers and the Federal Power," *Technology and Culture*, vol. 7, no. 1 (Winter), pp. 1–23.
Bush, Vannevar. (1945) *Science—The Endless Frontier*. Washington, DC: U.S. Government Printing Office.
Bush, Vannevar. (1946) *Endless Horizon*. Washington, DC: Public Affairs Press.
Bush, Vannevar. (1967) *Science Is Not Enough*. New York: William Morrow.
Cahill, Kevin M., ed. (2005) *Technology for Humanitarian Action*. New York: Fordham University Press and the Center for International Health and Cooperation.

Callahan, Daniel. (1998) *False Hopes: Overcoming the Obstacles to a Sustainable, Affordable Medicine*. New York: Simon & Schuster, 1998.

Cameron, Rondo, and A. J. Millard. (1985) *Technology Assessment: A Historical Approach*. Dubuque, IA: Kendall Hunt.

Campbell, Colin. (1987) *The Romantic Ethics and the Spirit of Modern Consumerism*. Oxford: Basil Blackwell.

Carley, Michael, and Ian Christie. (1993) *Managing Sustainable Development*. Minneapolis: University of Minnesota Press.

Carpenter, Stanley R. (1993) "When Are Technologies Sustainable?" In Larry A. Hickman and Elizabeth F. Porter, eds., *Technology and Ecology* (Carbondale, IL: Society for Philosophy and Technology, Department of Philosophy, Southern Illinois University), pp. 202–14. Also available with unnumbered pagination in the electronic journal of the Society for Philosophy and Technology: *Techné*, vol. 1, nos. 1–2 (Fall 1995).

Carson, Rachel. (1962) *Silent Spring*. Boston, MA: Houghton Mifflin.

Cartwright, Nancy. (1983) *How the Laws of Physics Lie*. Oxford: Oxford University Press.

Certeau, Michel de. (1980) *L'Invention du quotidien*, vol. 1: *Arts de faire*. Paris: Union Générale d'Editions. English version: *The Practice of Everyday Life*, trans. Steven Rendall (Berkeley: University of California Press, 1984).

Chesbrough, Henry. (2006) *Open Business Models: How to Thrive in the New Innovation Landscape*. Cambridge, MA: Harvard Business School Press.

Christensen, Steen Hyldgaard, Martin Meganck, and Bernard Delahousse, eds. (2007) *Philosophy in Engineering*. Aarhus: Academica.

Christensen, Steen Hyldgaard, Carl Mitcham, Li Bocong, and Yanming An, eds. (2012) *Engineering, Development and Philosophy: American, Chinese and European Perspectives*. Dordrecht: Springer.

Christensen, Steen Hyldgaard, Christelle Didier, Andrew Jamison, Martin Meganck, Carl Mitcham, and Byron Newberry, eds. (2015a) *Engineering Education and Practice in Context*, vol. 1: *International Perspectives on Engineering Education*. Dordrecht: Springer.

Christensen, Steen Hyldgaard, Christelle Didier, Andrew Jamison, Martin Meganck, Carl Mitcham, and Byron Newberry, eds. (2015b) *Engineering Education and Practice in Context*, vol. 2: *Engineering Identities, Epistemologies and Values*. Dordrecht: Springer.

Christiansen, Donald. (1972) "The New Professionalism," *IEEE Spectrum*, vol. 8, no. 6, p. 17.

Christie, A. G. (1922) "A Proposed Code of Ethics for All Engineers," *Annals of the American Academy of Political and Social Science*, vol. 101 (May), pp. 97–103.

Clarke, Jackie. (2001) "Engineering a New Order in the 1930s: The Case of Jean Coutrot," *French Historical Studies*, vol. 24, no. 1 (2001), pp. 63–86.

Clausen, H. W. (1922) "Procedure in Developing Ethical Standards Adopted by the American Association of Engineers," *Annals of the American Academy of Political and Social Science*, vol. 101 (May), pp. 89–94.

Coeckelbergh, Mark, Michael Funk, and Stefan Koller, eds. (2018) "Wittgenstein and Philosophy of Technology," theme issue, *Techné: Research in Philosophy and Technology*, vol. 22, no. 3, pp. 287–471.

Cohen, Bob. (1986) "Think Tank, Think Profits," *Newsweek* (April 28), p. 70.
Cohn, Carol. (1990) "Clean Bombs and Clean Language". In Elshtain and Tobias, eds., *Women, Militarism, and War: Essays in History, Politics, and Social Theory*. Savage, MD: Rowman & Littlefield.
Collins, Harry, and Trevor Pinch. (2002) *The Golem at Large: What You Should Know about Technology*. 2nd edition. Cambridge: Cambridge University Press.
Comité d'Etudes sur la Formation des Ingenieurs. (2000) "Engineering Education in France". http://cri.ensmp.fr/cefi/plaquet.html.
Committee on Science, Engineering, and Public Policy. (2007) *Rising above the Gathering Storm: Energizing and Employing America for a Brighter Economic Future*. Washington, DC: National Academies Press. (The select committee that prepared this report was the Committee on Prospering in the Global Economy of the 21st Century: An Agenda for American Science and Technology, Norman Augustine, chair.)
Conseil National des Ingenieurs et des Scientifiques de France (2001) *Charte d'ethique de l'Ingenieur*. Paris: CNISF.
Cooke, Morris Llewellyn. (1922) "Ethics and the Professional Engineer," *Annals of the American Academy of Political and Social Science*, vol. 101 (May), pp. 68–72.
Cooper, Barry. (1991) *Action into Nature: An Essay on the Meaning of Technology*. Notre Dame, IN: University of Notre Dame Press.
Coopersmith, Jennifer. (2010) *Energy, the Subtle Concept: The Discovery of Feynman's Blocks from Leibniz to Einstein*. Oxford: Oxford University Press.
Cross, Nigel. (1989) *Engineering Design Methods*. New York: John Wiley.
Cross, Wilbur. (1990) *The Code: An Authorized History of the ASME Boiler and Pressure Vessel Code*. New York: American Society of Mechanical Engineering.
Cuello Nieto, César. (1992) "Engineering and Ethics in the Dominican Republic". In Mitcham, *Engineering Ethics throughout the World: Introduction, Documentation, Commentary, and Bibliography* (University Park, PA: STS Press), pp. II2–II18.
Cuello Nieto, César, and Paul T. Durbin. (1993) "Sustainable Development and Philosophies of Technology." In Larry A. Hickman and Elizabeth F. Porter, eds., *Technology and Ecology* (Carbondale, IL: Society for Philosophy and Technology, Department of Philosophy, Southern Illinois University), pp. 215–39. Also available with unnumbered pagination in the electronic journal of the Society for Philosophy and Technology: *Techné*, vol. 1, nos. 1–2 (Fall 1995).
Cuny, Frederick C. (1983) *Disasters and Development*. New York: Oxford University Press.
Curd, Martin, and Larry May. (1984) *Professional Responsibility for Harmful Actions*. Dubuque, IA: Kendall Hunt.
Cutcliffe, Stephen H. (2000) *Ideas, Machines, and Values: An Introduction to Science, Technology, and Society Studies*. Lanham, MD: Rowman & Littlefield.
Davis, Michael. (1998) *Thinking Like an Engineer: Studies in the Ethics of a Profession*. New York: Oxford University Press.
Davis, Michael. (2002) *Profession, Code and Ethics*. Aldershot: Ashgate.
Davis, Michael. (2002) "Three Myths about Codes of Engineering Ethics," in his *Profession, Code, and Ethics* (Aldershot: Ashgate), pp. 121–31.
Davis, Michael, ed. (2005) *Engineering Ethics*. Aldershot: Ashgate.

Davis, Michael. (2010) "Distinguishing Architects from Engineers: A Pilot Study in Differences between Engineers and other Technologists". In Van de Poel and Goldberg, eds., *Philosophy and Engineering: An Emerging Agenda* (Dordrecht: Springer), pp. 15–30.

Davis, Michael. (2015) "Engineering as Profession: Some Methodological Problems in Its Study". In Christensen et al., eds., *Engineering Education and Practice in Context*, vol. 2: *Engineering Identities, Epistemologies and Values* (Dordrecht: Springer), pp. 65–79.

Davison, Aidan. (2001) *Technology and the Contested Meanings of Sustainability*. Albany: State University of New York Press.

Deforge, Yves. (1990) "Avatars of Design: Design before Design," *Design Issues*, vol. 6, no. 2 (Spring), pp. 43–50.

Dennett, Daniel. (1987) *The Intentional Stance*. Cambridge: MIT Press.

Dennett, Daniel. (1995) *Darwin's Dangerous Idea: Evolution and the Meanings of Life*. New York: Simon & Schuster.

Dessauer, Friedrich. (1927) *Philosophie der Technik: Das Problem der Realisierung* [Philosophy of technology: The problem of realization]. Bonn: F. Cohen.

Dessauer, Friedrich. (1956) *Streit um die Technik* [Controversy concerning technology]. Frankfurt am Main: J. Knecht. Abridged edition, Freiburg: Herder, 1959.

Devon, Richard. (1999) "Towards a Social Ethics of Technology: The Norms of Engagement," *Journal of Engineering Education*, vol. 88, no. 1 (January), pp. 87–92.

Dewey, John. (1922) "Education as Engineering," *New Republic*, vol. 32, no. 407 (20 September 1922), pp. 89–91.

Dewey, John. (1927) *The Public and Its Problems*. New York: Henry Holt.

DiBona, Chris, Sam Ockman, and Mark Stone, eds. (1999) *Open Sources: Voices from the Open Source Revolution*. Beijing: O'Reilly.

DiBona, Chris, Danese Cooper, and Mark Stone, eds. (2006) *Open Sources 2.0: The Continuing Evolution*. Beijing: O'Reilly.

Didier, Christelle. (1999) "Engineering Ethics in France: A Historical Perspective," *Technology in Society*, vol. 21, no. 4, pp. 471–86.

Didier, Christelle. (2000) "Engineering Ethics at the Catholic University of Lille (France): Research and Teaching in a European Context," *European Journal of Engineering Education*, vol. 25 (2000), pp. 325–35.

DiPrizio, Robert C. (2002) *Armed Humanitarians: U.S. Interventions from Northern Iraq to Kosovo*. Baltimore, MD: Johns Hopkins University Press.

Downey, Gary Lee. (2015) "Engineering Studies". In Wright, ed., *International Encyclopedia of the Social and Behavioral Sciences*, 2nd edition (Amsterdam: Elsevier), vol. 7, pp. 641–48.

Downey, Gary Lee, ed. (2015) Theme issue: "Liberal Studies in Engineering," *Engineering Studies*, vol. 7, nos. 2–3, pp. 99–220.

Downey, Gary Lee, and Juan C. Lucena. (1994) "Engineering Studies". In S. Jasanoff, G. E. Markle, J. C. Peterson, and T. Pinch, eds., *Handbook of Science and Technology Studies* (Newbury Park, CA: Sage), pp. 167–88. (Same pagination in revised edition, 2001.)

Downey, Gary Lee, Juan C. Lucena, and Carl Mitcham. (2007) "Engineering Ethics and Identity: Emerging Initiatives in Comparative Perspective," *Science and Engineering Ethics*, vol. 13, no. 4 (December), pp. 463–87.

Downey, Gary Lee, and Zhang Zhihui. (2015) "Nonlinear STS, Engineering Studies, and Dominant Images of Engineering Formation: An Interview with Professor Gary Downey," available at http://4sonline.org/files/Downey-2015-Nonlinear-STS-interview.pdf. This is the original English text of 盖瑞·唐尼，张志会"非线性STS,工程研究和工程形塑的主导性意象—对盖瑞•唐尼教授的访谈" *Fei xianxing STS, gongcheng yanjiu he gongcheng xingsu de zhudaoxing yixiang: dui Gairui Tangni jiaoshou de fangtan, Gongcheng yanjiu* [Journal of engineering studies], vol. 7, no. 4 (December), pp. 332–48.

Drews, Robert. (1993) *The End of the Bronze Age: Changes in Warfare and the Catastrophe ca. 1200 BC*. Princeton, NJ: Princeton University Press.

Dumont, Louis. (1986) *Essays on Individualism: Modern Ideology in Anthropological Perspective*. Chicago, IL: University of Chicago Press.

Dumont, Louis, with Christian Delacampagne. (1981) Interview: "Louis Dumont and the Indian Mirror," *Rain*, no. 43 (April 1981), pp. 4–7.

Dumouchel, Paul. (1992) "Gilbert Simondon's Plea for a Philosophy of Technology," *Inquiry: An Interdisciplinary Journal of Philosophy*, vol. 35, nos. 3–4 (September), pp. 407–21.

Dupuy, Jean-Pierre. (2009) "Critique of the Precautionary Principle and the Possibility for an 'Enlightened Doomsaying'". In Olsen et al., eds., *A Companion to the Philosophy of Technology* (Hoboken, NJ: Wiley-Blackwell), pp. 210–13.

Durbin, Paul T., ed. (1980) *A Guide to the Culture of Science, Technology, and Medicine*. New York: Free Press.

Durbin, Paul T., ed. (1991) *Critical Perspectives on Nonacademic Science and Engineering*. Bethlehem, PA: Lehigh University Press.

Durbin, Paul T. (1997) "Engineering Ethics and Social Responsibility: Reflections on Recent Development in the USA," *Bulletin of Science, Technology, and Society*, vol. 17, nos. 2–3 (April), pp. 77–83.

Durbin, Paul T. (2005) *Philosophy of Technology: In Search of Discourse Synthesis*. Privately published, 2005. Special issue, *Techné: Research in Philosophy and Technology*, vol. 10, no. 2 (Winter 2006), pp. 1–283.

Dusek, Val. (2006) *Philosophy of Technology: An Introduction*. Malden, MA: Blackwell.

Dyer, Frank Lewis, and Thomas Commerford Martin. (1929) *Edison: His Life and Inventions*. 2nd edition. 2 vols. New York: Harper.

Dym, Clive L. (1994) *Engineering Design: A Synthesis of Views*. New York: Cambridge University Press.

Dym, Clive L. (1998) "The Languages of Engineering Design: Representing Objects and Articulating Processes". Paper for a workshop on "The Empirical Turn in the Philosophy of Technology," Technical University Delft, April, 16–18.

Edsall, John T. (1975) *Scientific Freedom and Responsibility: A Report of the AAAS Committee on Scientific Freedom and Responsibility*. Washington, DC: American Association for the Advancement of Science.

Edwards, Clive A., Rattan Lal, Patrick Madden, Robert H. Miller, and Gar House. (1990) *Sustainable Agriculture*. Ankeny, IA: Soil and Water Conservation Society.
Einstein, Albert. (1960) *Einstein on Peace*. Eds. Otto Nathan and Heinz Norden. New York: Simon & Schuster.
Eisenstadt, Shmuel N., ed. (2012). *The Origins and Diversity of Axial Age Civilizations*. Albany, NY: SUNY Press.
Eliade, Mircea. (1959) *Cosmos and History: The Myth of the Eternal Return*, trans. Willard R. Trask. New York: Harper.
Ellul, Jacques. (1954) *La Technique ou l'enjeu du siècle*. Paris: A. Colin. English version: *The Technological Society*, trans. John Wilkinson (New York: Knopf, 1964).
Ellul, Jacques. (1962) "The Technological Order". Trans. John Wilkinson. In Stover, ed., *The Technological Order* (Detroit: Wayne State University Press), pp. 10–37.
English, J. Morley. (1975) Untitled review of Nicholas Georgescu-Roegen's *The Entropy Law and the Economic Process* (1971), *Engineering Economist*, vol. 20, no. 3, pp. 226–27.
Esteva, Gustavo, (1992) "Development." In Sacks, ed., *Development Dictionary* (London: Zed), pp. 6–25.
Ewen, Stuart. (1976) *Captains of Consciousness: Advertising and the Social Roots of the Consumer Culture*. New York: McGraw-Hill. 25th anniversary edition, with new preface, "Memoirs of a Commodity Fetishist," New York: Basic Books, 2001.
Ewen, Stuart. (1988) *All Consuming Images: The Politics of Style in Contemporary Culture*. New York: Basic Books.
Ewen, Stuart. (1996) *PR! A Social History of Spin*. New York: Basic Books.
Eyth, Max von. (1904) *Lebendige Krafte: Sieben Vortrage aus dem Gebiete der Technik* [Living powers: Seven lectures from the domain of technology]. Berlin: Julius Springer.
Ezrahi, Yaron. (1990) *The Descent of Icarus: Science and the Transformation of Contemporary Democracy*. Cambridge, MA: Harvard University Press.
Fealing, Kaye Husbands, Julia I. Lane, John H. Marburger III, and Stephanie S. Shipp, eds. (2011). *The Science of Science Policy: A Handbook*. Stanford, CA: Stanford University Press.
Feenberg, Andrew. (1991) *Critical Theory of Technology*. New York: Oxford University Press. Revised as *Transforming Technology: A Critical Theory Revisited* (New York: Oxford University Press, 2002).
Feenberg, Andrew. (1995a) *Alternative Modernity: The Technical Turn in Philosophy and Social Theory*. Berkeley: University of California Press.
Feenberg, Andrew. (1995b) "Subversive Rationalization: Technology, Power, and Democracy". In Feenberg and Hannay, eds., *Technology and the Politics of Knowledge* (Bloomington: Indiana University Press), pp. 3–22.
Feenberg, Andrew. (1999) *Questioning Technology*. New York: Routledge.
Feenberg, Andrew. (2017) *Technosystem: The Social Life of Reason*. Cambridge, MA: Harvard University Press.
Feller, Joseph, Brian Fitzgerlad, Scott A. Hissam, and Karim R. Lakhani, eds. (2005) *Perspectives on Free and Open Source Software*. Cambridge: MIT Press.
Feng, Patrick. (2000) "Rethinking Technology, Revitalizing Ethics: Overcoming Barriers to Ethical Design," *Science and Engineering Ethics*, vol. 6, no. 2, pp. 207–20.

Ferguson, Eugene S. (1992) *Engineering and the Mind's Eye*. Cambridge: MIT Press.
Ferré, Frederick. (1988) *Philosophy of Technology*. Englewood Cliffs, NJ: Prentice Hall. Reprinted Athens: University of Georgia Press, 1995.
Feynman, Richard. (1963) *The Feynman Lectures on Physics*. Vol. 1. Eds. Leighton and Sands. Reading, MA: Addison-Wesley.
Fielder, John H., and Douglas Dirsch, eds. (1992) *The DC-10 Case: A Study in Applied Ethics, Technology, and Society*. Albany: State University of New York Press.
Finch, James Kip. (1951) *Engineering and Western Civilization*. New York: McGraw-Hill.
Flores, Albert. (1977) "The National Project of Philosophy and Engineering Ethics to Bring Together Engineers, Philosophers," *Professional Engineer*, vol. 47, no. 8 (August), pp. 26–29.
Florman, Samuel C. (1976) *The Existential Pleasures of Engineering*. New York: St. Martin's Press. 2nd edition, 1994.
Florman, Samuel C. (1981) *Blaming Technology: The Irrational Search for Scapegoats*. New York: St. Martin's Press.
Florman, Samuel C. (1987) *The Civilized Engineer*. New York: St. Martin's Press.
Florman, Samuel C. (1994) *The Existential Pleasures of Engineering*. 2nd edition. New York: St. Martin's Press. (First published 1976.)
Florman, Samuel C. (1996) *The Introspective Engineer*. New York: St. Martin's Press.
Floyd, Juliet, and James E. Katz, eds. (2015) *Philosophy of Emerging Media: Understanding, Appreciation, Application*. New York: Oxford University Press.
Forman, Paul. (2007) "The Primacy of Science in Modernity, of Technology in Postmodernity, and of Ideology in the History of Technology," *History and Technology*, vol. 23, no. 1/2 (March/June), pp. 1–152.
Foucault, Michel. (1980) *Power/Knowledge: Selected Interviews and Other Writings, 1972–1977*. Ed. and trans. Colin Gordon. New York: Pantheon.
Frankena, William K. (1973) *Ethics*. 2nd edition. Englewood Cliffs, NJ: Prentice-Hall.
Franssen, Maarten. (2005) "Technical Functions". In Mitcham, ed., *Encyclopedia of Science, Technology, and Ethics* (Detroit: Macmillan Reference), vol. 4, pp. 1887–90.
Franssen, Maarten, Peter Kroes, Thomas A. C. Reydon, and Pieter E. Vermaas, eds. (2014) *Artefact Kinds: Ontology and the Human-Made World*. Dordrecht: Springer.
Franssen, Maarten, Gert-Jan Lokhorst, and Ibo van de Poel. (2018) "Philosophy of Technology". In Zalta, ed., *Stanford Encyclopedia of Philosophy* (Spring 2019 edition), https://plato.stanford.edu/archives/spr2019/entries/technology/.
Frascara, Jorge. (2019) "The Science of Design: Tomás Maldonado, Buenos Aires 1922 −Milano 2018," *Design Issues*, vol. 35, no. 2 (Spring), pp. 93–96.
Friedel, Robert. (2007) *A Culture of Improvement: Technology and the Western Millennium*. Cambridge: MIT Press.
Friedman, Batya, Peter H. Kahn Jr., and Alan Borning. (2006) "Value Sensitive Design and Information Systems". In Zhang and Galletta, eds., *Human-Computer Interaction in Management Information Systems: Foundations* (Armonk, NY: M. E. Sharpe), pp. 348–72.

Friends of the Earth. (1972) *The Stockholm Conference: Only One Earth, An Introduction to the Politics of Survival*. London: Earth Island.

Fritzsche, Albrecht, and Sascha Julian Oks, eds. (2018) *The Future of Engineering: Philosophical Foundations, Ethical Problems and Application Cases*. Dordrecht: Springer.

Frodeman, Robert, and Adam Briggle. (2016) *Socrates Tenured: The Institutions of Twenty-First-Century Philosophy*. London: Rowman & Littlefield International.

Frodeman, Robert, and Carl Mitcham, eds. (2002) "Toward a Philosophy of Science Policy," theme issue, *Philosophy Today*, vol. 48, no. 5 (supplement).

Fuller, R. Buckminster. (1969) *Operating Manual for Spaceship Earth*. New York: Simon and Schuster.

Galison, Peter. (1997) *Image and Logic: A Material Culture of Microphysics*. Chicago, IL: University of Chicago Press.

Galison, Peter. (2017) "The Pyramid and the Ring: A Physics Indifferent to Ontology." In B. Bensaude-Vincent et al., eds., *Research Objects in Their Technological Setting* (London: Routledge), pp. 15–26.

García de la Huerta, Marcos. (1991) "The Ethical Codes of Dictatorship: Ethics in Chile," *Perspectives on the Professions*, vol. 11, no. 1 (August), pp. 2–5.

García de la Huerta, Marcos, and Carl Mitcham. (2001) *La ética en la profesión de ingeniero: Ingeniería y ciudadanía* [Ethics in the profession of the engineer: Engineering and citizenship]. Santiago, Chile: Universidad de Chile.

Geerts, Robert-Jan, Bart Gremmen, Josette Jacobs, and Guido Ruivenkamp. (2014) "Towards a Philosophy of Energy," *Scientiae Studia*, vol. 12, special issue, pp. 105–27.

Gendron, Bernard. (1077) *Technology and the Human Condition*. New York: St. Martin's Press.

Georgescu-Roegen, Nicholas. (1971) *The Entropy Law and the Economic Process*. Cambridge, MA: Harvard University Press.

Georgescu-Roegen, Nicholas. (1975) "Energy and Economic Myths," *Southern Economic Journal*, vol. 41, no. 3 (January), pp. 347–81.

Giedion, Siegfried. (1948) *Mechanization Takes Command: A Contribution to Anonymous History*. New York: Oxford University Press.

Giovannitti, Len, and Fred Freed. (1965) *The Decision to Drop the Bomb*. New York: Coward-McCann.

Girard, René. (1965) *Deceit, Desire, and the Novel: Self and Other in Literary Structure*. Trans. Yvonne Freccero. Baltimore, MD: Johns Hopkins University Press.

Gispen, C. W. R. [Kees]. (1990). *New Profession, Old Order: Engineers and German Society, 1815–1914*. Cambridge: Cambridge University Press.

Glegg, Gordon L. *The Development of Design*. Cambridge: Cambridge University Press.

Glover, Jonathan. (1999) *Humanity: A Moral History of the Twentieth Century*. New Haven, CT: Yale University Press.

Goldberg, David E. (2010) "Why Philosophy? Why Now? Engineering Responds to the Crisis of a Creative Era". In Van de Poel et al., eds., *Philosophy and Engineering: An Emerging Agenda* (Dordrecht: Springer), pp. 255–63.

Goldberg, David E. (2013) "Is Engineering Philosophical Weak?" In Michelfelder et al., eds., *Philosophy and Engineering: Reflections on Practice, Principles and Process* (Dordrecht: Springer), pp. 391–405.

Goldberg, David E., and Mark Sommerville. (2014) *A Whole New Engineer: The Coming Revolution in Engineering Education.* (With Catherine Whitney.) Douglas, MI: Three Joy Associates.

Goldman, Steven L. (1984) "The *Techne* of Philosophy and the Philosophy of Technology," *Research in Philosophy and Technology*, vol. 7, pp. 115–44.

Goldman, Steven L. (1988) "The History of Engineering Education: Perennial Issues In the Supply and Training of Talent". Study commissioned by the U.S. Office of Technology Assessment in support of Congressional funding bills for engineering education in fiscal year 1989 and published in U.S. Office of Technology Assessment, *Educating Scientists and Engineers: Grade School to Grad School* (Springfield, VA: National Technical Information Service, March), pp. 244–305.

Goldman, Steven L. (1989) "Images of Technology in Popular Films: Discussion and Filmography," *Science, Technology, & Human Values*, vol. 14, no. 3 (Sumer), pp. 275–301.

Goldman, Steven L. (1990) "Philosophy, Engineering and Western Culture". In Durbin, ed., *Broad and Narrow Interpretations of Philosophy of Technology* (Dordrecht: Kluwer), pp. 125–52.

Goldman, Steven L. (1991) "The Social Captivity of Engineering". In Durbin, ed., *Critical Perspectives on Nonacademic Science and Engineering* (Bethlehem, PA: Lehigh University Press), pp. 121–45.

Goldman, Steven L. (1992) "No Innovation without Representation: Technological Action in a Democratic Society". In S. H. Cutcliffe, S. L. Goldman, M. Medina, and J. Sanmartín, eds., *New Worlds, New Technologies, New Issues* (Bethlehem, PA: Lehigh University Press, 1992), pp. 148–60.

Goldman, Steven L. (2004) "Why We Need a Philosophy of Engineering: A Work in Progress," *Interdisciplinary Science Reviews*, vol. 29, no. 2, pp. 163–67.

Goldman, Steven L. (2018) "Compromised Exactness and the Rationality of Engineering". In García-Díaz and Olaya, eds., *Social Systems Engineering: The Design of Complexity* (Hoboken, NJ: John Wiley), pp. 13–30.

Goodman, Paul. (1969) "Can Technology Be Humane?," *New York Review of Books*, vol. 13, no. 9 (November 20), pp. 27–34. Also available in his *New Reformation: Notes of a Neolithic Conservative* (New York: Random House, 1970), pp. 3–23.

Gorakhov, V. G. (1997) *Петр Климентьевич Энгельмейер: Инженер-механик и философ техники, 1855–1941* [Petr Kliment'evych Engel'meier: Inzhener-mekhanik i filosof tekhniki, 1855–1941 (Peter Klimentevych Engelmeier: Mechanical engineer and philosopher of technology, 1855–1941)]. Moscow: Nauka.

Gorman, Michael E., Matthew M. Mehalik, and Patricia H. Werhane. (2000) *Ethical and Environmental Challenges to Engineering*. Englewood Cliffs, NJ: Prentice Hall.

Goujon, Philippe, and Bertrand Hériard Dubreuil, eds. (2001) *Technology and Ethics: A European Quest for Responsible Engineering*. Leuven, Belgium: Peeters.

Graham, Loren R. (1993) *The Ghost of the Executed Engineer: Technology and the Fall of the Soviet Union*. Cambridge, MA: Harvard University Press.

Gravander, Jerry W. (1980) "The Origin and Implications of Engineers' Obligations to the Public Welfare," *PSA: Proceedings of the Biennial Meeting of the Philosophy of Science Association*, vol. 2, pp. 443–55.

Grayson, Lawrence P. (1980) "A Brief History of Engineering Education in the United States," *IEEE Transactions on Aerospace and Electronic Systems*, vol. AES-16, no. 3 (May), pp. 373–92.

Grayson, Lawrence P. (1993) *The Making of an Engineer: An Illustrated History of Engineering Education in the United States and Canada*. New York: John Wiley.

Greber, Henry. (1966) "The Philosophy of Engineering," *IEEE Spectrum*, vol. 3, no. 10 (October), pp. 112–15.

Grodzins, Morton, and Eugene Rabinowitch, eds. (1963) *The Atomic Age: Scientists in National and World Affairs—Articles from "The Bulletin of the Atomic Scientists," 1945–1962*. New York: Basic Books.

Gunn, Alastair S., and P. Aarne Vesilind. (1986) *Environmental Ethics for Engineers*. Chelsea, MI: Lewis Publishers.

Gunn, Alastair S., and P. Aarne Veslind. (2003) *Hold Paramount: The Engineer's Responsibility to Society*. Pacific Grove, CA: Thomson Brooks Cole.

Haas, Peter M. (2005) "Science and International Environmental Governance." In Peter Dauvergne, ed., *Handbook of Global Environmental Politics* (Northampton, MA: Edward Elgar), pp. 383–401.

Habermas, Jürgen. (1983) "Modernity—An Incomplete Project". In Foster, ed., *The Anti-Aesthetic: Essays on Postmodern Culture* (Port Townsend, WA: Bay Press, 1983), pp. 3–15.

Hacking, Ian. (1975) *Why Does Language Matter to Philosophy?* Cambridge: Cambridge University Press.

Haraway, Donna (1991) "A Cyborg Manifesto: Science, Technology, and Socialist-Feminism in the Late Twentieth Century," in her *Simians, Cyborgs and Women: The Reinvention of Nature* (New York: Routledge, 1991), pp. 149–81.

Harding, C. Francis, and Donald T. Canfield. (1936) *Legal and Ethical Phases of Engineering*. New York: McGraw-Hill. Revised by Canfield and J. H. Bowman as *Business, Legal and Ethical Phases of Engineering* (New York: McGraw-Hill, 1948).

Harre, Rom. (1970) *The Principles of Scientific Thinking*. Chicago, IL: University of Chicago Press.

Harris, Charles E., Jr., Michael S. Pritchard, and Michael J. Rabins. (1995) *Engineering Ethics: Concepts and Cases*. Belmont, CA: Wadsworth. 2nd edition, 2000. 3rd edition, 2005. 4th edition, 2008. 5th edition, 2013.

Harrison, Andrew. (1978) *Making and Thinking: A Study of Intelligent Activities*. Indianapolis, IN: Hackett.

Harrison, Lawrence E. (1985) *Underdevelopment Is a State of Mind: The Latin American Case*. Lanham, MD: University Press of America, 1985.

Hayek, F. A. (1967) "The Results of Human Action but Not of Human Design". In his *Studies in Philosophy, Politics, and Economics* (Chicago, IL: University of Chicago Press), pp. 96–105.

Hayes, Denis. (1978) *Repairs, Reuse, Recycling—First Steps toward a Sustainable Society*, Worldwatch Paper 23 (September).

Heidegger, Martin (1954) "Die Frage nach der Technik" [The question concerning technology]. In *Vorträge und Aufsätze* (Pfullingen: Neske), pp. 13–44.
Herf, Jeffrey. (1986) *Reactionary Modernism: Technology, Culture, and Politics in Weimar and the Third Reich*. Cambridge: Cambridge University Press.
Hering, Carl. (1922) "Ethics of the Electrical Engineer," *Annals of the American Academy of Political and Social Science*, vol. 101 (May), pp. 86–89.
Herkert, Joseph R. (2001). "Future Directions in Engineering Ethics Research: Microethics, Macroethics, and the Role of Professional Societies," *Science and Engineering Ethics*, vol. 7, no. 3, pp. 403–14.
Herkert, Joseph R. (2002) "Continuing and Emerging Issues in Engineering Ethics Education," *The Bridge*, vol. 32, no. 3 (Fall), pp. 8–13.
Hesse, Mary B. (1966) *Models and Analogies in Science*. Notre Dame, IN: University of Notre Dame Press.
Hilpinen, Risto. (1992) "Artifacts and Works of Art," *Theoria*, vol. 58, no. 1, pp. 58–82.
Hively, William. (1988) "Profile: Union of Concerned Scientists," *American Scientist*, vol. 76, no. 1 (January-February), pp. 18–20.
Hodges, Michael. (1980) "Means/Ends and the Nature of Engineering," *PSA: Proceedings of the Biennial Meeting of the Philosophy of Science Association*, vol. 2, pp. 456–63.
Hofstadter, Richard. (1963) *Anti-Intellectualism in American Life*. New York: Knopf.
Hollander, Rachelle D. (1983) "Conference Report: Engineering Ethics," *Science, Technology, and Human Values*, vol. 8, no. 1 (Winter), pp. 25–29.
Hollander, Rachelle D., and Nicholas H. Steneck. (1990) "Science- and Engineering-Related Ethics and Values Studies: Characteristics of an Emerging Field of Research," *Science, Technology, & Human Values*, vol. 15, no. 1 (Winter), pp. 84–104.
Hösle, Vittorio, and Christian Illies, eds. (2005) *Darwinism and Philosophy*. Notre Dame, IN: University of Notre Dame Press.
Houkes, Wybo. (2009) "The Nature of Technological Knowledge". In Meijers, ed., *Philosophy of Technology and Engineering Sciences* (Amsterdam: Elsevier), 309–50.
Howard, Albert. (1927) *The Development of Indian Agriculture*. London: Oxford University Press.
Howard, Albert. (1940) *An Agricultural Testament*. London: Oxford University Press.
Hughes, Thomas P. (1980) "National Socialist Ideology and German Engineers". In Friedlander and Milton, eds., *The Holocaust: Ideology, Bureaucracy, and Genocide* (Millwood, NY: Kraus International Publishers), pp. 165–81.
Hughes, Thomas P. (1989) *American Genesis: A Century of Invention and Technological Enthusiasm, 1870–1970*. New York: Viking.
Huizinga, Johan. (1955) *Homo Ludens: A Study of the Play-Element in Culture*. Boston, MA: Beacon.
Huning, Alois, and Carl Mitcham. (1993) "The Historical and Philosophical Development of Engineering Ethics in Germany," *Technology in Society*, vol. 15, no. 4, pp. 427–39.
Ignatieff, Michael. (2007) *The Rights Revolution*. 2nd edition. Toronto: Anansi Press.

Ihde, Don. (1986) *Consequences of Phenomenology*. Albany: State University of New York Press.
Ihde, Don. (1990) *Technology and the Lifeworld: From Garden to Earth*. Bloomington: Indiana University Press.
Ihde, Don. (1993) *Introduction to Philosophy of Technology*. New York: Paragon House.
Ihde, Don. (2008) *Ironic Technics*. Copenhagen: Automatic Press/VIP.
Illich, Ivan. (1967) "The Seamy Side of Charity," *America*, vol. 116, no. 3 (January 21), pp. 88–91.
Illich, Ivan. (1973) *Tools for Conviviality*. New York: Harper and Row.
Illich, Ivan. (1974) *Energy and Equity*. New York: Harper and Row.
Illich, Ivan. (1978) *Toward a History of Needs*. New York: Pantheon.
Illich, Ivan. (1983) *Gender*. New York: Pantheon.
Illich, Ivan. (2018) *The Powerless Church and Other Selected Writings, 1955–1985*. Eds. Valentina Borremans and Sajay Samuel. University Park: Pennsylvania State University Press.
International Union for Conservation of Nature and Natural Resources. (1980) *World Conservation Strategy: Living Resource Conservation for Sustainable Development*. Gland, Switzerland: International Union for Conservation of Nature and Natural Resources, United Nations Environment Programme, and World Wildlife Fund.
Institution of Civil Engineers (ICE). (1828) Royal Charter. Available in *Charter, Supplemental Charters, By-Laws, and List of Members of the Institution of Civil Engineers*. London: Institution of Civil Engineers, 1908.
Institution of Civil Engineers (ICE). *Power for the Use of Man*. London: Institution of Civil Engineers, 1978.
Institution of Civil Engineers (ICE). (2008) Code of Professional Conduct. Available at www.ice.org.uk.
Israel, Paul. (1998) *Edison: A Life of Inventions*. New York: John Wiley.
Jackson, Barbara Ward, and René Dubos. (1972) *Only One Earth: The Care and Maintenance of a Small Planet*. New York: W. W. Norton.
Jackson, Wes, Wendell Berry, and Bruce Colman, eds. (1984) *Meeting the Expectations of the Land*. San Francisco, CA: North Point Press.
Jakobsen, Bernhard F. (1955) *Ethics and the American Society of Civil Engineers: A Contribution to the Engineering Profession*. Los Angeles: n.p.
Jasanoff, Sheila. (1990) *The Fifth Branch: Science Advisers as Policymakers*. Cambridge, MA: Harvard University Press.
Jaspers, Karl. (1949) *Vom Ursprung und Ziel der Geschichte*. Zurich: Artemis Verlag. English version: *Origin and Goal of History*, trans. Michael Bullock (London: Routledge & Kegan Paul, 1953).
Jesiek, Brent K., and Leah H. Jamieson. (2017) "The Expansive (Dis)Integration of Electrical Engineering Education," *IEEE Access*, vol. 5, pp. 4561–73.
Johnson, Deborah G., ed. (1991) *Ethical Issues in Engineering*. Englewood Cliffs, NJ: Prentice Hall.

Johnson, Deborah G. (2001) "Nature and Technology: Virtual Violations". Presidential address, 12th Biennial Conference of the Society for Philosophy and Technology, University of Aberdeen, Aberdeen, Scotland, July 11.

Johnson, Deborah G., and Jameson M. Wetmore. (2008) "STS and Ethics: Implications for Engineering Ethics". In E. J. Hackett, O. Amsterdamska, M. E. Lynch, and J. Wajcman, eds., *The Handbook of Science and Technology Studies*, 3rd edition (Cambridge, MA: MIT Press), pp. 567–81.

Johnson, Robert R. (1998) *User-Centered Technology: A Rhetorical Theory for Computers and Other Mundane Artifacts*. Albany: State University of New York Press.

Johnston, Stephen, Alison Lee, and Helen McGregor. (1996) "Engineering as Captive Discourse," *Society for Philosophy and Technology Electronic Journal*, vol. 1, nos. 3–4.

Johri, Aditya, and Barbara M. Olds, eds. (2015) *Cambridge Handbook of Engineering Education Research*. Cambridge: Cambridge University Press.

Jonas, Hans. (1984) *The Imperative of Responsibility: In Search of an Ethics for the Technological Age*. Trans. Jonas and David Herr. Chicago, IL: University of Chicago Press.

Jones, Stacy V. (1973) *Inventions Necessity Is Not the Mother of*. New York: Quadrangle.

Joy, Bill. (2000) "Why the Future Doesn't Need Us," *Wired*, vol. 8, no. 4 (April), pp. 238–62.

Kaag, John, and Sujata K. Bhatia. (2014) "Fools for Tools: Why Engineers Need to Become Philosophers," *Chronicle Review* (November 28), pp. B13–B15.

Kaasgaard, Klaus. (2000) *Software Design and Usability: Talks with Bonnie Nardi, Jakob Nielsen, David Smith, Austin Henderson and Jed Harris, Terry Winograd, and Stephanie Rosenbaum*. Copenhagen: Copenhagen Business School Press.

Kandachar, Prabhu, and Minna Halme, eds. (2008) *Sustainability Challenges and Solutions at the Base of the Pyramid: Business, Technology and the Poor*. Sheffield: Greenleaf.

Kapp, Ernst. (1877) *Grundlinien einer Philosophie der Technik: Zur Entstehungsgeschichte dur Cultur aus neuen Gesichtspunkten*. Braunschweig: Westermann.

Kawashima, Kazuhiko, Syunsuke Ikeda, Akihiko Hirotani, and Kozo Katayama. (2004) "Roles of the Japan Society of Civil Engineering on the Continuing Education for Engineering Ethics," *Proceedings of the 9th World Conference on Continuing Engineering Education*, Tokyo, May 15–20, pp. 101–6.

Kelty, Christopher M. (2008) *Two Bits: The Cultural Significance of Free Software*. Durham, NC: Duke University Press.

Kemper, Bart. (2004) "Evil Intent and Design Responsibility," *Science and Engineering Ethics*, vol. 10, no. 2, pp. 303–9.

Kendall, Henry W. (2000) *A Distant Light: Scientists and Public Policy*. New York: Springer-Verlag.

Kennedy, David. (2004) *The Dark Sides of Virtue: Reassessing International Humanitarianism*. Princeton, NJ: Princeton University Press.

Kevles, Daniel J. (1978) *The Physicists: The History of a Scientific Community in Modern American*. New York: Knopf.

Kierkegaard, Søren. (1980) *The Sickness unto Death: A Christian Psychological Exposition for Upbuilding and Awakening*. Trans. H. V. Hong and E. H. Hong. Princeton, NJ: Princeton University Press. Original publication, 1849.

Kitcher, Philip. (2003) *Science, Truth, and Democracy*. New York: Oxford University Press.

Kitcher, Philip. (2011) *Science in a Democratic Society*. Amherst, NY: Prometheus Books.

Kline, Stephen J. (1965) *Similitude and Approximation Theory*. New York: McGraw-Hill.

Koen, Billy Vaughn. (2003) *Discussion of the Method: Conducting the Engineer's Approach to Problem Solving*. New York: Oxford University Press.

Kondo, Dorinne K. (1990) *Crafting Selves: Power, Gender, and Discourses of Identity in a Japanese Workplace*. Chicago, IL: University of Chicago Press.

Kranzberg, Melvin. (1986) "Technology and History: 'Kranzberg's Laws'," *Technology and Culture*, vol. 27, no. 3, pp. 544–60.

Kroes, Peter. (2012) *Technical Artifacts: Creations of Mind and Matter*. Dordrecht: Springer.

Ladd, John. (1980) "The Quest for a Code of Professional Ethics: An Intellectual and Moral Confusion". In Rosemary A. Chalk, Mark S. Frankel, and Sallie Birket Chafer, eds., *AAAS Professional Ethics Project: Professional Ethics Activities in the Scientific and Engineering Societies* (Washington, DC: American Association for the Advancement of Science), pp. 154–59.

Ladenson, Robert F., James Choromokos, Ernest d'Anjou, Martin Pimsler, and Howard Rosen. (1980) *A Selected Annotated Bibliography of Professional Ethics and Social Responsibility in Engineering*. Chicago: Center for the Study of Ethics in the Professions, Illinois Institute of Technology.

Lakoff, Sanford A., ed. (1980) *Science and Ethical Responsibility*. Reading, MA: Addison-Wesley.

Lasswell, Harold D. (1927) *Propaganda Technique in the World War*. New York: Peter Smith.

Lasswell, Harold D. (1930) *Psychopathology and Politics*. Chicago, IL: University of Chicago Press.

Lasswell, Harold D. (1936) *Politics: Who Gets What, When, How*. New York: McGraw-Hill.

Lasswell, Harold D. (1948) *Power and Personality*. New York: Norton.

Lasswell, Harold D. (1951) "The Policy Orientation". In Lerner and Lasswell, eds., *The Policy Sciences: Recent Developments in Scope and Method* (Stanford, CA: Stanford University Press), pp. 3–15.

Lasswell, Harold D. (1969) *Political Communication: Public Language of Political Elites in India and the United States*. New York: Holt, Rinehart and Winston.

Latouche, Serge. (2010) *Farewell to Growth*. Cambridge: Polity Press. (French original, Petit traité de la décroissance sereine, 2007.)

Latour, Bruno. (1988) *Science in Action: How to Follow Scientists and Engineers through Society*. Cambridge, MA: Harvard University Press.

Layton, Edwin T., Jr. (1962) "Veblen and the Engineers," *American Quarterly*, vol. 14, no. 1 (Spring), pp. 64–72.

Layton, Edwin T., Jr. (1971a) *The Revolt of the Engineers: Social Responsibility and the American Engineering Profession*. Cleveland, OH: Case Western Reserve University Press. 2nd edition, with new preface, Baltimore, MD: Johns Hopkins University Press, 1986.

Layton, Edwin T., Jr. (1971b) "Mirror-Image Twins: The Communities of Science and Technology in 19th-Century America," *Technology and Culture*, vol. 12, no. 4 (October), pp. 562–80.

Layton, Edwin T., Jr. (1974) "Technology as Knowledge," *Technology and Culture*, vol. 15, no. 1 (January), pp. 31–41.

Layton, Edwin T., Jr. (1976) "Engineering Ethics and the Public Interest: A Historical View". *American Society of Mechanical Engineers*, 76-WA/TS-9 (Technical Paper). Reprinted in Baum and Flores (1980), vol. 1, pp. 26–29.

Layton, Edwin T., Jr. (1977) "Conditions of Technological Development". In Spiegel-Rösing and De Solla Price, eds., *Science, Technology and Society: A Cross-Disciplinary Perspective* (Beverly Hills, CA: Sage, 1977), pp. 197–222.

Lélé, Sharachchandra M. (1991) "Sustainable Development: A Critical Review," *World Development*, vol. 19, no. 6 (June), pp. 607–21.

Lélé, Sharachchandra M. (1991) "Sustainable Development: A Critical Review," *World Development*, vol. 19, no. 6 (June), pp. 607–21.

Lemons, John, and Donald A. Brown, eds. (1995) *Sustainable Development: Science, Ethics, and Public Policy*. Boston: Kluwer.

Lenk, Hans, and Günter Ropohl, eds. (1987) *Technik und Ethik* [Technology and ethics]. Stuttgart: Philipp Reclam.

Levinson, Mark. (2013) *Particle Fever*. [documentary film] New York: Anthos Media.

Lewis, C.S. (1947) *The Abolition of Man*. New York: Macmillan.

Li Bocong 李伯聪. (1993) "我造物，故我在 *Wo zao wu, gu wo zai* [I create therefore I am]," *Studies in Dialectics of Nature*, vol. 9, no. 12, pp. 9–19.

Li Bocong. (2002) 工程哲学导论 *Gongcheng zhexue daolun* [Introduction to philosophy of engineering]. Zhengzhou: Daxiang Press.

Li Bocong. (2002) 工程哲学引论：我造物故我在 *Gongcheng zhexue yin lun: Wo zaowu guwo za* [Introduction to philosophy of engineering: I create therefore I am]. Zhengzhou: Daxiang Press.

Li, Bocong. (2008) 绝对命令伦理学和协调伦理学—四谈工程伦理学 *Juedui mingling lunlixue he xietiao lunlixue: si tan gongcheng lunlixue* [Categorical imperative ethics and coordination ethics: the 4th discussion on engineering ethics], *Lunlixue yanjiu* [Studies in ethics], vol. 5, pp. 42–48.

Li Bocong. (2010) "The Rise of Philosophy of Engineering in the East and the West." In Van de Poel and Goldberg, eds., *Philosophy and Engineering: An Emerging Agenda* (Dordrecht: Springer), pp. 31–40.

Li Bocong. (2012) "From a Micro-Macro Framework to a Micro-Meso-Macro Framework." In Christensen et al., eds., *Engineering, Development and Philosophy: American, Chinese and European Perspectives* (Dordrecht: Springer), pp. 23–36.

Li Bocong. (2015) "Engineering Action in Micro-, Meso-, and Macro-contexts". In Christensen et al., eds., *Engineering Identities, Epistemologies and Values: Engineering Education and Practice in Context*, vol. 2 (Dordrecht: Springer), pp. 369–379.

Li Bocong. (2018) "On Relationships between the History and Philosophy of Engineering." In Mitcham et al., eds., *Philosophy of Engineering, East and West* (Dordrecht: Springer), pp. 3–12.

Li Bocong, ed. (2010) 工程社会学导论：工程共同体研究 *Gongcheng shehui xue daolun: Gongcheng gongtongti yanjiu* [Introduction to sociology of engineering: Research on the engineering community]. Hangzhou: Zhejiang University Press.

Lieberman, Henry, Fabio Paternò, and Volker Wulf, eds. (2006) *End User Development*. Dordrecht: Springer.

Light, Andrew, and David Roberts. (2000) "Toward New Foundations in Philosophy of Technology: Mitcham and Wittgenstein on Descriptions," *Research in Philosophy and Technology*, vol. 19, pp. 125–47.

Lindinger, Herbert, ed. (1991) *Ulm Design: The Morality of Objects*. Trans. David Britt. Cambridge, MA: MIT Press.

Lessig, Lawrence. (1999) *Code: And Other Laws of Cyberspace*. New York: Basic Books.

Locke, John. (1954) *Essays on the Law of Nature: The Latin Text with a Translation, Introduction and Notes, together with Transcripts of Locke's Shorthand in the Journal for 1676*. Edited by W. von Leyden. Oxford: Oxford University Press.

Lonergan, Bernard J.F. (1957) *Insight: A Study of Human Understanding*. New York: Philosophical Library.

Losonsky, Michael. (2006) *Linguistic Turns in Modern Philosophy*. Cambridge: Cambridge University Press.

Lovins, Amory B. (1975) *World Energy Strategies: Facts, Issues, and Options*. San Francisco, CA: Friends of the Earth.

Lovins, Amory B. (1976) "Energy Strategy: The Road Not Taken?," *Foreign Affairs*, vol. 55, no. 1 (October), pp. 65–96.

Lovins, Amory B. (1977) *Soft Energy Paths: Toward a Durable Peace*. San Francisco, CA: Friends of the Earth.

Lucena, Juan, ed. (2013) *Engineering Education for Social Justice*. Dordrecht: Springer.

Lucena, Juan, Jen Schneider, and Jon A. Leydens. (2010) *Engineering and Sustainable Community Development*. San Rafael, CA: Morgan and Claypool.

Ludwig, Karl-Heinz. (1974) *Technik und Ingenieure in Dritten Reich* [Technology and engineering in the Third Reich]. Düsseldorf: Droste Verlag.

Luegenbiehl, Heinz C. (2010) "Ethical Principles for Engineers in a Global Environment". In Van de Poel and Goldberg, eds., *Philosophy and Engineering: An Emerging Agenda* (Dordrecht: Springer), pp. 147–59.

Luegenbiehl, Heinz C., and Rockwell F. Clancy. (2017) *Global Engineering Ethics*. Oxford: Butterworth-Heinemann.

Luegenbiehl, Heinz C., and Jun Fudano. (2014) "Japanese Perspectives". In Holbrook and Mitcham, eds., *Ethics, Science, Technology, and Engineering* (Detroit: Macmillan Reference), vol. 2, pp. 615–19.

Lugo, Elena. (1985) *Ética professional para la ingeniería*. Mayagüez, Puerto Rico: Librería Universal.
Lusher, D.S. (1986) "The Changing Role of the Professional Institutions in Hong Kong," *Journal of the Hong Kong Institution of Engineers*, vol. 14, no. 2 (February), pp. 39–40.
Lyotard, Jean-François. (1979) *La condition postmoderne: Rapport sur le savoir*. Paris: Editions de Minuit. English version: *The Postmodern Condition: A Report on Knowledge,* trans. Goeff Bennington and Brian Massumi (Minneapolis: University of Minnesota Press, 1984).
Madhavan, Guru. (2015) *Applied Minds: How Engineers Think*. New York: W. W. Norton.
Madhavan, Guru. (2016) *Think Like an Engineer: Inside the Minds That are Changing Our Lives*. London: Oneworld.
Malone, Thomas F., and Robert Corell. (1989) "Mission to Planet Earth Revisited: An Update on Studies of Global Change," *Environment*, vol. 31, no. 3 (April), pp. 6–11.
Margolin, Victor. (1989) "Postwar Design Literature: A Preliminary Mapping". In Margolin, ed., *Design Discourse: History, Theory, Criticism* (Chicago, IL: University of Chicago Press), pp. 265–287.
Margolin, Victor and Richard Buchanan, eds. (1995). *The Idea of Design*. Cambridge, MA: MIT Press.
Marshall, E. D., E. O. Pfrang, E. V. Leyendecker, K. A. Woodward, R. P. Reed, M. B. Kasen, and T. R. Shives. (1982) *Investigation of the Kansas City Hyatt Regency Walkways Collapse*. (National Bureau of Standards Building Science Series 143.) Washington, DC: U.S. Department of Commerce, National Bureau of Standards.
Martin, Mike W., and Roland Schinzinger. (1983) *Ethics in Engineering*. New York: McGraw Hill. 2nd edition, 1989. 3rd edition, 1996. 4th edition, 2005.
Martínez Contreras, Jorge, Raúl Gutiérrez Lombardo, and Paul Durbin, eds. (1997) *Tecnología, desarrollo económico y sustentabilidad*. Mexico City: Centro de Estudios Filosóficos, Políticos y Sociales Vicente Lombardo Toledano. Special no. 2 of the Centro publication *Ludus Vitalis*. Proceedings of a Society for Philosophy and Technology conference, Puebla, México, 1996.
Marx, Leo. (1997) "*Technology*: The Emergence of a Hazardous Concept," *Social Research*, vol. 64, no 3 (Fall 1997), pp. 965–988. Revised version: "*Technology*: The Emergence of a Hazardous Concept," *Technology and Culture*, vol. 51, no. 3 (2010), pp. 561–77.
Maslow, Abraham. (1954) *Motivation and Personality*. New York: Harper and Row.
McCarthy, Natasha. (2009) *Engineering: A Beginner's Guide*. Oxford: One World.
McCarthy, Natasha. (2017) "Interdisciplinarity, Incommensurability and Engineering in Policymaking: Using Philosophy to Understand Challenges at the Engineering-Policy Interface". In Michelfelder et al., eds., *Philosophy and Engineering: Exploring Boundaries, Expanding Connections* (Dordrecht: Springer), pp. 139–150.
McClean, G.F. (1993) "Integrating Ethics and Design," IEEE Technology and Society *Magazine*, vol. 12, no. 3, pp. 19–30.

McClelland, E. H. (1917) *Engineering Ethics*. Pittsburgh: Carnegie Library of Pittsburgh. (*Bibliographies*, vol. 1), pp. 3–17.
McDonald, Allan J., with James R. Hansen. (2009) *Truth, Lies, and O-Rings: Inside the Space Shuttle Challenger Disaster*. Gainesville: University Press of Florida.
McFadden, Robert D. (2010) "Samuel T. Cohen, Neutron Bomb Inventor, Dies at 89," *New York Times* (December 1).
McMahon, Michal A. (1984) *The Making of a Profession: A Century of Electrical Engineering in America*. New York: Institute of Electrical and Electronic Engineers.
McNeill, William H. (1982) *The Pursuit of Power: Technology, Armed Force, and Society since A.D. 1000*. Chicago, IL: University of Chicago Press.
Mead, Daniel W. (1916) *Contracts, Specifications and Engineering Relations*. New York: McGraw-Hill. 2nd edition, 1933. 3rd edition, 1956.
Mead, Margaret, ed. (1953) *Cultural Patterns and Technical Change*. Paris: UNESCO.
Meadows, Donella H., Dennis L. Meadows, Jørgen Randers, and William W. Behrens III. (1972) *The Limits to Growth: A Report for The Club of Rome's Project on the Predicament of Mankind*. New York: Universe Books.
Meadows, Donella H., Dennis L. Meadows, and Jørgen Randers. (1992) *Beyond the Limits: Confronting Global Collapse, Envisioning a Sustainable Future*. Post Mills, VT: Chelsea Green.
Meadows, Donella H., Jørgen Randers, and Dennis L. Meadows. (2005) *Limits to Growth: The Thirty Year Update*. London: Earthscan.
Meijers, Anthonie. (2009) *Philosophy of Technology and Engineering Sciences (Handbook of the Philosophy of Science*, vol. 9). Amsterdam: Elsevier.
Melzer, Arthur M., Jerry Weinberger, and M. Richard Zinman, eds. (1993) *Technology in the Western Political Tradition*. Ithaca, NY: Cornell University Press.
Mendelsohn, Everett, Merritt Roe Smith, and Peter Weingart, eds. (1988) *Science, Technology, and the Military*. Sociology of the Sciences: A Yearbook, no. 12, 2 vols. Dordrecht, Netherlands: Kluwer Academic.
Merchant, Carolyn. (1980) *The Death of Nature: Women, Ecology, and the Scientific Revolution*. San Francisco, CA: Harper and Row.
Merton, Robert K. (1938) "Science and the Social Order," *Philosophy of Science*, vol. 5, pp. 321–37. Reprinted in his *The Sociology of Science: Theoretical and Empirical Investigations* (Chicago, IL: University of Chicago Press, 1973), pp. 254–266.
Merton, Robert K. (1942) "Science and Technology in a Democratic Order," *Journal of Legal and Political Sociology*, vol. 1, pp. 115–26. Reprinted as "The Normative Structure of Science" in his *The Sociology of Science: Theoretical and Empirical Investigations* (Chicago, IL: University of Chicago Press, 1973), pp. 267–78.
Mesarovic, Mihailo and Eduard Pestel. (1974) *Mankind at the Turning Point: The Second Report to the Club of Rome*. New York: Dutton.
Metzger, Robert Charles. (2003) *Debugging by Thinking: A Multidisciplinary Approach*. Boston, MA: Digital Press.

Michelfelder, Diane P, Natasha McCarthy, and David E. Goldberg, eds. (2013) *Philosophy and Engineering: Reflections on Practice, Principles and Process.* Dordrecht: Springer.

Michelfelder, Diane P., Byron Newberry, and Qin Zhu, eds. (2017) *Philosophy and Engineering: Exploring Boundaries, Expanding Connections.* Dordrecht: Springer.

Middendort, William H. (1986) *Design of Devices and Systems.* New York: Dekker.

Milbrath, Lester W. (1989) *Envisioning a Sustainable Society: Learning Our Way Out.* Albany: State University of New York Press.

Millikan, Ruth. (1984) *Language, Thought, and Other Biological Categories.* Cambridge, MA: MIT Press.

Mills, C. Wright. (1963) "Man in the Middle: The Designer". In Horowitz, ed., *Power, Politics, and People: The Collected Essays of C. Wright Mills* (New York: Oxford University Press), pp. 374–86.

Minear, Larry, ed. (2002) *The Humanitarian Enterprise: Dilemmas and Discoveries.* Bloomfield, CT: Kumarina Press.

Mitcham, Carl. (1978) "Types of Technology," *Research in Philosophy and Technology*, vol. 1, pp. 229–94.

Mitcham, Carl. (1987) "Schools for Whistle Blowers: Educating Ethical Engineers," *Commonweal*, vol. 114, no. 7 (April 10), pp. 201–5.

Mitcham, Carl. (1989) "Science, Technology, and the Theory of Progress". In Goldman, ed., *Science, Technology, and Social Progress* (Bethlehem, PA: Lehigh University Press), pp. 240–52.

Mitcham, Carl. (1991) "Engineering as Productive Activity: Philosophical Remarks". In Durbin, ed., *Critical Perspectives on Non-Academic Science and Engineering* (Bethlehem, PA: Lehigh University Press), pp. 80–117.

Mitcham, Carl. (1992) *Engineering Ethics throughout the World: Introduction, Documentation and Bibliography.* A collection of engineering ethics codes and related materials with introductory analysis and supplementary annotated bibliography. Version 1.0. University Park, PA: STS Press.

Mitcham, Carl. (1994a) *Thinking through Technology: The Path between Engineering and Philosophy.* Chicago, IL: University of Chicago Press.

Mitcham, Carl. (1994b) "Engineering Design Research and Social Responsibility". In Shrader-Frechette, *Ethics of Scientific Research* (Lanham, MD: Rowman & Littlefield), pp. 153–68.

Mitcham, Carl. (1997a) *Thinking Ethics in Technology: Hennebach Lectures and Papers, 1995–1996.* Golden: Colorado School of Mines Press.

Mitcham, Carl. (1997b) "The Sustainability Question". In Gottlieb, ed., *The Ecological Community: Environmental Challenges for Philosophy, Politics, and Morality* (New York: Routledge), pp. 359–79.

Mitcham, Carl. (1999) "Why the Public Should Participate in Technical Decision Making". In Von Schomberg, ed., *Democratising Technology: Theory and Practice of a Deliberative Technology Policy* (Hengelo, Netherlands: International Centre for Human and Public Affairs), pp. 39–50.

Mitcham, Carl. (2001) "*Dasein* versus Design: The Problematics of Turning Making into Thinking," *International Journal of Technology and Design Education*, vol. 11, no. 1, pp. 27–36.

Mitcham, Carl. (2018) "Speeding Things Down." In Volker Friedrich, ed., *Technik denken: Philosophische Annäherungen* (Stuttgart: Franz Steiner Verlag), pp. 31–39.

Mitcham, Carl, and R. Shannon Duval. (2000) *Engineer's Toolkit: Engineering Ethics*. Upper Saddle River, NJ: Prentice Hall.

Mitcham, Carl, and Erik Fisher. (2012) "Ethics and Policy". In Chadwick, ed., *Encyclopedia of Applied Ethics*, 2nd edition. (San Diego: Academic Press), vol. 2, pp. 165–72.

Mitcham, Carl, and Robert Frodeman. (2002) "The Plea for Balance in the Public Funding of Science," *Technology in Society*, vol. 24, pp. 83–92.

Mitcham, Carl, and Robert Frodeman. (2006) "Science Policy". In Borchert, ed., *Encyclopedia of Philosophy*, 2nd edition (Detroit: Macmillan Reference), vol. 8, pp. 673–76.

Mitcham, Carl, Li Bocong, Byron Newberry, and Zhang Baichun, eds. (2018) *Philosophy of Engineering, East and West*. Dordrecht: Springer.

Mitcham, Carl, and Robert Mackey, eds. (1972) *Philosophy and Technology: The Philosophical Problems of Technology*. New York: Free Press.

Mitcham, Carl, and David Muñoz. (2010) *Humanitarian Engineering*. San Rafael, CA: Morgan & Claypool Publishers.

Mitcham, Carl, Barbara M. Olds, and Ron L. Miller. (2001) "A Plea for Pursuing New Dimensions of Assessment in the Teaching and Learning of Research Integrity". http://ori.dhhs.gov/multimedia/acrobat/papers/mitcham.pdf.

Mokyr, Joel. (2002) *The Gifts of Athena: Historical Origins of the Knowledge Economy*. Princeton, NJ: Princeton University Press.

Mokyr, Joel. (2009) *The Enlightened Economy: An Economic History of Britain, 1700–1850*. New Haven, CT: Yale University Press.

Morarity, Gene. (2008) *The Engineering Project: Its Nature, Ethics, and Promise*. University Park, PA: Penn State University Press.

Morison, George S. (1895) "Address at the Annual Convention," *Transactions of the ASCE*, vol. 33 (June), p. 483.

Morris-Suzuki, Tessa. (1994) *The Technological Transformation of Japan: From Seventeenth to the Twenty-first Century*. Melbourne: Oxford University Press.

Moulakis, Athanasios. (1994) *Beyond Utility: Liberal Education for a Technological Age*. Columbia: University of Missouri Press.

Mulder, Karel. (2006) *Sustainable Development for Engineers: A Handbook and Resource Guide*. Abingdon, UK: Routledge.

Mumford, Lewis. (1967) *The Myth of the Machine*. Vol. 1: *Technics and Human Development*. New York: Harcourt, Brace, Jovanovich.

Mumford, Lewis. (1970) *The Myth of the Machine*. Vol. 2: *The Pentagon of Power*. New York: Harcourt, Brace, Jovanovich.

Murphy, Colleen, Paolo Gardoni, Hassan Bashir, Charles E. Harris, Jr., and Eyad Masad, eds. (2015) *Engineering Ethics for a Globalized World*. Dordrecht: Springer.

Murphy, Glenn. (1950) *Similitude in Engineering*. New York: Ronald.

Myers, Norman, and Julian L. Simon. (1994) *Scarcity or Abundance: A Debate on the Environment*. New York: W. W. Norton.

Nash, Hugh, ed. (1979) *The Energy Controversy: Soft Path Questions and Answers.* San Francisco, CA: Friends of the Earth.

Nash, Roderick Frazier. (1989) *The Rights of Nature: A History of Environmental Ethics.* Madison: University of Wisconsin Press.

National Academy of Engineering (NAE). (2002) *Engineering and Environmental Challenges: Technical Symposium on Earth Systems Engineering.* Washington, DC: National Academies Press.

National Academy of Engineering (NAE). (2004) *The Engineer of 2020: Visions of Engineering in the New Century.* Washington, DC: National Academies Press.

National Academy of Engineering (NAE). (2005) *Educating the Engineering of 2020: Adapting Engineering Education to the New Century.* Washington, DC: National Academies Press.

National Society of Professional Engineers (NSPE). (2007) "NSPE Code of Ethics for Engineers". Available at www.nspe.org.

Newberry, Byron. (2004) "The Dilemma of Ethics in Engineering Education," *Science and Engineering Ethics*, vol. 10, pp. 343–51.

Newberry, Byron. (2005) "Engineering Globalization: Oxymoron or Opportunity?," *IEEE Technology and Society Magazine* (Fall), pp. 8–15.

Newberry, Byron. (2007) "Are Engineers Instrumentalists?," *Technology in Society*, vol. 29, no. 1 (January), pp. 107–19.

Newberry, Byron. (2013) "Engineered Artifacts". In Michelfelder et al., eds., *Philosophy and Engineering: Reflections on Practice, Principles and Process* (Dordrecht: Springer), pp. 165–176.

Newberry, Byron. (2015a) "The Dialectics of Engineering". In Christensen et al., eds., *Engineering Identities, Epistemologies and Values* (Dordrecht: Springer), pp. 9–22.

Newberry, Byron. (2015b) "The Engineering Adjective, and Why It Might Matter," *Engineering Studies*, vol. 7, nos. 2–3, pp. 142–44.

Newberry, Byron. (2015c) "Efficiency Animals: Efficiency as an Engineering Value". In Christensen et al., eds., *Engineering Identities, Epistemologies and Values* (Dordrecht: Springer), pp. 199–214.

Newell, Frederick Haynes. (1922) "Ethics of the Engineering Profession," *Annals of the American Academy of Political and Social Science*, vol. 101 (May), pp. 76–85.

Noble, David F. (1977) *America by Design: Science, Technology, and the Rise of Corporate Capitalism.* New York: Knopf.

Nobel, David F. (1984) *Forces of Production: A Social History of Industrial Automation.* New York: Knopf.

Nordmann, Alfred. (2011) "The Age of Technoscience". In Nordmann, et al., eds., *Science Transformed? Debating Claims of an Epochal Break* (Pittsburg: University of Pittsburgh Press), pp. 18–30.

Nordmann, Alfred, Hans Radder, and Gregor Schiemann, eds. (2011) *Science Transformed? Debating Claims of an Epochal Break.* Pittsburgh, PA: University of Pittsburgh Press.

Nordmann, Alfred. (2017) "Vanishing Friction Events and the Inverted Platonism of Technoscience". In B. Bensaude-Vincent, et al., eds., *Research Objects in their Technological Setting* (London: Routledge), pp. 56–69.

Norman, Donald, and Stephen Draper, eds. (1986) *User-Centered System Design: New Perspectives on Human-Computer Interaction*. Hillsdale, NJ: Lawrence Erlbaum.

Nozick, Robert. (1974) *Anarchy, State, Utopia*. Cambridge, MA: Harvard University Press.

NRC (National Research Council). (1989) *Alternative Agriculture*. Washington, DC: National Academies Press.

Nussbaum, Bruce. (2005) "Annual Design Awards 2005," *Business Week* (July 4, 2005), pp. 62–63.

Nye, David E. (1990) *Electrifying America: Social Meanings of a New Technology, 1880–1940*. Cambridge, MA: MIT Press.

Nye, David E. (1995) *American Technological Sublime*. Cambridge, MA: MIT Press.

Nye, David E. (1998) *Consuming Power: A Social History of American Energies*. Cambridge, MA: MIT Press.

Odagiri, Hiroyuk. (1998) "Education as a Source of Network, Signal, or Nepotism: Managers and Engineers during Japan's Industrial Development". In W. Mark Fruin, ed., *Networks, Markets, and the Pacific Rim: Studies in Strategy* (New York: Oxford University Press), pp. 141–53.

Ohashi, Hideo. (2001) "Re-Engineering of Engineering Profession in Japan," *JSME News* (Japan Society of Mechanical Engineers), vol. 12, no. 1 (June), pp. 2–5.

Olsen, Jan Kyrre Berg, Stig Andur Pedersen, and Vincent F. Hendricks, eds. (2009) *A Companion to the Philosophy of Technology*. Malden, MA: Wiley-Blackwell.

Olson, Richard G. (2016) *Scientism and Technocracy in the Twentieth Century: The Legacy of Scientific Management*. Lanham, MD: Lexington Books.

Oppenheimer, J. Robert. (1947) *Physics in the Contemporary World*. Cambridge, MA: MIT Press.

Ortega y Gasset, José. (1939) *Meditación de la técnica*. In *Obras completas*, first edition (Madrid: Revista de Occidente, 1945–1947), vol. 5, pp. 317–375.

Papanek, Victor. (1983) *Design for Human Scale*. New York: Van Nostrand.

Papanek, Victor. (1984) *Design for the Real World: Human Ecology and Social Change*. 2nd edition. New York: Van Nostrand Reinhold.

Parsons, Maurice G. (1914) "The Philosophy of Engineering," *Transactions of the American Society of Civil Engineers*, vol. 77, Paper No. 1282 (December), with Discussion by Lewis M. Haupt, Charles Kirby Fox, A. H. Markwart, Morgan Cilley, and the author, pp. 38–63.

Paxton, Roland. (2008) "Thomas Telford". In *Oxford Dictionary of National Biography*, available online at https://doi.org/10.1093/ref:odnb/27107.

Peccei, Aurelio. (1977) *The Human Quality*. Oxford: Pergamon.

Perrow, Charles. (1984) *Normal Accidents: Living with High-Risk Technologies*. New York: Basic Books.

Petersen, James C., and Dan Farrell. (1986) *Whistleblowing: Ethical and Legal Issues in Expressing Dissent*. Dubuque, IA: Kendall Hunt.

Petroski, Henry. (1985) *To Engineer Is Human: The Role of Failure in Successful Design*. New York: St. Martin's Press.

Petroski, Henry. (1994) *Design Paradigms: Case Histories of Error and Judgment in Engineering*. New York: Cambridge University Press.

Petroski, Henry. (1995) *Engineers of Dreams: Great Bridge Builders and the Spanning of America*. New York: Knopf.

Petroski, Henry. (1996) *Invention by Design: How Engineers Get from Thought to Thing*. Cambridge, MA: Harvard University Press.

Petroski, Henry. (1997) *Remaking the World: Adventures in Engineering*. New York: Knopf.

Petroski, Henry. (2003) *Small Things Considered: Why There Is No Perfect Design*. New York: Knopf.

Petroski, Henry. (2004) *Pushing the Limits: New Adventures in Engineering*. New York: Knopf.

Petroski, Henry. (2006) *Success through Failure: The Paradox of Design*. Princeton, NJ: Princeton University Press.

Petroski, Henry. (2010) *The Essential Engineer: Why Science Alone Will Not Solve Our Global Problems*. New York: Knopf.

Petroski, Henry. (2012) *To Forgive Design: Understanding Failure*. Cambridge, MA: Harvard University Press.

Pfatteicher, Sarah K. A. (2003) "Depending on Character: ASCE Shapes Its First Code of Ethics," *Journal of Professional Issues in Engineering Education and Practice*, vol. 129, no. 1 (January), pp. 21–31.

Pielke, Roger, Jr. (2007) *The Honest Broker: Making Sense of Science in Policy and Politics*. Cambridge: Cambridge University Press.

Pinkus, Rosa Lynn B., Larry J. Shuman, Norman P. Hummon, and Harvey Wolfe. (1997) *Engineering Ethics: Balancing Cost, Schedule, and Risk—Lessons Learned from the Space Shuttle*. Cambridge: Cambridge University Press.

Pitt, Joseph C. (2000) *Thinking about Technology: Foundations of the Philosophy of Technology*. New York: Seven Bridges Press.

Pitt, Joseph C. (2011) *Doing Philosophy of Technology: Essays in a Pragmatist Spirit*. Dordrecht: Springer.

Pitt, Joseph C. (2013) "Fitting Engineering into Philosophy". In Michelfelder et al., eds., *Philosophy and Engineering: Reflections on Practice, Principles and Process* (Dordrecht: Springer), pp. 91–101.

Polanyi, Karl. (1944) *The Great Transformation*. New York: Farrar and Rinehart.

Polanyi, Karl. (1957) "Aristotle Discovers the Economy". In K. Polanyi, C. M. Arensberg, and H. Pearson, eds. *Trade and Market in the Early Empires:* Economics in History and Theory (Glencoe, IL: Free Press), pp. 64–94.

Polanyi, Michael. (1962) "The Republic of Science: Its Political and Economic Theory," *Minerva*, vol. 1, pp. 54–74.

Popper, Karl. (1945) *The Open Society and Its Enemies*. London: Routledge. Revised and expanded editions: rev. (1950), 2nd rev. (1952), 3rd edition rev. (1957), 4th edition rev. (1962), and 5th edition (1966).

Pörksen, Uwe. (1989) *Plastik Wörter: Die Sprache einer Internationalendiktatur*. Stuttgart: Klett-Cotta. English version: *Plastic Words: The Tyranny of Modular Language*,

trans. Jutta Mason and David Cayley (University Park, PA: Penn State University Press, 1995).
Press Release. (2006) "Maurice L. Albertson, a Founder of the Peace Corps, to Receive Honorary Degree at Colorado State May 12". Department of Public Relations, Colorado State University, May 10.
Rapp, Friedrich, ed. (1974) *Contributions to a Philosophy of Technology: Studies in the Structure of Thinking in the Technological Sciences*. Dordrecht: D. Reidel.
Rapp, Friedrich. (1981) *Analytical Philosophy of Technology*. Trans. Stanley R. Carpenter and Theodor Langenbruch. Dordrecht: D. Reidel. German original, *Analytische Technikphilosophie* (Freiburg: Karl Alber, 1974).
Raven-Hansen, Peter, ed. (1987) *First Use of Nuclear Weapons: Under the Constitution, Who Decides?* New York: Greenwood Press (for Federation of American Scientists).
Raymo, Chet. (2008) *When God Is Gone, Everything Is Holy: The Making of a Religious Naturalist*. Notre Dame, IN: Ave Maria Press.
Raymond, Eric S. (1999) *The Cathedral and the Bazaar: Musings on Linux and Open Source by an Accidental Revolutionary*. Beijing: O'Reilly.
Redclift, Michael. (1994) "Sustainable Development: Economics and Environment". In Redclift and Sage, eds., *Strategies for Sustainable Development: Local Agendas for the Southern Hemisphere* (New York: John Wiley), pp. 17–34.
Redclift, Michael and Colin Sage, eds. (1994) *Strategies for Sustainable Development: Local Agendas for the Southern Hemisphere*. New York: John Wiley.
Rice, Calvin W. (1922) "The Ethics of the Mechanical Engineers," *Annals of the American Academy of Political and Social Science*, vol. 101 (May), pp. 72–76.
Rieff, David. (2002) *A Bed for the Night: Humanitarianism in Crisis*. New York: Simon and Schuster.
Riessen, Hendrik van. (1949) *Filosofie en techniek* [Philosophy and technology]. Kampen: J. H. Kok.
Riley, Donna. (2008) *Engineering and Social Justice*. San Rafael, CA: Morgan and Claypool.
Rodale, Robert. (1972) *Sane Living in a Mad World: A Guide to the Organic Way of Life*. Emmaus, PA: Rodale Press.
Rodale, Robert. (1981) *Our Next Frontier: A Personal Guide for Tomorrow's Lifestyle*. Emmaus, PA: Rodale Press.
Roemischer, Jessica. (n.d.) "A New Axial Age: Karen Armstrong on the History—and the Future—of God," www.enlightennext.org/magazine/j31/armstrong.asp?page=1.
Rogers, C. Thomas. (1980) "The End-Use Problem in Engineering Ethics," *PSA: Proceedings of the Biennial Meeting of the Philosophy of Science Association*, vol. 2, pp. 464–80.
Rogers, G. F. C. (1983) *The Nature of Engineering: A Philosophy of Technology*. London: Macmillan.

Roland, Alex. (2009) "War and Technology," *Footnotes: Newsletter of Foreign Policy Research Institute Wachman Center*, vol. 14, no. 2 (February). Available at fpri.org/footnotes/1402.200902.roland.wartechnology.html.
Rolt, L. T. C. (1958) *Thomas Telford*. London: Longmans.
Ropohl, Günter. (2002) "Mixed Prospect of Engineering Ethics," *European Journal of Engineering Education*, vol. 27, no. 2, pp. 149–55.
Rorty, Richard M., ed. (1992) *The Linguistic Turn: Essays in Philosophical Method, with Two Retrospective Essays*. Chicago, IL: University of Chicago Press. (First published, 1967.)
Rosen, Stanley. (1988) *The Quarrel between Philosophy and Poetry*. New York: Routledge.
Rosen, Stanley. (1993) "*Technē* and the Origins of Modernity". In Arthur M. Melzer, Jerry Weinberger, and M. Richard Zinman, eds., *Technology in the Western Political Tradition* (Ithaca, NY: Cornell University Press), pp. 69–84.
Rosenberg, Nathan. (1982) *Inside the Black Box: Technology and Economics*. New York: Cambridge University Press.
Rostow, Walt W. (1952) *The Process of Economic Growth*. New York: W. W. Norton. 2nd edition, 1962.
Rostow, Walt W. (1956) "The Take-Off into Self-Sustained Growth," *Economic Journal*, vol. 50, no. 4 (October), pp. 150–200.
Rostow, Walt W. (1960) *The Stages of Economic Growth: A Non-Communist Manifesto*. New York: Cambridge University Press. Third edition, 1991.
Rostow, Walt W., ed. (1963) *The Economics of Take-Off into Sustained Growth*. New York: St. Martin's Press.
Rostow, Walt W. (1990) *Theories of Economic Growth from David Hume to the Present: With a Perspective on the Next Century*. New York: Oxford University Press.
Rotblat, Joseph. (1972) *Scientists in the Quest for Peace: A History of the Pugwash Conferences*. Cambridge, MA: MIT Press.
Rotblat, Joseph, ed. (1982) *Scientists, the Arms Race and Disarmament: A UNESCO/Pugwash Symposium*. London: Taylor & Francis.
Rotblat, Joseph. (1999) "A Hippocratic Oath for Scientists," *Science*, vol. 286, whole no. 5444 (19 November), p. 1475.
Rotblat, Joseph. (2000) "Taking Responsibility," *Science*, vol. 289, whole no. 5480 (4 August), p. 729.
Royal Academy of Engineering (RAE). (2010) *Philosophy of Engineering*. Vol. 1 of the Proceedings of a Series of Seminars. London: Royal Academy of Engineering.
Royal Academy of Engineering (RAE). (2011) *Philosophy of Engineering*. Vol. 2 of the Proceedings of a Series of Seminars. London: Royal Academy of Engineering.
Royal Academy of Engineering (RAE). (2014). *Public Projects and Procurement in the UK: Sharing Experience and Changing Practice*. London: Royal Academy of Engineering.
Rüdiger, Mogens, ed. (2008) *The Cultures of Energy*. Newcastle, UK: Cambridge Scholars Publishing.

Rupp, Stephanie. (2013) "Considering Energy: E = mc^2 = (magic • culture)^2," in Strauss et al., eds., *Cultures of Energy: Power, Practices, Technologies* (Walnut Creek, CA: Left Coast Press), pp. 79–95.

Ruse, Michael. (1994) "Sustainability". In Skirbekk, ed., *The Notion of Sustainability and Its Normative Implications* (Oslo, Norway: Scandinavian University Press), pp. 7–27.

Russell, Bertrand. (1912) *The Problems of Philosophy*. Oxford: Oxford University Press.

Ruttan, Vernon W. (1988) "Sustainability Is Not Enough," *American Journal of Alternative Agriculture*, vol. 3, nos. 2–3 (Spring-Summer), pp. 128–30.

Sabbagh, Karl. (1996) *Twenty-First-Century Jet: The Making and Marketing of the Boeing 777*. New York: Scribners.

Sachs, Wolfgang, ed. (1989) "A Critique of Ecology," *New Perspectives Quarterly* (Spring), pp. 16–19.

Sachs, Wolfgang, ed. (1992) *The Development Dictionary: A Guide to Knowledge as Power*. London: Zed Books.

Sachs, Wolfgang, ed. (1993) *Global Ecology: A New Area of Political Conflict*. London: Zed Books.

Safina, Carl. (1998) "To Save the Earth, Scientists Should Join Policy Debates," *Chronicle of Higher Education* (November 6), p. A80.

Sagoff, Mark. (1985) *Risk-Benefit Analysis in Decisions Concerning Public Safety and Health*. Dubuque, IA: Kendall Hunt.

Salomon, Jean-Jacques. (1973) *Science and Politics*. Trans. Noël Lindsay. Cambridge, MA: MIT Press. (French original, 1970.)

Sarewitz, Daniel, and Roger A. Pielke, Jr. (2007) "The Neglected Heart of Science Policy: Reconciling Supply of and Demand for Science," *Environmental Science and Policy*, vol. 10, pp. 5–16.

Sassower, Raphael. (1997) *Technoscientific Angst: Ethics and Responsibility*. Minneapolis: University of Minnesota Press.

Schatzberg, Eric. (2018) *Technology: Critical History of a Concept*. Chicago, IL: University of Chicago Press.

Schillinger, George. (1984) "Man's Enduring Technological Dilemma: Prometheus, Faust, and Other Macro-Engineers," *Technology in Society*, vol. 6, no. 1, pp. 59–71.

Schmidgen, Henning. (2012) "Inside the Black Box: Simondon's Politics of Technology," *SubStance* #129, vol. 41, no. 3, pp. 16–31.

Schneider, Jen, Juan Lucena, and Jon A. Leydens. (2009) "Engineering to Help: The Value of Critique in Engineering Service," *IEEE Technology and Society Magazine*, vol. 28, no. 4 (Winter), pp. 42–48.

Schumacher, E. F. (1973) *Small Is Beautiful: Economics as If People Mattered*. New York: Harper and Row.

Schumpeter, Joseph A. (1942) *Capitalism, Socialism and Democracy*. New York: Harper.

Scrivener, Stephen A. R., Linden J. Ball, and Andrée Woodcock, eds. (2000) *Collaborative Design: Proceedings of CoDesigning 2000*. London: Springer.

Seffah, Ahmed, Jan Gulliksen, and Michel C. Desmarais, eds. (2005) *Human-Centered Software Engineering: Integrating Usability in the Software Development Lifecycle*. Dordrecht: Springer.
Sen, Amartya. (1981) *Poverty and Famine: An Essay on Entitlement and Deprivation*. New York: Oxford University Press.
Şen, Zekâi. (2014) *Philosophical, Logical and Scientific Perspectives in Engineering*. Dordrecht: Springer.
Serres, Michel. (1995) *Conversations on Science, Culture, and Time*. With Bruno Latour. Trans. Roxanne Lapidus. Ann Arbor: University of Michigan Press.
Shibata, Masako. "Controlling National Identity and Reshaping the Role of Education: The Vision of State Formation in Meiji Japan and the German *Keiserreich*," *History of Education*, vol. 33 (2004), pp. 75–85.
Shrader-Frechette, K. S. (1991) *Risk and Rationality: Philosophical Foundations for Populist Reforms*. Berkeley: University of California Press.
Simon, Herbert A. (1968) *The Sciences of the Artificial*. Cambridge, MA: MIT Press. 2nd edition, 1981. 3rd edition, 1996.
Simon, Julian L. (1981) *The Ultimate Resource*. Princeton, NJ: Princeton University Press.
Simon, Julian L., ed. (1995) *The State of Humanity*. Oxford: Blackwell.
Simon, Julian L., and Herman Kahn, eds. (1984) *The Resourceful Earth: A Response to Global 2000*. Oxford: Blackwell.
Simondon, Gilbert. (1958) *Du mode d'existence des objets techniques*. Paris: Méot. 2nd, Paris: Aubier, 1989. English trans. Cécile Malaspina and John Rogove: *On the Mode of the Existence of Technical Objects* (Minneapolis, MN: Univocal, 2017).
Sitarz, Daniel, ed. (1994) *Agenda 21: The Earth Summit Strategy to Save Our Planet*. Boulder, CO: EarthPress.
Sloterdijk, Peter. (2013) *In the World Interior of Capital: Towards a Philosophical Theory of Globalization*. Cambridge: Polity Press.
Smil, Václav. (2006) *Energy: A Beginner's Guide*. Oxford: One World.
Smil, Václav. (2017) *Energy and Civilization: A History*. Cambridge, MA: MIT Press.
Smiles, Samuel. (1867) *The Life of Thomas Telford, Civil Engineer: With an Introductory History of Roads and Travelling in Great Britain*. London: John Murray.
Smith, Alice Kimbrall. (1965) *A Peril and a Hope: The Scientists Movement in America, 1945–1947*. Chicago, IL: University of Chicago Press.
Smith, Cecil O. Jr. (1990) "The Longest Run: Public Engineers and Planning in France," *American Historical Review*, vol. 95, no. 3 (June), pp. 657–92.
Smith, Douglas K., and Robert C. Alexander. (1988) *Fumbling the Future: How Xerox Invented, Then Ignored, the First Personal Computer*. New York: Morrow.
Smith, Merrit Roe, ed. (1985) *Military Enterprise and Technological Change: Perspectives on the American Experience*. Cambridge, MA: MIT Press.
Smith, Ralph J. (1956) *Engineering as a Career*. New York: McGraw-Hill. 2nd edition, 1962. 3rd edition, 1969. 4th edition, with Blaine Butler and William Kerns Lebold, 1983.

Smyser, W. R. (2003) *The Humanitarian Conscience: Caring for Others in the Age of Terror*. New York: Palgrave Macmillan.

Snow, C. P. (1959) *The Two Cultures and the Scientific Revolution*. Cambridge: Cambridge University Press. 2nd edition, *The Two Cultures: And a Second Look* (New York: Oxford University Press, 1963).

Sowell, Thomas. (1972) *Say's Law: An Historical Analysis*. Princeton, NJ: Princeton University Press.

Sparke, Penny. (1986) *An Introduction to Design and Culture in the Twentieth Century*. London: Allen and Unwin.

Spiegel-Rosing, Ina, and Derek de Solla Price. (1977) *Science, Technology and Society: A Cross-Disciplinary Perspective*. London: Sage.

Sprat, Thomas. (1667) *The History of the Royal-Society of London for the Improving of Natural Knowledge*. London: T. R. for J. Martyn. Early English Books Online, a project of the Text Creation Partnership. Available at http://name.umdl.umich.edu/A61158.0001.001.

Staudenmaier, John M. (1985) *Technology's Storytellers: Reweaving the Human Fabric*. Cambridge, MA: MIT Press.

Steinmetz, Charles P., Harold W. Buck, and Schuyler Skaats Wheeler. (1908) "Proposed Code of Ethics: American Institute of Electrical Engineers Committee on Code of Ethics," *Transactions of the American Institute of Electrical Engineers*, vol. 26, part 2, p. 1422.

Stone, Jeremy J. (1999). *"Every Man Should Try": Adventures of a Public Interest Activist*. New York: Public Affairs Press.

Strauss, Leo. (1975) "The Three Waves of Modernity". In Gildin, ed., *Political Philosophy: Six Essays by Leo Strauss* (Indianapolis, IN: Bobbs-Merrill), pp. 81–98.

Strauss, Sarah, Stephanie Rupp, and Thomas Love, eds. (2013) *Cultures of Energy: Power, Practices, Technologies*. Walnut Creek, CA: Left Coast Press.

Strickland, Donald A. (1968) *Scientists in Politics: The Atomic Scientists Movement, 1945–1946*. Lafayette, IN: Purdue University Press.

Strong, Maurice ed. (1973) *Who Speaks for the Earth?* New York: W. W. Norton.

Sutz, Judith. (1993) "The Social Implications of Information Technologies: A Latin American Perspective". In Mitcham, ed., *Philosophy of Technology in Spanish Speaking Countries* (Dordrecht: Springer), pp. 297–308.

Tang, Xiaofeng, and Dean Nieusma. (2017) "Contextualizing the Code: Ethical Support and Professional Interests in the Creation and Institutionalization of the 1974 IEEE Code of Ethics," *Engineering Studies*, vol. 9, no. 3, pp. 166–194.

Taylor, Frederick W. (1911) *The Principles of Scientific Management*. New York: Harper.

Telford, Thomas. (1838) *The Life of Thomas Telford, Civil Engineer*. Edited by John Rickman. London: Payne and Foss.

Terry, Fiona. (2002) *Condemned to Repeat? The Paradox of Humanitarian Action*. Ithaca, NY: Cornell University Press.

Thackara, John, ed. (1988) *Design after Modernism: Beyond the Object*. New York: Thames and Hudson.
Tierney, Thomas F. (1993) *The Value of Convenience: A Genealogy of Technical Culture*. Albany: State University of New York Press.
Tisdell, Clem. (1988) "Sustainable Development: Differing Perspectives of Ecologists and Economists, and Relevance," *World Development*, vol. 16, no. 3 (March), pp. 373–84.
Thibodeau, Francis R., and Hermann H. Field, eds. (1984) *Sustaining Tomorrow: A Strategy for World Conservation and Development*. Hanover, NH: University Press of New England.
Thorstein, Veblen (1921) *The Engineers and the Price System*. New York: B. W. Huebsch.
Tijmes, Pieter, and Reginald Luijf. (1995) "The Sustainability of Our Common Future: An Inquiry into the Foundations of an Ideology," *Technology in Society*, vol. 17, no. 3, pp. 327–36.
Torvalds, Linus. (1992) "LINUX's History". July 31, 1992. (A collection of emails with narrative notes available on a number of web sites. Accessed from www.cs.cmu.edu/~awb/linux.history.html.)
Torvalds, Linus, with David Diamond. (2001) *Just for Fun: The Story of an Accidental Revolutionary*. New York: HarperCollins.
Toulmin, Stephen. (1982) "How Medicine Saved the Life of Ethics," *Perspectives in Biology and Medicine*, vol. 25, no. 4 (Summer), pp. 736–50.
Turkle, Sherry. (1984) *The Second Self: Computers and the Human Spirit*. New York: Simon & Schuster.
Turner, Fred. (2006) *From Counterculture to Cyberworld: Stewart Brand, the Whole Earth Network, and the Rise of Digital Utopianism*. Chicago, IL: University of Chicago Press.
Ubbelohde, A. R. (1955) *Man and Energy*. New York: Penguin.
UNESCO (United Nations Educational, Scientific and Cultural Organization). (2010) *Engineering: Issues, Challenges and Opportunities for Development*. Paris: UNESCO.
Unger, Stephen H. (1982) *Controlling Technology: Ethics and the Responsible Engineer*. New York: Holt, Rinehart and Winston. 2nd edition, New York: John Wiley, 1994.
United States Atomic Energy Commission. *In the Matter of J. Robert Oppenheimer: Transcript of Hearing before Personnel Security Board*. Washington, DC: U.S. Government Printing Office.
Vallor, Shannon. (2016) *Technology and the Virtues: A Philosophical Guide to a Future Worth Wanting*. New York: Oxford University Press.
Van Creveld, Martin. (1989) *Technology and War: From 2000 BC to the Present*. New York: Free Press.
Van de Poel, Ibo, and David Goldberg, eds. (2010) *Philosophy and Engineering: An Emerging Agenda*. (With associate eds. Michael Davis, Carl Mitcham, and Aarne Viesland.) Dordrecht: Springer.

Van de Poel, Ibo, and Lambèr Royakkers. (2011) *Ethics, Technology, and Engineering: An Introduction*. Malden, MA: Wiley-Blackwell.

Van den Bergh, Jeroen C. J. M., and Jan van der Straaten, eds. (1994) *Toward Sustainable Development: Concepts, Methods, and Policy*. Washington, DC: Island Press.

Vanderburg, Willem H. (1995) "Can a Technical Civilization Sustain Human Life?," *Bulletin of Science, Technology, and Society*, vol. 15, nos. 2–3 (April), pp. 92–98.

Vanderburg, Willem H. (2000) *The Labyrinth of Technology: A Preventive Technology and Economic Strategy as a Way Out*. Toronto: University of Toronto Press.

Vanderburg, Willem H. (2005) *Living in the Labyrinth of Technology*. Toronto: University of Toronto Press.

Vanderburg, Willem H. (2016) *Our Battle for the Human Spirit: Scientific Knowing, Technical Doing, and Daily Living*. Toronto: University of Toronto Press.

Vandivier, Kermit. (1972) "The Aircraft Brake Scandal," *Harper's*, vol. 244 (April), pp. 45–52.

Vandivier, Kermit. (1975) "Engineers, Ethics, and Economics," American Society of Civil Engineering, *Proceedings of the Conference on Engineering Ethics* (New York: ASCE), pp. 20–24.

Vattimo, Gianni. (1988) *The End of Modernity: Nihilism and Hermeneutics in Postmodern Culture*. Trans. Jon R. Snyder. Baltimore: Johns Hopkins University Press.

Vaughan, Diane. (1996) *The Challenger Launch Decision: Risky Technology, Culture, and Deviance at NASA*. Chicago, IL: University of Chicago Press.

Veblen, Thorstein. (1921) *The Engineers and the Price System*. New York: B. W. Huebsch.

Venturi, Robert. (1977) *Complexity and Contradiction in Architecture*, 2nd edition. New York: Museum of Modern Art.

Verbeek, Peter-Paul. (2005) *What Things Do: Philosophical Reflections on Technology, Agency, and Design*. Trans. Robert P. Crease. University Park: Pennsylvania State University Press.

Verbeek, Peter-Paul. (2011) *Moralizing Technology: Understanding and Designing the Morality of Things*. Chicago, IL: University of Chicago Press.

Verbeek, Peter-Paul. (2009) "Let's Make Things Better: A Reply to My Readers," *Human Studies*, vol. 32, no. 2 (June), pp. 251–61.

Verein Deutscher Ingenieure (VDI). (2002) *Ethische Grundsätze des Ingenieurberufs* [Ethical Foundations of the Engineering Profession]. Dusseldorf: VDI.

Vermaas, Pieter. (2007) "Fiat Functions in Engineering". Paper presented at the Society for Philosophy and Technology Biannual Meeting, Charlestown, SC, July 11.

Vermaas, Pieter. (2010) "Focussing Philosophy of Engineering: Analysis of Technical Functions and Beyond." In Van de Poel and Goldberg, eds., *Philosophy and Engineering: An Emerging Agenda* (Dordrecht: Springer), pp. 61–73.

Vesilind, P. Aarne, ed. (2005) *Peace Engineering: When Personal Values and Engineering Careers Converge*. Woodsville, NH: Lakeshore Press.

Vesilind, P. Aarne, (1995) "Evolution of the American Society of Civil Engineers Code of Ethics," *Journal of Professional Issues in Engineering Education and Practice*, vol. 121, pp. 4–10.
Vesilind, P. Aarne, and Alastair S. Gunn. (1998) *Engineering, Ethics, and the Environment*. New York: Cambridge University Press.
Vincenti, Walter G. (1990) *What Engineers Know and How They Know It: Analytical Studies from Aeronautical History*. Baltimore, MD: Johns Hopkins University Press.
Vincenti, Walter G. (1991) "Introduction." In Hedy E. Sladovich, ed., *Engineering as a Social Enterprise* (Washington, DC: National Academy Press), pp. 1–4.
Vlot, Ad. (2000) "Toward a Juridical Turn for the Ethics of Technology? An Aerospace Case". In Kroes and Meijers, eds., *The Empirical Turn in the Philosophy of Technology, Research in Philosophy and Technology*, vol. 20, pp. 207–18.
Vogel, Ezra F. (1971) *Japan's New Middle Class: The Salary Man and His Family in a Tokyo Suburb*. Berkeley: University of California Press.
Vowe, Gerhard. (2008) "Politics, Polity, Policy." In Lynda Lee Kaid and Christina Holtz-Bacha, eds., *Encyclopedia of Political Communication* (Thousand Oaks, CA: Sage), pp. 620–21.
Vries, Marc J. de, Nigel Cross, and D. P. Grant, eds. (1993) *Design Methodology and Relationships with Science*. Boston, MA: Kluwer.
Vrin, Hélène, and Irina Gouzévitch. (2011) "The Rise of the Engineering Profession in Eighteenth Century Europe: An Introductory Overview," *Engineering Studies*, vol. 3, no. 3 (December), pp. 153–69.
Wallace, William A. (1984) "The Intelligibility of Nature: A Neo-Aristotelian View," *Review of Metaphysics*, vol. 38, no. 1, whole no. 149 (September), pp. 33–56.
Ward, John K. (1989) "The Future of an Explosion," *American Heritage of Invention and Technology*, vol. 5, no. 1 (Spring-Summer), pp. 58–63.
Ward, Lester F. (1883) *Dynamic Sociology*, vol. 2. New York: Appleton.
Watson, Garth. (1988) *The Civils: The Story of the Institution of Civil Engineers*. London: Thomas Telford.
Weber, Max. (1946) "Politics as a Vocation". In H. H. Gerth and C. Wright Mills, trans. and eds., *From Max Weber: Essays in Sociology* (New York: Oxford University Press), 77–128. (German original 1919.)
Weber, Steven. (2004) *The Success of Open Source*. Cambridge, MA: Harvard University Press.
Weil, Vivian. (1980) *Report of the Workshops on Ethical Issues in Engineering*. Chicago: Center for the Study of Ethics in the Professions, Illinois Institute of Technology.
Weil, Vivian. (1983) *Beyond Whistleblowing: Defining Engineers' Responsibilities*. Chicago: Center for the Study of Ethics in the Professions, Illinois Institute of Technology. (Proceedings of the Second National Conference on Ethics in Engineering, March 1982.)
Weil, Vivian. (1984) "The Rise of Engineering Ethics," *Technology in Society*, vol. 6, no. 4, pp. 341–45.
Weiser, Mark. (1991) "The Computer for the 21st Century," *Scientific American*, vol. 265, no. 3 (September), pp. 94–95, 98–102, and 103.

Welin, Stellan. (1992) "Engineering Ethics and the Fofors Affair". In Mitcham, *Engineering Ethics throughout the World: Introduction, Documentation, Commentary, and Bibliography* (University Park, PA: STS Press), pp. II19–II27.

Wells, H. G. (1928) *The Open Conspiracy: Blueprints for a World Revolution.* London: Gollancz.

Wells, H. G. (1934) *Experiment in Autobiography: Discoveries and Conclusions of a Very Ordinary Brain (since 1866).* New York: Macmillan.

Wells, Paula, Hardy Jones, and Michael Davis. (1986) *Conflicts of Interest in Engineering.* Dubuque, IA: Kendal Hunt.

Weszkalnys, Gisa. (2013) "Oil's Magic: Contestation and Materiality," in Strauss, Rupp, and Love, eds., *Cultures of Energy: Power, Practices, Technologies* (Walnut Creek, CA: Left Coast Press), pp. 267–83.

White, Leslie A. (1959) *The Evolution of Culture: The Development of Civilization to the Fall of Rome.* New York: McGraw-Hill.

Whitehead, Alfred North. (1925) *Science and the Modern World.* New York: Macmillan.

Wiener, Norbert. (1947) "A Scientist Rebels". *Atlantic Monthly*, vol. 179, no. 1 (January), p. 46.

Wilgus, William J. (1922) "Shall Corporations Be Authorized to Practice Engineering?," *Annals of the American Academy of Political and Social Science*, vol. 101 (May), pp. 94–96.

Williams, Rosalind. (2002) *Retooling: A Historian Confronts Technological Change.* Cambridge, MA: MIT Press.

Wimsatt, William C. (2007) *Re-Engineering Philosophy for Limited Beings: Piecewise Approximations to Reality.* Cambridge, MA: Harvard University Press.

Winner, Langdon. (1980) "Do Artifacts Have Politics?" *Daedalus*, vol. 109, no. 1 (Winter), pp. 121–136. Reprinted in his *The Whale and the Reactor: A Search for Limits in an Age of High Technology* (Chicago, IL: University of Chicago Press, 1986), pp. 19–39.

Winner, Langdon. (1986) "Technologies as Forms of Life". In his *The Whale and the Reactor: A Search for Limits in an Age of High Technology* (Chicago, IL: University of Chicago Press), pp. 3–18.

Winner, Langdon. (1991) "Artifact/Ideas and Political Culture," *Whole Earth Review*, no. 73 (Winter), pp. 18–24.

Winograd, Terry, and Fernando Flores. (1987) *Understanding Computers and Cognition: A New Foundation for Design.* Reading, MA: Addison-Wesley.

Winther, Tanya. (2013) "Space, Time, and Sociomaterial Relationships: Moral Aspects of the Arrival of Electricity in Rural Zanzibar," in Strauss et al., eds., *Cultures of Energy: Power, Practices, Technologies* (Walnut Creek, CA: Left Coast Press), pp. 164–76.

Wisnioski, Matthew. (2012) *Engineers for Change: Competing Visions of Technology in 1960s America.* Cambridge, MA: MIT Press.

Wittgenstein, Ludwig. (1953) *Philosophical Investigations.* Oxford: Blackwell.

Wood, Barbara. (1984) *E. F. Schumacher: His Life and Thought.* New York: Harper and Row.

World Commission on Environment and Development. (1987) *Our Common Future*. New York: Oxford University Press.
Worster, Donald. (1993) *The Wealth of Nature: Environmental History and the Ecological Imagination*. New York: Oxford University Press.
Worster, Donald. (1994) *Nature's Economy: A History of Ecological Ideas*. 2nd edition. Cambridge: Cambridge University Press.
Wulf, William A. (2004) "Keynote Address". In *Emerging Technologies and Ethical Issues in Engineering*, Papers from a Workshop, October 14–15, 2003 (Washington, DC: National Academy Press), pp. 1–6.
Xenos, Nicholas. (1989) *Scarcity and Modernity*. New York: Routledge.
Yin Dengxiang 殷登祥. (1997) 当代中国科学技术和社会的发展 *Dangdai zhongguo kexue jishu he shehui de fazhan* [The development of science, technology, and society in contemporary China]. Wuhan: Hubei Press.
Yin Ruiyu 殷瑞钰, Wang Yingluo 汪应洛, and Li Bocong. (2007) 工程哲学 *Gongcheng zhexue* [Philosophy of engineering]. Beijing: Higher Education Press.
Yin Ruiyu, Wang Liheng 王礼恒, Wang Yingluo, and Li Bocong, eds. (2007) 工程与哲学 *Gongcheng yu zhexue* [Engineering and philosophy]. Vol. 1. Beijing: Beijing Institute of Technology Press.
Yin Ruiyu, Sun Yongfu 孙永福, Wang Yingluo, Li Bocong, and Qiu Lianghui 丘亮辉 eds. (2018) 工程与哲学 *Gongcheng yu zhexue* [Engineering and philosophy]. Vol. 2. Xi'an: Xidian University Press.
Zhu, Qin, and Brent K. Jesiek. (2017) "Engineering Ethics in Global Context: Four Fundamental Approaches," ASEE Annual Conference & Exposition, Columbus, Ohio, June 25–28, Paper ID #19721.
Zimmerli, Walther Ch. (1986) "Who Is to Blame for Data Pollution? On Individual Moral Responsibility with Information Technology." In Carl Mitcham and Alois Huning, eds., *Philosophy and Technology II: Information Technology and Computers in Theory and Practice*, Boston Studies in the Philosophy of Science, vol. 90 (Boston, MA: D. Reidel), pp. 291–305.

Index

AAAS (American Association for the Advancement of Science), 91, 98–101, 353
AAES (American Association of Engineering Societies), 165
ABET (Accreditation Board for Engineering and Technology), 57–58, 59–60, 61, 164–65, 209–10, 211, 221, 222, 226, 261, 262, 283, 299, 302, 383–84
ABET EC 2000, 210, 211, 214. *See also* "Engineering Criteria 2000" (ABET)
Academy of Military Engineering, 156
Accreditation Board for Engineering and Technology (ABET). *See* ABET (Accreditation Board for Engineering and Technology)
Achterhuis, Hans, 5, 206n43
"ACM (Association for Computing Machinery) Code of Ethics and Professional Conduct," 127
Adams, Henry, 1–2, 22
Agenda 21, 188, 189
agricultural engineering, 193, 198, 351
Agricultural Productivity Research component of Food Security Act, 194
agricultural science, 193
Aiken, William, 162
Albertson, Maurice (Maury), 251–52
Alder, Ken, 229
Alexander, Jennifer, 162
Allenby, Branden, 360
alternative agriculture, 193, 194
Alternative Agriculture (report), 195
alternative energy, 193, 331, 356
alternative technology, 121, 130, 134, 178, 194, 250
America by Design (Noble), 11
American Association for the Advancement of Science (AAAS), 91, 98–101, 353
American Association of Engineering Societies (AAES), 165
American Bar Association, 158, 293
American Friends Service Committee, 247
American Genesis (Hughes), 74
American Institute of Electrical Engineers (AIEE), 158, 159, 165, 208, 293, 298
American Journal of Alternative Agriculture, 194
American Medical Association, 158, 293
American Society for Engineering Education (ASEE), 209, 211
American Society of Civil Engineers (ASCE), 158–59, 167, 169, 170, 208, 236, 264, 293, 297–98

American Society of Mechanical Engineers (ASME), 158, 159, 162, 171, 208
Ammianus Marcellinus, 156
Anders, Günther, 21
Anderson, Scott, 250
Annals of the American Academy of Political and Social Science, 160
Annan, Kofi, 249
Anthropocene, 2, 286
Applied Minds: How Engineers Think (Madhavan), 17
appropriate technology, 130
Aquinas, Thomas. *See* Thomas Aquinas
Arendt, Hannah, 45, 85, 192, 197, 358, 359–60
architect/architecture, 41, 228, 371, 373–75; and art/artist, 43, 47, 76; books on, 372; and engineer/engineering, 15, 210, 235, 264, 339, 349, 369; hydraulic, 369, 382; postmodern, 69
De Architectura (Vitruvius), 264, 372–74; principles of good building, 374
Aristotle, 48, 61, 69, 147, 154, 155, 309, 314, 330, 349, 351
Arnett, Jeffrey, 285
art/artist, 41, 43, 47, 49, 74–76, 210, 228, 307, 339
Arthur, Brian, 19
ASC (Atomic Scientists of Chicago), 95
Asociación Dominicano de Ingenieros y Arquitectos (AIDA), 235
Aspen Institute, 113–14
Association for the Promotion of Technical Activity in Prussia, 232
Association of German Engineers (*Verein Deutscher Ingenieure* VDI), 221, 231, 232, 233, 295–96
Association of Professional Engineers (Ontario, Canada), 301
atomic bomb, 21, 29, 90, 91, 95, 179
atomic energy, 92, 95, 222, 229, 317
Atomic Scientists of Chicago (ASC), 95
Augustine, Norman, 217

Austin, J. L., 145
autodidacts, 125, 295
automatic train control (ATC) system, 172, 174, 176
Auyang, Sunny, 17, 355–56
Axial Age, 280–81, 285–86, 354

Babbage, Thomas, 154
Bacon, Francis, 30, 63, 85, 106–7, 368, 370, 375
balance of nature, 204n2
Baptist World Aid, 247
Barrett, Michael, 253
Barsoux, Jean-Louis, 228
Barton, Clara, 246
Bashir, Hassan, 240
"The Basic Principles of International Scientific Exchange" (Japanese Science Council), 223
Bataille, George, 331
Bauhaus, 349
Baum, Robert J., 212–13, 297
Bay Area Rapid Transit (BART) system, as engineering disaster, 169, 172–73, 174, 176, 177, 212
Bayet, Albert, 154
Bayles, Michael, 262
Beder, Sharon, 180n8
Bekenntnis des Ingenieurs (Confession of Engineers) (VDI), 295
Bentham, Jeremy, 308, 309, 375
Bergson, Henri, 330
Bernays, Edward, 21–22, 350
Beuth, Christian Peter, 232
Bhopal, India industrial disaster, 214
Bijker, Wiebe, 143
Bildung, 106, 108, 109, 110, 111, 231
Bildungsroman, 108
"black box" opening, 61, 70n19, 380
Blankenzee, Max, 172
Block, Jan, 32
Blockley, David, 17
Blue, Ethan, 37
Bofors, 234–35
Boisjoly, Roger, 173, 174, 214, 299

Boot, Max, 27, 28, 33
Borgmann, Albert, 46, 133, 348, 352, 359, 378–79, 380, 381
Bowen, Richard, 22, 37
Boyle, Charles, 30
Branscomb, Lewis, 318
Bratman, Michael, 350
Bratt, Ingbar, 234–35
Briggle, Adam, 350
Brindley, James, 373
Brooks, Harvey, 316, 317, 319
Broome, Taft H., Jr., 4, 7, 8, 13, 14, 262, 269–70, 271, 342
Brown, Lester, 195
Bruder, Robert, 172
Brundtland, Gro, 189–90
Brunel, Isambard Kingdom, 372
Bucciarelli, Louis, 7, 8, 43, 64, 144
Buchanan, Richard, 43, 44, 79
Bugliarello, George, 178
Building a Sustainable Society (Brown), 195
Bulletin of the Atomic Scientists, 91, 93, 94–95, 97, 102, 103
Bunge, Mario, 3, 11, 142, 338, 342
Bureaux des Dessinateurs du Roi, 156
Burns, Robert, 374
Bush, Vannevar, 33, 280, 355
business, as compared to medicine, law, military, and engineering, 275
Business Week, 74

California Society of Professional Engineers (CSPE), 172
Cambridge Handbook of Engineering Education Research (Johri and Olds), 20
Campbell, Donald, 339, 340
Canfield, Donald, 210
"Canons of Ethics" (American Bar Association), 158
"Can Technology Be Humane?" (Goodman), 97
CARE (Cooperative Action for American Relief Everywhere), 247

Carpenter, Stanley R., 201
Carson, Rachel, 89–90
Carter, Jimmy, 193, 251, 252
The Cathedral and the Bazaar (Raymond), 127
Catholic Medical Mission Board, 247
Cattell, James McKeen, 98
Center for Engineering Ethics and Society (National Academy of Engineering), 286
Center for Science, and Technology Policy Research (University of Colorado, Boulder), 215
Center for the Study of Ethics in the Professions (Illinois Institute of Technology), 212, 297
A Century of Innovation (National Academy of Engineering), 279
Certeau, Michel de, 82, 131
Challenger, as engineering disaster, 64, 169, 173, 174, 214, 223, 299
Charles I (King), 370
Charles II (King), 370
Charte d'Ethique de l'Ingénieur (Charter of Ethics of the Engineer) (CNISF), 221, 222, 228, 301
Charter of the Institution of Civil Engineers (ICE), 259, 365–68
Chesbrough, Henry, 119
Chile, engineering ethics as alternative development in, 236–37, 238
China: movement from ethics to politics in, 306; and philosophy of engineering in, 303–4
Chinatown (film), 170
Chinese Academy of Engineering (CAE), 8, 286
Chinese Society for Philosophy of Engineering, 8
Christensen, Steen, 7, 303
Churchill, Winston, 180n11, 185
Churchill principle, 359
Cicero, 75
civil engineer: Tredgold's description of, 368–69; use of term, 15, 62, 157, 264

civil engineering, defined, 264–65
civilian engineering, emergence of, 157–58
Clancy, Rockwell, 240
Clausewitz, Carl von, 243, 249
Club of Rome, 186–87, 193
CNISF (National Council of Engineers and Scientists of France, *Conseil National des Ingénieurs et Scientifiques de France*), 221, 228, 301
"Code of Conduct" (European Federation of National Engineering Associations, FEANI), 302
"Code of Ethics for Civil Engineers" (Japanese Society of Civil Engineers), 226–27
"Code of Ethics for Engineers" (IEEE), 298
"Code of Ethics for Engineers" (NSPE), 265
"Code of Honour" (Engineering Society of Finland), 301
"Code of Medical Ethics" (American Medical Association), 158
"Code of Professional Conduct" (ICE), 265
"Code of Professional Ethics of the Engineers of the Colegio de Ingenieros de Chile (Association of Engineers of Chile)," 236
codes of ethics. *See* ethics, codes of
Cohen, Samuel, 29
Cohn, Carol, 30
Colegio de Ingenieros de Chile (Association of Engineers of Chile), 236, 301
Colegio Dominicano de Ingenieros, Arquitectos y Agrimensores (CODIA or Dominican Association of Engineers, Architects, and Surveyors), 235, 236
Collins, Harry, 143, 337
Colorado School of Mines (CSM), program in humanitarian engineering, 114–15

Commission des Titres d'Ingenieurs (Engineering titles commission), 230
Committee for Scientific Freedom and Responsibility (CSFR), 91, 98–101
Committee on Science, Engineering, and Public Policy, 217
commodity, 46, 133, 348, 377, 378, 379
Common Crisis (Independent Commission on International Development Issues), 189
Common Security (Independent Commission on Disarmament and Security Issues), 189
Companion to the Philosophy of Technology (Olsen, Pedersen, and Hendricks), 303
Comte, Auguste, 63
Condorcet, Marquis de, 186
conduct, codes of: of European Federal of National Engineering Associations (FEANI), 126–27; of Institution of Civil Engineers (ICE), 265
conflicts of interest, 117, 169, 212, 237, 266
Confucius, 280, 304, 308, 309, 351
Conseil National des Ingénieurs et Scientifiques de France (CNISF or National Council of Engineers and Scientists of France), 221, 228, 301
consequentialism, 45, 46, 308, 355
conservation, 188–89, 192, 193, 320
Consortium for Science, Policy, and Outcomes (Arizona State University), 215
Contracts, Specifications and Engineering Relations (Mead), 210
Contribution to a Philosophy of Technology (Rapp), 3
Controlling Technology: Ethics and the Responsible Engineer (Unger), 213
convenience: distractions of, 378–80; paradox of, 380–81; use and convenience, 365–66, 368, 369, 374, 375–76, 379, 381, 382, 383, 384

conviviality: program or technical conviviality, 120, 132–33, 134; pursuit of, 136; use of term, 131
convivial software, 121, 134–35
convivial technology, 130, 131
convivial tools, 130, 132
Cooke, Morris L., 162
Coopersmith, Jennifer, 330
Corps des Ponts et Chaussees, 229
Corps of Royal Engineers, 372
Coutrot, Jean, 229
Crafting Selves: Power, Gender, and Discourses of Identity in a Japanese Workplace (Kondo), 223
Critical Perspectives on Nonacademic Science and Engineering (Durbin), 6–7, 11, 12, 143
CRRNs (cyber-weapons, remote-controlled weapons, robot weapons, and nano-weapons), 37
CSPE (California Society of Professional Engineers), 172
Cuello Nieto, César, 201, 235
Cueva Badallo, Ana, 11
The Culture of Energy (Rudiger), 329
cultured engineer project, 116
"Cultures of Energy Initiative" (Rice University), 329
Cultures of Energy (Strauss et al.), 329
"the culture wars," 61
Cuny, Frederick (Fred), 250–51
Cutcliffe, Stephen, 7, 9

Daley, Herman, 201
Darwin, Charles, 75–76
Dasein, 73–75
Davidson, Donald, 143
Davis, Michael, 6, 14, 159, 208, 210, 212, 260, 261–62, 292, 342
Davy, Humphrey, 193
DC-10 airline disasters, 64, 169, 212, 223
The Death of Nature (Merchant), 30
Debugging by Thinking (Metzger), 83
Deforge, Yves, 41

democracy: according to Aristotle, 309; according to Bentham and Mill, 309; according to Churchill, 185; limitations of, 34; public involvement as promoting, 127; Telford as tinctured with fashionable doctrines of, 375; undermining of, 126, 169
Denmark, and philosophy of engineering in, 303
Dennett, Daniel, 84, 339
Derrida, Jacques, 48
Descartes, René, 30, 63, 85, 107
design: authenticity in engineering design, 86–87; from *Dasein* to, 73–75; engineering design as turning of making into thinking, 79–81; and engineering drawing, 41; ethics of, 44–46; etymology of, 75–77; existence of, 40–43; failures of, 64, 65–66, 83; origin of word, 39; as proactivity, 41; social dimensions of modern design, 43–44; toward an inner ethics of, 48–50; two versions of ethics in, 46–47; unintended consequences of, 84, 85; what constitutes good design/better design, 64; use of term, 39. *See also* engineering design; nanotechnological design; technological design
Designing Engineers (Bucciarelli), 7, 144
Dessauer, Friedrich, 11, 56, 110, 294
development, conflicts of, 199
device/device paradigm, 16, 46, 378, 379, 383
Devon, Richard, 215
Dewey, John, 21, 109, 143, 167, 293, 316
Dialogues Concerning Natural Religion (Hume), 75
Dickens, Charles, 89
Didier, Christelle, 6
Diesel, Eugene, 32
Diesel, Rudolf, 32
Diogo, Maria Paula, 10

disasters (in engineering), 64, 65–66, 169–73, 174, 176, 177, 212, 214, 223, 299
Discussion of the Method: Conducting the Engineer's Approach to Problem Solving (Koen), 110
Doctors without Borders (*Médecins sans Frontieres*), 248, 252–53
Dominican Association of Engineers, Architects, and Surveyors (*Colegio Dominicano de Ingenieros, Arquitectos y Agrimensores,* CODIA), 235, 236
Dominican Republic, as engineering ethics failure, 235–36, 238
Downey, Gary Lee, 10, 20, 221–40
Dual Natures Program, 4, 18–19
Dubreuil, Bertrand Hériard, 305
Dumont, Louis, 21, 361
Dunant, Henri, 246
Dupuy, Jean-Pierre, 21
Durbin, Paul, 6–7, 11–12, 143, 201, 296, 299, 300
Durkheim, Émile, 63
Dusek, Val, 3
duty *plus respicere*, 81, 82–84, 87, 126, 177–78, 344, 349, 350, 351
Dyer, Henry, 225
Dym, Clive, 56, 65, 66

Earth Summit (1992), 188
Eaton, Cyrus, 95
Eckhart, Meister, 106
École des Mines, 156, 230
École des Ponts et Chaussées, 156, 230
École Polytechnique, 156–57, 229, 230
"The Economodernist Manifesto" (Simon), 203
ECRA (Engineering College Research Association), 211
Edison, Thomas, 121–22, 123, 125, 128, 129, 133
Edsall, John, 99
Edsall report, 99, 102
efficiency, as ideal in engineering ethics, 156, 161–63, 269

Einstein, Albert, 91–92, 93, 94, 102, 330
Eisenhower, Dwight, 96
Eisenstadt, Shmuel, 354
Elizabeth II (Queen), 366
Ellul, Jacques, 3, 202
Emerging Technology and Ethical Issues in Engineering (workshop), 215
Endless Horizon (Bush), 355–56
energy: alternative energy, 193, 331, 356; anthropologies of, 327–29; constraints of, 327–34; ethics of, 331–34; philosophies of, 329–31; Type I versus Type II energy ethics, 331–34
Energy and Equity (Illich), 332
Engelmeier, Peter, 55, 110, 294
engineering: agricultural engineering, 193, 198, 351; ambivalent social role of, 217; can it be philosophical? 105–18; as civilian profession, 243; as compared to medicine and law, 267–68, 273, 275; as core of technology, 358; defined, 260–63; definition of (according to Davis, 292; according to Greber, 383; according to Rogers, 383; according to Tredgold, 55, 58, 62, 125–26, 292, 298, 381–82; according to Vincenti, 383; *McGraw-Hill Encyclopedia of Science and Technology*, 259); as detaching from military context, 241, 264, 292; disasters in (*see* disasters (in engineering)); duty *plus respicere* (*see* duty *plus respicere*); emergence of, 61–62; end goal of, 155; engineering philosophy for, 381–82; English-speaking engineering, 17, 18, 208, 215, 243, 244, 260, 287, 348, 349, 361, 365, 366, 369, 375, 382, 383, 384; and ethics, 63–66(*see also* engineering ethics); ethics into, 211–14; European engineering, characterization of, 17–18; fundamental obligation of, 111; genetic engineering, 86, 103,

193, 204, 341, 358; humanitarian engineering, 241–58, 282, 305, 353, 354; ideals of, 126, 153–81; importance of philosophy to, 53–69; invention of, 125; as originating in the military, 35, 109, 156–57, 292; origins of, 153; origins/meaning of in English, 14–18; philosophical engineering, 111, 113, 117–18; philosophical engineering ethics, 298, 300, 307; philosophical inadequacy in, 259–77; philosophy for (as distinct from philosophy of, 8, 111; engineering philosophy for engineering, 381–82); philosophy of (according to Durbin, 11–12; according to Goldman, 13; according to Li Bocong, 9; according to Simondon, 290; in China *vs.* the West, 9; comparing approaches to, 139–50; as defined discourse, 1; as distinct from philosophy for, 8, 111; and ethics, 18–19; as not just for engineers or philosophers, 358); post-engineering, 216–19; preventive engineering, 202; question of, 341–43; reenvisioning of, 284–87; safety engineering, 284; as social experimentation, 270; sociological inadequacy of, 273–77; techno-engineering, 30, 31, 35, 198; toward a linguistic philosophy of, 144–48; toward a political philosophy of, 308–10; true grand challenge for, 279–90; use of knowledge in, 263–65; use of self-assessments in, 217; as way of life, 358

Engineering: A Beginner's Guide (McCarthy), 17

engineering agency, 1, 2, 20, 354, 355, 379, 382, 383

Engineering and War (Blue, Levine, and Nieusma), 37–38

Engineering—An Endless Frontier (Auyang), 17

"Engineering as a Social Enterprise" (symposium), 177

Engineering College Research Association (ECRA), 211

"Engineering Criteria 2000" (ABET), 59. *See also* ABET EC 2000

engineering design: authenticity in, 86–87; metaphysics of, 84–86; problematics of, 81–82; as turning of making into thinking, 79–81

engineering disasters. *See* disasters (in engineering)

engineering education: on bringing philosophy into, 20; challenges for humanitarian engineering education programs, 254; content requirements of, 57–58, 59, 61; as cutting intellectual exploration short, 285; ethical component of, 265–66; ethics into, 211–14; initial institutions of, 156–57; in Japan, 226; justification for humanities in, 282–84; professional ethics as topic in, 207–19; reconfiguration of, 113

Engineering Education and Practice in Context (Christensen et al.), 306

Engineering Education (periodical), 211. *See also Journal of Engineering Education* (JEE)

engineering epistemology and metaphysics, 338–41

Engineering Epoch, 2

engineering ethics: as alternative development in Chile, 236–37; Dominican Republic as failure in, 235–36; emergence of, 125; enforcement of, 172; from engineering ethics to politics, 291–310; globalization of, 300–304; in global perspectives, 221–40; history of key ideas in, 208–10; integration of environmental ethics into, 167; philosophical engineering ethics, 298, 300, 307; quantitative observations about, 210–11;

questions of, 343–44; as social reform in Sweden, 234–35; use of term, 292, 293
Engineering Ethics: Concepts and Cases (Harris, Pritchard, and Rabins), 213, 215, 299
Engineering Ethics for a Globalized World (Murphy et al.), 240
engineering ethos, 115, 125, 155, 163, 357
Engineering: Issues, Challenges and Opportunities for Development (UNESCO), 280
Engineering Philosophy (Bucciarelli), 7, 144
engineering philosophy of technology, 110, 111
engineering policy: classics in science policy, 316–17; normative arguments for, 319–25; and science, technology, and engineering, 318–19; what is policy? 313–16
Engineering Society of Finland, 301
Engineering Studies (English journal), 10, 20
Engineering Studies: Interdisciplinary Perspectives (Chinese journal), 8, 10
engineering use and convenience, 365–84
The Engineer of 2020: Visions of Engineering in the New Century (National Academy of Engineering), 111, 168, 217, 218, 284
engineer-philosophers, 55, 115, 268, 290, 305
engineer(s): defined, 264; as having duty *plus respicere* (*see* duty *plus respicere*); many autodidacts as, 125; origins/meaning of in English, 14–18; Professional Engineers (PEs), 165
The Engineers and the Price System (Veblen), 162, 209, 269
Engineers' Council for Professional Development (ECPD), 57, 164, 209
Engineers for Change (Wisnioski), 164

Engineers without Borders (EWB), 251, 253
An Enquiry Concerning the Principles of Morals (Hume), 375
"An Enquiry into Modes of Existence" (Latour), 22–23
The Entropy Law and the Economic Process (Georgescu-Roegen), 200
Environmental Ethics (journal), 201
environmentalism, 167–68
environmentalists *vs.* developmentalists, 205n14
Envisioning a Sustainable Society: Learning Our Way Out (Milbrath), 195
epistemology, 14, 18, 20, 31, 54, 55, 62, 63, 139, 144, 145, 155, 274, 275, 294, 338–41, 344, 347, 352, 356, 366
The Essential Engineer: Why Science Alone Will Not Solve Our Global Problems (Petroski), 319–20
Estates School of Engineering, 156
Esteva, Gustavo, 205n17
Ethical Problems in Engineering (Alger, Christensen, Olmsted), 209
Ethical Problems in Engineering (Baum and Flores), 212–13
ethics: codes of (American Bar Association, 158, 293; American Institute of Electrical Engineers (AIEE), 158, 159, 293; American Medical Association, 158, 293; American Society of Civil Engineers (ASCE), 158–59, 169, 170; American Society of Mechanical Engineers (ASME), 158, 159; Association of Engineers of Chile, 236; Association of Professional Engineers (Ontario, Canada), 301; *Colegio de Ingenieros de Chile*, 301; first explicit professional engineering ethics codes, 126; French state engineers, 228; Institute of Electrical and Electronic Engineers (IEEE), 166, 167, 298; Institution of Civil

Engineers (ICE), 265; Institution of Engineers (Australia), 301; Institution of Engineers (India), 301; Japanese Society of Civil Engineers, 226–27; National Society for Professional Engineers (NSPE), 265; *Unión Panamericana de Asociaciones de Ingenieros* (UPADI or Pan American Federation of Engineering Societies), 301–2; World Federation of Engineering Organizations (WFEO), 302); described, 39; of designing, 44–46; in engineering, use of term, 292; engineering and, 63–66; into engineering education, 211–14; engineering ethics education in the American context, 207–19; for engineers as not same as ethics in engineering, 292–93; historical dialectics of ethics and engineering, 155–74; integration of environmental ethics into engineering ethics, 167; introduction of into design studies, 40; macro-ethics, 215, 305, 309; in medical profession, 128; and philosophy of engineering, 18–19; policy turn, 214–16; social ethics of technology, 215; "soft" ethics, 178; tendency for engineers to adopt some form of military ethics, 126; toward an inner ethics of design, 48–50; two versions of in design, 46–47. *See also* engineering ethics
Ethics, Technology, and Engineering (van de Poel and Royakkers), 303
Ethics and Values in Science and Technology (EVIST) program (National Science Foundation), 213, 297, 298
Ethics in Engineering (Martin and Schinzinger), 213, 298–99
"The Ethics of the Professions and of Business" (*Annals of the American Academy of Political and Social Science*), 160

Ethische Grundsätze des Ingenieurberufs (Ethical Foundations of the Engineering Profession) (VDI), 221, 222, 231
Europe, movement from ethics to politics in, 305–6
European Federation of National Engineering Associations (FEANI), 302
European Recovery Program (Marshall Plan), 247
Ewen, Stuart, 348
The Existential Pleasures of Engineering (Florman), 141, 282, 362
experiential gap, 381
experimental modeling, 41–42
Eyth, Max von, 233

FAS Public Interest Report, 94
Faust II (Goethe), 59
Fédération Européenne d'Associations Nationales d'Ingénieurs (FEANI or European Federation of National Engineering Associations), 238
Federation of American Scientists (FAS), 91, 93–94, 97, 102, 103
Federation of Atomic Scientists, 93
Feenberg, Andrew, 19–20, 22, 123, 131
Feng, Patrick, 87
Ferré, Frederick, 3
Feynman, Richard, 330
Fichte, Johann Gottlieb, 232
Filosofie en techniek (van Riessen), 79
First Use of Nuclear Weapons: Under the Constitution, Who Decides? (FAS), 94
Flores, Albert, 212–13, 297
Flores, Fernando, 71n36
Florman, Samuel, 11, 16, 17, 110, 122, 141, 268–69, 271, 282, 362
focal engineering, 381
focal things and practices, 378–79
folk philosophy, 293, 318
Food, Agriculture, Conservation, and Trade Act (1990), 195

Food Security Act (1985), 194
Forces of Production (Noble), 11
Ford, Dan, 97
Ford Pinto, as engineering disaster, 169, 212
Foreign Affairs, 192
Forum on Philosophy, Engineering, and Technology (fPET), 4–5, 20
4TU Centre for Ethics and Technology, 286, 303
France, engineers, progress, and rational state in, 228–30, 238
Franck, James, 94
Franck report, 94–95
Franco Battle, Orlando, 235–36
Frankenstein (Shelly), 279
Franklin Institute, 171
Frascara, Jorge, 349
free and open source software, 119–37
free and open source software (FOSS), 119
free *libre* open source software (FLOSS), 119
Free Software Foundation, 135
Freud, Sigmund, 31–32
Fritzsche, Albrecht, 5
Frodeman, Robert, 350
Fujian Shipping School, 303
Fuller, Buckminster, 3, 198
The Future of Engineering (Fritzsche and Oks), 5

Gadamer, Hans Georg, 68
Galilei, Galileo, 41, 63
Galison, Peter, 324, 356–57
Galt, John, 242
García de la Huerta, Marcos, 236
Gardoni, Paolo, 240
Geerts, Robert-Jan, 331
Gendron, Bernard, 360
genetic engineering, 86, 103, 193, 204, 341, 358
George IV (King), 15, 365, 368
Georgescu-Roegen, Nicholas, 200–201
Germany: engineering and *Bildung* in, 231–34, 238; initiating engineering-philosophical discussions in, 294–96; and philosophy of engineering in, 302
Gilson, Etienne, 144
Girard, René, 237
Global Engineering Ethics (Luegenbiehl and Clancy), 240
Global Grant Challenges Summits, 280
globalization: of engineering ethics, 300–304; globalized diversity, 237–39; impact of on engineering ethics, 221–40
Glover, Jonathan, 256, 257–58
Goethe, Wolfgang, 59, 108–9
Goldberg, David, 5, 18, 20, 273–77, 354
Goldman, Steven, 7, 11, 12–13, 128, 168, 217, 269
Gongcheng yu zhexue 工程与哲学 (Engineering and philosophy) (Yin Ruiyu et al.), 8
Gongcheng zhexue daolun 工程哲学导论 (Introduction to philosophy of engineering) (Li Bocong), 8
good building, principles of, 374
Goodman, Paul, 97
Gorgias (Plato), 109–10
"The Gospel of Global Efficiency" (Sachs), 198
Goujon, Philippe, 305
Gouzévitch, Irina, 17
Grand Challenges (NAE), 276, 279, 280, 284, 286, 321, 354
Grayson, Lawrence, 352
The Great Instauration (Bacon), 106–7
Greber, Henry, 383
Gropius, Walter, 349
Grundlinien einer Philosophie der Technik (Kapp), 110
Guide to the Culture of Science, Technology, and Medicine (Durbin), 296
Gunn, Alastair, 65, 167, 168, 272

Habermas, Jürgen, 22, 43
hacker culture, 124, 125

Hacking, Ian, 147–48
The Handbook of Science and Technology Studies, 10
Haraway, Donna, 69
Harding, Francis, 210
Harris, Charles E., Jr., 65, 213, 215, 240, 299
Hayek, Friedrich A. von, 51n3
Haynes, Denis, 195
Hedrich, Richard, 296
Hegel, G. W. F., 110, 179, 181n17, 294, 336
Heidegger, Martin, 21, 54–55, 68, 73–74, 287, 289, 348, 363
Hendricks, Vincent F., 303
Herbst, Clarence, 113, 114
Herbst Program of Humanities for Engineers (University of Colorado), 113–14
Herder, Johann Gottfried, 244
Herf, Jeffrey, 233
Herkert, Joseph, 20, 215, 305, 309
heuristics, 110, 115, 142, 336–37, 339, 344, 345
Heywood, John, 20
Hickam, Homer, Jr. (a.k.a. Sonny), 73
Hickman, Larry, 143
"A Historical Definition of Engineering" (Layton), 14
History of the Royal Society (Sprat), 370
Hjortsvang, Holger, 172, 174, 176
Hobbes, Thomas, 30, 34, 107, 375
Holbrook, J. Britt, 73–87
Hollander, Rachelle, 297
homo faber, 121, 358, 359, 362
homo politicus, 359
The Honest Broker (Pielke), 322
Hong Kong, engineering ethics as institutional protection in, 227–28, 238
Hong Kong Institution of Engineers (HKIE), 227, 238
honour, code of, Engineering Society of Finland, 301
Hoover, Herbert, 247, 251, 252
Howard, Albert, 193

Hughes, Thomas P., 74, 231
Huizinga, Johan, 48
The Human Condition (Arendt), 358
humanism, 115, 244, 284, 349
humanitarian engineering, 241–58, 282, 305, 353, 354
humanitarianism: described, 115–16; in history, 243–46; phases in modern humanitarianism, 246–50
Humanitarianism in Question (Barrett and Weiss), 253
humanitarian movement, 243, 246, 249, 255, 256
humanitarian software, 120
humanities, rationale for study of, 282–84
Humanity (Glover), 256
human rights, 94, 98, 100–101, 115, 217, 231, 244–45, 247, 251, 253–54
human sustainability, 196, 202
Humboldt, Wilhelm von, 109
Hume, David, 31, 75, 245, 330, 369, 370, 375
Humphrey, Hubert, 252
Hydrolevel Corporation, as conflict of interest case, 169, 171, 173–74

Ichisuke, Fujioka, 225
idealism: free and open source movement as heir to traditional of professional-ethical idealism, 120; professional idealism among scientists and engineers, 89–103; professional technoscientific idealism, 91, 101, 102; scientific-engineering idealism, 91
ideals, in engineering ethics, simplified, 153–81. *See also* efficiency; loyal obedience; participation; public safety, health, and welfare
IEEE Technology and Society Magazine, 213
Ignatieff, Michael, 132
Ihde, Don, 3, 141, 348, 362, 380
Illich, Ivan, 130–32, 255–56, 332, 333
Imperial College of Engineering (Japan), 225

In Context (journal), 196
Independent Commission on Disarmament and Security Issues, 189
Independent Commission on International Development Issues, 189
industrial revolution, 21, 28, 75, 76, 89, 136, 208, 241, 374, 376
informed consent, 99, 128, 168, 269, 270, 271, 299, 331
Ingegnería Senza Frontiere, 251
Ingeniería Sin Fronteras, 251
Ingénieurs Assistance Internationale, 251
Ingénieurs Sans Frontires, 251
Ingeniører uden grænser, 251
Ingenjörer och Naturvetare utan Gräner, 251
innovation policy, 313
Insight: A Study of Human Understanding (Lonergan), 144
Institute of Electrical and Electronic Engineers (IEEE), 158, 165–66, 172–73, 293, 298
Institute of Radio Engineers (IRE), 165, 298
Institution of Civil Engineers (ICE), 125, 153, 208, 241, 261, 264, 265, 292, 300–301, 365–68, 382–83
Institution of Engineers (India), 301
Institution of Mechanical Engineers, 125
Institutions of Engineers (Australia), 301
intermediate technology, 130, 194
International Committee of the Red Cross/Red Crescent (ICRC), 246–47
International Handbook on Chemical Weapons Proliferation (FAS), 94
International Network of Engineering Studies (INES), 10
International Union for Conservation of Nature and Natural Resources, 187, 188–89
In the World Interior of Capital (Sloterdijk), 352
An Introduction to Design and Culture in the Twentieth Century (Sparke), 77–78

invention: method of, 121, 123, 125, 133; patenting of, 153
IRE (Institute of Radio Engineers), 165, 298

Jakobsen, Bernhard F., 169–71
James, William, 293
Japan, engineering and profession as household, 222–27, 237
Japan Accreditation Board for Engineering Education (JABEE), 221, 222, 223, 226
Japanese Science Council (JSC), 222, 223
Japanese Society of Civil Engineers, 226
Japan Federation of Engineering Societies, 225
Jasanoff, Sheila, 314
Jaspers, Karl, 280, 363
Jefferson, Thomas, 156
Jesiek, Brent, 20, 240
Johnson, Deborah, 143, 214
Johnson, Robert, 120
Jonas, Hans, 21, 45
Journal of Engineering Education (JEE), 211
Joy, Bill, 103

Kan, F. Y., 227
Kant, Immanuel, 186, 245, 294, 308, 338
Kantaro, Suzuki, 225
Kapp, Ernst, 55, 110
Kass, Leon, 112
Kemper, Bart, 82
Kendall, Henry, 97, 102
Kennedy, David, 249
Kennedy, John F., 252, 319
Kierkegaard, Søren, 86–87
Kitcher, Philip, 344
Koen, Billy Vaughn, 11, 23, 110, 115, 142, 282, 337
Kondo, Dorinne, 223–24, 225
Kouchner, Bernard, 248
Kranakis, Eda, 13

Kranzberg, Melvin, 216
Kuhn, Thomas, 324
Kundera, Milan, 74

The Labyrinth of Technology (Vanderburg), 202
Ladd, John, 215, 305
Large Hadron Collider, 285
Lasswell, Harold, 316–17, 319
Latouche, Serge, 201
Latour, Bruno, 3–4, 21, 22, 143, 288, 348, 349, 352, 359
law, as compared to medicine and engineering, 267–68, 273, 275
Layman, Ronald, 11
Layton, Edwin T., Jr., 10, 13–14, 153, 161, 164, 293, 342
Lebentige Krafte (Eyth), 233
Legal and Ethical Phases of Engineering (Harding and Canfield), 210
Lenk, Hans, 13, 296
Lessig, Lawrence, 119
Levine, Michael, 37
Levinson, Mark, 285
Leydens, Jon, 20, 256
Li Bocong, 5, 6, 8, 9, 10, 11, 13, 306, 307, 309
Liebig, Justus von, 193
lifeworld, 1, 22, 68, 123, 318, 348, 380. *See also* techno-lifeworld
Light, Andrew, 146
Limited Nuclear Test Ban Treaty, 93, 95
The Limits of Growth (Club of Rome), 186, 187, 196, 200
limits-to-growth argument, 195, 197, 200
linguistic philosophy, of engineering, 144–48
Linux, 124
Lives of Engineers, 375
Locke, John, 375, 376
logic, 18, 39, 40, 54, 62, 83, 108, 142, 338
Lonergan, Bernard, 144

Los Angeles County Flood Control District dam, as engineering disaster, 170
Los Angeles Daily News, 170
Los Angeles Times, 170
Lovins, Amory, 192–93, 195
low-input sustainable agriculture (LISA), 194–95
loyal obedience: as ideal in engineering ethics, 156–61; Weil's pamphlet on, 212
Lucena, Juan, 10, 221–40, 256
Luegenbiehl, Heinz, 6, 240, 342
Lyotard, François, 69, 146

Machen, Bernie, 285
Machiavelli, Niccolò, 30, 106, 107
MacIntyre, Alasdair, 144
Mackey, Robert, 139–50
macro-ethics, 215, 305, 309
Madhavan, Guru, 17
making, turning of into thinking, 77, 79–87
Malafouris, Lambros, 362
Maldonado, Tomás, 349
The Man Who Tried to Save the World (Anderson), 250
management: Earth management, 198, 206n40; of human beings, 298; input-output management, 197; management captivity, 12; scientific management, 161, 162
Manhattan Project, 92, 93, 96
Mankind at the Turning Point (Club of Rome), 187
The Mantra of Efficiency (Alexander), 162
Margolin, Julius, 79
Maritain, Jacques, 144
Marshall Plan, 247
Martin, Mike, 66, 168, 213, 269–70, 271, 298, 299
Marx, Karl, 63, 107, 110, 122, 243, 348, 374
Marx, Leo, 2

Masad, Eyad, 240
Maslow, Abraham, 253
McCarthy, Natasha, 5, 17, 18, 324–25
McCloskey, Deirdre, 376
McDonnell and Miller, Inc. (M&M), 171
McNeill, William H., 27, 28
Mead, Daniel, 210
Mead, Margaret, 328
Mechanical Engineering, 164
Médecins sans Frontieres (MSF or Doctors without Borders), 248, 252–53
Media Lab (MIT), 68
medicine, as cvompared to law and engineering, 267–68, 273, 275
Meditación de la técnica (Ortega y Gasset), 43
Meijers, Anthonie, 5
Mendelsohn, Everett, 27
Mensch und Technik (Humanity and Technology) (VDI), 295
Merchant, Carolyn, 30
The Merchant of Venice (Shakespeare), 244
Merton, Robert, 120, 154
metaphysics, 338–41
Metzger, Robert Charles, 83, 84
Michelfelder, Diane, 5
Microsoft, 124
Milbrath, Lester W., 195
military: as compared to medicine, law, business, and engineering, 275; engineering as detaching from, 241, 264, 292; engineering as originating in, 109, 156–57, 292; relationships with science, engineering, and technology, 27–38
Military Academy at West Point, 156, 157
Military Enterprise and Technological Change (Smith), 27
military-industrial complex, 96
Mill, John Stuart, 45, 308, 309
Millikan, Ruth, 267
MIT, Media Lab at, 68
Mitcham, Carl, 3, 5, 9, 294, 296

modeling paradox, 176–77
Mokyr, Joel, 376
Monju fast-breeder reactor high-technology accident, 223
Morgan, Granger, 319
Morgan, Lewis H., 327
Morison, George S., 161, 168
Morton, Timothy, 203
Morton-Thiokol, 173
Moulakis, Athanasios, 114
Moulton, Forest Ray, 98
MSF (*Médecins sans Frontieres*), 248, 252–53
Mulder, Karel, 202
"Multiple Facets of Philosophy and Engineering" (Durbin), 11
Mumford, Lewis, 31, 342, 348
Muñoz, David, 241–58
Murphy, Colleen, 240
The Myth of the Machine (Mumford), 31

nanotechnological design, 86, 116
Nansen, Fridtjof, 247
Napoleon Bonaparte, 156–57
Nash, Roderick Frazier, 180n6
National Academy of Engineering (NAE), 111, 167–68, 177, 216–17, 279, 284, 286, 313, 321, 354
National Council of Engineers and Scientists of France (*Conseil National des Ingénieurs et Scientifiques de France,* CNISF), 221, 228, 301
National Endowment for the Humanities (NEH), 212–13, 296, 297, 298
National Project on Philosophy and Engineering Ethics (NEH), 297, 298
National Renewable Energy Laboratory, 193
National Science Foundation (NSF), 33, 114, 213, 296
National Society of Professional Engineers (NSPE), 165, 209, 261, 265, 297
natural environment, protection of, 167–68

Nersessian, Nancy, 20
Netherlands, and philosophy of engineering and technology, 303
Newberry, Byron, 5, 14, 16, 115, 342, 343, 353
New Farm (journal), 193
New Frontiers for American Youth: Perspective on the Peace Corps (Albertson), 252
"The New Science Scheme: Science for Society and the Fusion of Humanity and Natural Sciences" (JSC), 223
Newton, Isaac, 63, 75
The New Yorker, 89–90
Nietzsche, Friedrich, 30, 31, 48, 378
Nieusma, Dean, 37, 180n4, 181n15
Nieuwentyt, Bernard, 75
Nightingale, Florence, 246
Nobel, Alfred, 32
Noble, David, 11, 293
Nordmann, Alfred, 356, 357
Nouveau Christianisme (Saint-Simon), 161
nuclear power plants: monitoring of, 97; safe operation of, 160
nuclear weapons, development of, 32, 54, 91–96, 97, 98, 102, 103, 242, 298, 381. *See also* atomic bomb
Nye, David, 328–29

Oberrealschulen, 232, 233
October Sky (film), 73, 74
Ohashi, Hideo, 225, 226
Oks, Sascha Julian, 5
Oldenburg, Henry, 154
Olsen, Jan Kyrre Berg, 303
Olson, Richard, 162
On the Mode of the Existence of Technical Objects (Simondon), 287
open source movements, 119–37
Oppenheimer, J. Robert, 90, 95
organic farming, 193–94
Organic Gardening and Farming (journal), 193

Organization for Economic Cooperation and Development (OECD), 247
Ortega y Gasset, José, 43, 121, 123–24, 286
Our Common Future (World Commission on Environment and Development), 187, 189–90
Owen, Robert, 243
OXFAM, 247

Palchinsky, Peter, 59
Paley, William, 75
Palmer, Henry Robinson, 367
Palo Alto Research Center (PARC), 68
Pan-American Federation of Engineering Societies (*Unión Panamericana de Asociaciones de Ingenieros/União Panamericana de Associações de Engenheiros*, UPADI), 238, 301–2
paradox: of convenience, 380–81; of information technology, 176; modeling paradox, 176–77
Parson, Talcott, 180n3
participation, as ideal in engineering ethics, 168
Particle Fever (film), 285
Paul, Saint, 245
Paul VI (pope), 199
Payne, James H., 169–71
Peace Corps, 247, 251–52
Peace of Westphalia, 370
Peccei, Aurelio, 186
Pedersen, Stig Arthur, 303
Petroski, Henry, 16, 65, 282, 319
Pevsner, Nikolaus, 372
Pfatteicher, Sarah, 158–59
Der Phänomenologie des Geistes (Hegel), 110
phenomenology, 142, 287–88, 348, 359, 362, 378, 381
Philosophical, Logical and Scientific Perspectives in Engineering (Şen), 20
philosophical engineering, 111, 113, 117–18

philosophical engineering ethics, 298, 300, 307
philosophical humanitarianism, 244
Philosophical Investigations (Wittgenstein), 145
Philosophical Transactions (Royal Society), 153, 154
philosophy: branches of/types of, 139, 141–44; can it be engineering? 335–45; common/uniting elements of, 62–63; core areas of, 63; currents in contemporary philosophy, 140; emergence of according to Aristotle, 61; folk philosophy, 293, 318; importance of philosophy to engineering, 53–69; linguistic philosophy of engineering, 144–48; policy turn in, 215; of science, 21; self-defense and, 54–56, 67; self-interest and, 56–60, 67; self-knowledge and, 67–69; significance of, 63
Philosophy and Engineering: An Emerging Agenda (*P&E*) (van de Poel and Goldberg), 5, 12, 20, 22, 23
Philosophy and Engineering: Exploring Boundaries, Expanding Connections (Michelfelder, Newberry, and Qa Zhu), 5
Philosophy and Engineering: Reflections on Practice, Principles and Process (Michelfelder, McCarthy, and Goldeberg), 5
Philosophy and Technology (journal), 3
Philosophy and Technology: Technology as a Philosophical Problem (Mitcham and Mackey), 2–3
Philosophy in Engineering (Christensen), 7, 303
"Philosophy of Engineering and Technology" book series (Vermaas), 5
Philosophy of Engineering, East and West (Mitcham, Li Bocong, and Zhang Baichun), 5, 6

Philosophy of Science Association, 297
Philosophy of Technology and Engineering Sciences (Meijers), 5
Pielke, Roger, Jr., 322–24
Pinch, Trevor, 143, 337
Pitt, Joseph, 3, 143
planned making, 39
Plato, 31, 41, 49, 109–10, 147, 309, 314, 349
Polanyi, Karl, 43, 351
Polanyi, Michael, 68, 154, 316
policy: defined, 313–14; engineering policy, 313–25; innovation policy, 313
politics: from engineering ethics to, 291–310; movement from politics to policy, 316; questions of, 343–44
Politics (Aristotle), 314
Politics: Who Gets What, When, How (Lasswell), 316
Pope, Alexander, 185
Popper, Karl, 120
Pörksen, Uwe, 196
post-engineering, 216–19
Pound, Ezra, 47, 108
Prevention (journal), 193
preventive engineering, 202
principal-agent problems, 169
The Principles of Scientific Management (Taylor), 162
Pritchard, Michael, 66, 213, 215, 299
The Problems of Philosophy (Russell), 150
Professional Engineers (PEs), 165
professional organizations (for engineering profession), 164–65, 208–9, 227, 238, 261, 292, 293. *See also specific organizations*
professional technoscientific idealism, 91, 101, 102
program conviviality, 120, 134
public disclosure, principle of, 173–74
public safety, health, and welfare, as ideal in engineering ethics, 156, 163–68, 215–16, 243, 256, 270

Pugwash Conferences on Science and
    World Affairs, 95
Pugwash movement, 91, 95–96, 97,
    102, 103
Pulteney, William, 375
Puritans, 49
The Pursuit of Power (McNeill), 27

Qiu Lianghui, 8
Quine, Willard Van Orman, 143, 338,
    340

Rabins, Michael, 65, 66, 213, 215, 299
Rapp, Friedrich, 3
Raymo, Chet, 338, 344
Raymond, Eric, 127
Reagan, Ronald, 97
Reden an die deutsche Nation (Lectures
    to the German nation) (Fichte), 232
*Re-Engineering Philosophy for Limited
    Beings* (Wimsatt), 335, 337
*Renewable Agriculture and Food
    Systems* (journal), 194
renewable energy, 193, 329, 333
Republic (Plato), 31, 314
Rerum gestarum (Ammianus
    Marcellinus), 156
research and development (R&D),
    122–23
Research Center for Engineering and
    Society (RCES), 8, 9
responsible conduct of research
    (RCR), 154
responsible research and innovation
    (RRI), 154
The Revolt of the Engineers (Layton),
    293
Rieff, David, 249
The Rights Revolution (Ignatieff), 132
Riley, Donna, 37
Rising Above the Gathering Storm
    (Committee on Science, Engineering,
    and Public Policy), 217
Roberts, David, 146
Rocket Boys (Hickam), 73

Rodale, J. I., 193
Rodale, Robert, 193, 194, 195, 197
Rodale Institute, 193
Rogers, C. Thomas, 66
Rogers, G. F. C., 383
Roosevelt, Franklin D., 91–92
Ropohl, Günther, 12, 13, 296
Rorty, Richard, 144–45
Rosen, Stanley, 33
Rosenberg, Nathan, 70n19
Rostow, Walt W., 191–92, 198, 243
Rotblat, Joseph, 95, 96, 102
Rousseau, Jean-Jacques, 30, 31
Royakkers, Lambèr, 303
Royal Academy of Engineering, 4, 313,
    324
Royal Institute of Technology
    (Stockholm), 234
Royal Society, 153–54, 368, 370
Russell, Bertrand, 91, 92–93, 149–50
Ryle, Gilbert, 145

Sabbagh, Karl, 81
Sachs, Wolfgang, 198
safety engineering, 284
Safina, Carl, 103
Saint-Simon, Henri de, 161
Sakharov, Andrei, 94
Salomon, Jean-Jacques, 316, 317, 319
Sanmartín, José, 201
SARE (Sustainable Agriculture
    Research and Education)
    Program, 195
Sarewitz, Daniel, 322, 360
Sartre, Jean-Paul, 244
Sassower, Raphael, 91
Save the Children, 247
scenario planning, 217–18
Schatzberg, Eric, 2
Scheler, Max, 378
Schinzinger, Roland, 65, 66, 168, 213,
    269–70, 271, 298, 299
Schneider, Jen, 256
Schumacher, E. F., 130, 194, 250
Schumpeter, Joseph, 192

Schweitzer, Albert, 245
science: agricultural science, 193; criticism of, 91, 93; ethos of according to Merton, 154–55; origins of, 153; philosophy of, 21; relationships between science, engineering, technology, and military, 27–38; rising emphasis on questions concerning relations between science and society, 98. *See also* science, technology, and society (STS); science, technology, engineering, and mathematics (STEM); science policy; technoscience
Science (magazine), 196
Science, Technology, & Human Values (journal), 3
science, technology, and engineering relations, 6
science, technology, and society (STS), 3, 9, 12, 91, 103, 216, 283, 303, 304, 337, 350
Science, Technology, and the Military (Mendelsohn, Smith, and Weingart), 27
science, technology, engineering, and mathematics (STEM), 17, 36–37
Science and Engineering Ethics (journal), 211
Science and Human Rights Coalition, 353–54
Science and Politics (Salomon), 317
Science and the Modern World (Whitehead), 74
Science in Action (Latour), 3–4
Science Is Not Enough (Bush), 280
The Science of Science Policy (Fealing et al.), 319
science policy, 30, 32, 36, 215, 280, 313, 316–17, 318, 319, 321, 322, 325, 355, 356
Science: The Endless Frontier (Bush), 355
scientific freedom, 99, 100, 101
scientific modeling *vs.* projective and technological modeling, 42

scientific revolution, 217, 218
scientific-engineering idealism, 91
Scientists in the Quest for Peace (Rotblat), 95
Seacole, Mary, 246
Second Axial Age, 280–81, 285–86, 354
Second Treatise of Government (Locke), 375
Sein und Zeit (Heidegger), 73–74
self-defense, and philosophy, 54–56, 67
self-interest, and philosophy, 56–60, 67
self-knowledge: and philosophy, 67–69; as true grand challenge for engineering, 279–90
Sellars, Wilfrid, 143
Sen, Amartya, 198–99
Şen, Zekâi, 20
Serres, Michel, 91
Shakespeare, William, 156, 244, 372–73
Shelley, Percy Bysshe, 279
Shelly, Mary, 279
Shrader-Frechette, K. S., 180n9, 204n2
Silent Spring (Carson), 90
Simon, Herbert, 55, 68, 73, 339
Simon, Julian, 197, 203
Simondon, Gilbert, 287–90
Sinclair, George, 383
Sloterdijk, Peter, 349, 352
Small Is Beautiful (Schumacher), 130, 194, 250
Smeaton, John, 15, 153, 264, 292, 373
Smelser, Neil, 180n3
Smiles, Samuel, 371, 375
Smith, Adam, 369, 374
Smith, Alice Kimball, 93
Smith, Cecil, 229–30
Smith, Jessica, 327–34
Smith, Merritt Roe, 27
Smith, Ralph J., 16
Snow, C. P., 21, 89, 90, 91, 102, 281
"The Social Captivity of Engineering" (Goldman), 12
Social Implications of Technology (IEEE interest group), 213

Society for Philosophy and Technology (SPT), 3, 4, 12, 20, 201
Society for the Dialectics of Nature, 8–9
Society for the History of Technology (SHOT), 10, 61
Society for the Promotion of Engineering Education (SPEE), 211
Society for the Social Implications of Technology (IEEE), 213
Society of Civil Engineers, 125, 153, 292, 373
Socrates, 110, 111–12, 113, 116, 183, 200, 280, 286
Socrates Society Seminars, 114
Soddy, Frederick, 327
Soft Energy Paths, 192
software: argument for not updating of, 136; convivial software, 121, 134–35; creation of, 121; free and open source software, 119–37; humanitarian software, 120
"Software Engineering Code of Ethics and Professional Practice" (ACM), 126–27
Solar Energy Research, Development, and Demonstration Act, 193
Solar Energy Research Institute, 193
Sparke, Penny, 77–78
special (personal) interests, 169
speed trap, 136–37
Sprat, Thomas, 370
SPT (Society for Philosophy and Technology), 3, 4, 12, 20, 201
Sputnik, 73, 85
The Stages of Economic Growth: A Non-Communist Manifesto (Rostow), 191, 243
"Statement on Atomic Research in Japan" (Japanese Science Council), 222
The State of Humanity (Simon), 197
State of the World reports (Worldwatch Institute), 195
Staudenmaier, John, 61

STEM (science, technology, engineering, and mathematics), 17, 36–37
Stephenson, George, 372
Stimson, Henry L., 94, 95
Stone, Jeremy, 94
Strategic Defense Initiative, 34, 94, 97
Strauss, Leo, 30
Strawson, P. F., 145
Strickland, Donald, 93
structural differentiation, 180n3
STS (science, technology, and society), 3, 9, 12, 91, 103, 216, 283, 303, 304, 337, 350
Student Pugwash, 96
Sun Yongfu, 8
sustainability: criticisms of, 196–99; environmentalism and, 167–68; historical and philosophical background for, 184–86; immediate origins of concept of sustainable development, 186–91; origins and ambivalences of, 183–206; and some near neighbors, 191–96
Sustainable Agriculture Research and Education (SARE) Program, 195
Sustainable Development for Engineers: A Handbook and Resource Guide (Mulder), 202
sustainable energy, 193, 195, 331
Sutz, Judith, 237
Sweden, engineering ethics as social reform in, 234–35, 238
systems architect, 324–25
Szilard, Leo, 91–92

"The Take-off into Sustained Growth" (Rostow), 191
Tang, Xiaofeng, 180n4, 181n15
Taylor, Frederick W., 162
Techné: Research in Philosophy and Technology (journal), 3, 201
technical conviviality, 120, 132–33, 134
technical decision-making, involvement of public in, 117, 127–28, 306

technical self-education, 125
Technik, 110, 232, 233, 234, 294
Technikbewertung (Technology evaluation), 234
Technische Hochschulen, 232
technocratic efficiency, as ideal in engineering ethics, 161–63. *See also* efficiency
techno-engineering, 30, 31, 35, 198
techno-human condition, 286, 351, 358–63, 381
The Techno-Human Conditions (Allenby and Sarewitz), 360
techno-lifeworld, 44, 45, 46, 47, 50, 53, 318, 378, 380
technological design, 45, 74, 77–79, 81, 84, 85, 87, 128
technological modeling *vs.* scientific modeling, 42
The Technological Society (Ellul), 3
technology: alternative technology, 121, 130, 134, 178, 194, 250; appropriate technology, 130; convivial technology, 130, 131; engineering as core of, 358; engineering philosophy of, 110, 111; history of, 121–25; intermediate technology, 130, 194; paradox of information technology, 176; philosophy of (engineering as lacking prominent role in, 2–4; philosophy of engineering as another name for, 1); relationships of to warfare, 27–38; social ethics of, 215. *See also* science, technology, and society (STS); science, technology, engineering, and mathematics (STEM); technology assessment; technology policy; technology transfer
Technology and Culture (*T&C*) (journal), 10, 13
Technology and Ethics: A European Quest for Responsible Engineering (Goujon and Dubreuil), 305–6
Technology and War (Van Creveld), 27

technology assessment, 84, 99, 212, 234, 238, 295, 296, 302
Technology: Critical History of a Concept (Schatzberg), 2
Technology in the Western Political Tradition (Melzer, Weinberger, and Zinman), 33
technology policy, 32, 36, 313, 318, 321
Technology's Storytellers (Staudenmaier), 61
technology transfer, 122, 125, 128–29, 131, 133, 134, 135
technoscience: criticism of, 98; science, engineering, and technology as having merged into, 101; scientific intellectual criticism of, 91, 93; scientists serving as moral critics of, 97; technoscientific criticism of, 102; use of term, 31
technoscientific activism, 96
Technoscientific Angst (Sassower), 91
technoscientific change, kinds of harms from, 97
technoscientific weapons, pursuit of, 98
Telford, Thomas: architect, 372; death of, 375; early years of, 369–70, 375; first ICE president, 366, 367; stone mason to engineer, 370–75
Thayer, Sylvanus, 157
Thayer School of Engineering (Dartmouth College), 157
Theses on Feuerbach (Marx), 107
think globally, act locally, 177
Thinking through Technology (Mitcham), 350
Think Like an Engineer: Inside the Minds That Are Changing Our Lives (Madhavan), 17
Thomas Aquinas, 75, 385
Three Mile Island nuclear accident, 97, 212
"The Three Principles for the Peaceful Use of Atomic Energy," 222, 223
Tierney, Thomas, 375, 377–78, 383
Tocqueville, Alexis de, 158

Tokaimura reactor, high-technology accident at, 223
Tools for Conviviality (Illich), 130
Torvalds, Linus, 124
Toulmin, Stephen, 299
A Treatise of Human Nature (Hume), 375
Tredgold, Thomas, 55, 58, 62, 110, 125–26, 259, 260, 264–65, 292, 298, 366, 368, 375, 381
Truman, Harry S., 129, 179, 190, 242–43, 252, 323
Tu Weiming, 354
21st-Century Jet: The Making and Marketing of the Boeing 777 (Sabbagh), 81
"two cultures" argument/problem, 21, 89, 90, 281–82
"2000-watt society," 321

Ubbelhode, A. R., 328
Ulm School, 349
The Ultimate Resource (Simon), 197
The Unbearable Lightness of Being (Kundera), 74
"Underdevelopment Is a State of Mind" (Harrison), 129
Unger, Stephen H., 65, 66, 213
unintended consequences, 32, 35, 84, 90, 256, 320
Union of Concerned Scientists (UCS), 91, 96–97, 102, 103
Unión Panamericana de Asociaciones de Ingenieros/União Panamericana de Associações de Engenheiros (UPADI or Pan-American Federation of Engineering Societies), 238, 301–2
United Nations (UN): Conference on Sustainable Development (UNCSD), 201; creation of, 247; Educational, Scientific and Culture Organization (UNESCO), 247, 279–80; High Commissioner for Refugees (UNHCR), 247; International Children's Emergency Fund (UNICEF), 247; "Millennium Declaration," 248–49; Millennium Project, 249; Sustainable Development goals, 201
United States: Department of Agriculture, 193; Environmental Protection Agency, 90; initiating engineering-philosophical discussions in, 296–300
Universal Declaration of Human Rights, 247
University of Chinese Academy of Sciences (UCAS), 8
University of Tokyo, engineering program at, 225
use and convenience. *See* convenience

Vallor, Shannon, 349
Van Creveld, Martin, 27, 28
van de Poel, Ibo, 5–6, 9, 303
Vanderburg, Willem, 196, 202
van Riessen, Hendrick, 79, 81
Vasari, Giorgio, 41
Veblen, Thorstein, 161–62, 209, 269, 342
Venturi, Robert, 69
Verbeek, Peter-Paul, 120, 348, 363
Verein Deutscher Ingenieure (VDI or Association of German Engineers), 221, 231, 232, 233, 295–96
Vermaas, Pieter, 5, 18–19, 268
Vesilind, Aarne, 65, 167, 168, 272
Vietnam War, opposition to, 96–97, 212
Vincenti, Walter, 6, 177, 267, 342–43, 382, 383
Vitruvius, 373
Vlot, Ad, 81
Von Braun, Wernher, 242
Vowe, Gerhard, 313–14
Vrin, Hélène, 17

Wang Liheng, 8
Wang Nan, 291–310
Wang Yingluo, 8

warfare and technology, 27–38
War Made New (Boot), 27
Watergate, impact of, 212
Watt, James, 372
The Wealth of Nations (Smith), 369, 374
weapons of mass destruction (WMDs), 29, 34, 82
Weber, Max, 63, 314, 378
Weber, Steven, 119
Weil, Vivian, 210, 211, 212, 297
Weingart, Peter, 27
Weiser, Mark, 68
Weiss, Thomas G., 253
Wells, H. G., 154
What Engineers Know and How They Know It (Vincenti), 267, 342–43
whistle-blowing, 100, 173–74, 212, 213, 216, 235, 300, 305
Whitbeck, Caroline, 214
White, Leslie, 327–28, 330
Whitehead, Albert North, 74, 121
Why Does Language Matter to Philosophy? (Hacking), 147
"Why the Future Doesn't Need Us" (Joy), 103
Wiener, Norbert, 34, 290
Wilhelm Meister's Apprenticeship (Goethe), 108–9
William and Flora Hewlett Foundation, 114
Williams, Rosalind, 17, 218
Wimsatt, William, 335–41, 342, 343–45, 356, 357
Winner, Langdon, 22, 44, 128, 168, 305
Winograd, Terry, 71n36

Wisnioski, Matthew, 164, 298
Wittgenstein, Ludwig, 18, 143, 145, 146
Workshop on Philosophy and Engineering (WPE), 4–5
World Commission on Environment and Development, 187, 189–90
World Conference on Science (1999), 102
World Conservation Strategy (International Union for Conservation of Nature and Natural Resources), 187, 188–89
World Federation of Engineering Organizations (WFEO), 238–39, 302
World Food Programme (WPG), 247
Worldwatch Institute, 195
Worster, Donald, 180n6, 204n3
Wo zaowu, guwo zai 我造物，故我在 (I create therefore I am) (Li Bocong), 8
Wu, Chyuan-Yuan, 10
Wulf, William, 215

X-Crise (think tank), 229
Xerox Palo Alto Research Center (PARC), 68

Yamao, Yozo, 225
Yin Dengxiang, 9
Yin Ruiyu, 8

Zadig, Cecilia, 235
Zhang Baichun, 5
Zhang Zhuhui, 10
Zhu, Qin, 5, 240
Zimmerli, Walther Ch., 181n184

About the Author

Carl Mitcham is International Distinguished Professor of Philosophy of Technology at Renmin University of China and Professor Emeritus of Humanities, Arts, and Social Sciences at the Colorado School of Mines. His publications include *Thinking through Technology* (1994) and *Ethics and Science: An Introduction* (2012, with Adam Briggle).

www.ingramcontent.com/pod-product-compliance
Lightning Source LLC
Chambersburg PA
CBHW021813300426
44114CB00009BA/161